Neuronal Dynamics

Neuronal Dynamics

Edited by:

Stefano Spezia

ARCLER PRESS

www.arclerpress.com

Neuronal Dynamics

Stefano Spezia

Arcler Press
2010 Winston Park Drive,
2nd Floor
Oakville, ON L6H 5R7
Canada
www.arclerpress.com
Tel: 001-289-291-7705
001-905-616-2116
Fax: 001-289-291-7601
Email: orders@arclereducation.com

ISBN: 978-1-77361-124-2 (Hardcover)

DECLARATION

ABOUT THE EDITOR

STEFANO SPEZIA

Stefano Spezia is Ph.D. holder in Applied Physics at the University of Palermo since April 2012. His major research experience is in noise-induced effects in nonlinear systems, especially in the fields of modeling of complex biological systems and simulation of semiconductor spintronic devices. Associate member of the Italian Physical Society and European Physical Society.

TABLE OF CONTENTS

SECTION IV: NEURONAL NETWORKS DYNAMICS

SECTION V: DYNAMICS OF COGNITION

LIST OF CONTRIBUTORS

Spezia Stefano

Aidan C. Daly
Department of Computer Science

David J. Gavaghan
Department of Computer Science

Chris Holmes
Department of Statistics, University of Oxford, Oxford, UK

Jonathan Cooper
Department of Computer Science

Michael D. Forrest
Department of Computer Science, University of Warwick, Coventry, CV4 7AL, UK
Institute for Adaptive and Neural Computation, School of Informatics, University of Edinburgh, Edinburgh, EH1 2QL, UK

Florian Aspart
Department of Software Engineering and Theoretical Computer Science, Technische Universität Berlin, Berlin, Germany
Bernstein Center for Computational Neuroscience Berlin, Berlin, Germany

Josef Ladenbauer
Department of Software Engineering and Theoretical Computer Science, Technische Universität Berlin, Berlin, Germany
Bernstein Center for Computational Neuroscience Berlin, Berlin, Germany

Klaus Obermayer
Department of Software Engineering and Theoretical Computer Science, Technische Universität Berlin, Berlin, Germany

Bernstein Center for Computational Neuroscience Berlin, Berlin, Germany

Vladimir Kornijcuk
Center for Electronic Materials, Korea Institute of Science and Technology, Seoul, South Korea
Department of Materials Science and Engineering, Seoul National University of Science and Technology, Seoul, South Korea

Hyungkwang Lim
Center for Electronic Materials, Korea Institute of Science and Technology, Seoul, South Korea
Department of Materials Science and Engineering, Seoul National University, Seoul, South Korea

Jun Yeong Seok
Center for Electronic Materials, Korea Institute of Science and Technology, Seoul, South Korea
Department of Materials Science and Engineering, Seoul National University, Seoul, South Korea

Guhyun Kim
Center for Electronic Materials, Korea Institute of Science and Technology, Seoul, South Korea
Department of Materials Science and Engineering, Seoul National University, Seoul, South Korea

Seong Keun Kim
Center for Electronic Materials, Korea Institute of Science and Technology, Seoul, South Korea

Inho Kim
Center for Electronic Materials, Korea Institute of Science and Technology, Seoul, South Korea

Byung Joon Choi
Department of Materials Science and Engineering, Seoul National University of Science and Technology, Seoul, South Korea

Doo Seok Jeong
Center for Electronic Materials, Korea Institute of Science and Technology, Seoul, South Korea

Frederike Kneer
Institut für Theoretische Physik, Technische Universität Berlin, Hardenbergstrasse 36, D-10623 Berlin, Germany
Institut für Softwaretechnik und Theoretische Informatik Technische Universität Berlin, Marchstr. 23, D-10587 Berlin, Germany

Eckehard Schöll
Institut für Theoretische Physik, Technische Universität Berlin, Hardenbergstrasse 36, D-10623 Berlin, Germany

Markus A Dahlem
Institut für Physik, Humboldt Universität zu Berlin, Robert-Koch-Platz 4, D-10115 Berlin, Berlin, Germany

Ri-Qi Su
School of Electrical, Computer, and Energy Engineering, Arizona State University, Tempe, AZ 85287, USA

Ying-Cheng Lai
School of Electrical, Computer, and Energy Engineering, Arizona State University, Tempe, AZ 85287, USA
Department of Physics, Arizona State University, Tempe, AZ 85287, USA

Xiao Wang
School of Biological and Health Systems Engineering, Arizona State University, Tempe, AZ 85287, USA

André C. Marreiros
The Wellcome Trust Centre for Neuroimaging, Institute of Neurology, University College London, UCL, 12 Queen Square, London, UK WC1N 3BG, UK

Stefan J. Kiebel
The Wellcome Trust Centre for Neuroimaging, Institute of Neurology, University College London, UCL, 12 Queen Square, London, UK WC1N 3BG, UK

Karl J. Friston
The Wellcome Trust Centre for Neuroimaging, Institute of Neurology, University College London, UCL, 12 Queen Square, London, UK WC1N 3BG, UK

E. C. Y. Ho
Department of Neuroscience & Institute for Brain Science, Brown University, Providence, USA

Wilson Truccolo
Department of Neuroscience & Institute for Brain Science, Brown University, Providence, USA
U.S. Department of Veterans Affairs, Center for Neurorestoration and Neurotechnology, Providence, USA

Gustavo Deco
Institució Catalana de Recerca i Estudis Avançats (ICREA), Universitat Pompeu Fabra, Department of Technology, Computational Neuroscience, Barcelona, Spain

Viktor K. Jirsa
Theoretical Neuroscience Group, Institut Sciences de Mouvement, Marseille, France
Center for Complex Systems and Brain Sciences, Department of Physics, Florida Atlantic University, Boca, Florida, United States of America

Peter A. Robinson
School of Physics, University of Sydney, Sydney, New South Wales, Australia
Brain Dynamics Center, Westmead Millennium Institute, Westmead Hospital and University of Sydney, Westmead, New South Wales, Australia
Faculty of Medicine, University of Sydney, Sydney, New South Wales, Australia

Michael Breakspear
School of Psychiatry, University of New South Wales, Sydney, and The Black Dog Institute, Randwick, New South Wales, Australia
School of Physics, University of Sydney, Sydney, New South Wales, Australia

Karl Friston
Wellcome Trust Centre for Neuroimaging, University College London, London, United Kingdom

PREFACE

Neuronal Dynamics is a field of knowledge that creates models of individual neurons and biological neural networks of any part of the nervous system. Ongoing research efforts of spiking neural networks attempts to gain a better understanding of the brain and/or realize its electronic replicas that partially imitate brain functionalities such as learning and memory. In particular, the cortex is a complex system, characterized by its dynamics and architecture, which underlie many functions such as action, perception, learning, language, and cognition. Despite its structural architecture has been studied for more than a hundred years, its dynamics is not entirely understood.

The book begins with a brief introduction that provides some basilar concepts about the neurophysiology of the neurons. In particular, the morphology of both neurons and synapsis, the action potentials and electrical properties of the cell membrane have been considered. Section 1 focuses on the most influential and enduring cellular model: the Hodgkin–Huxley action potential model, which historically was constructed for the squid giant axon and persists to this day. Section 2 discusses of integrate-and-fire neuron models which present the advantage to be well applicable to study the dynamics of large neuronal populations, due to their computational efficiency and analytical tractability. Section 3 present recent works about the FitzHugh-Nagumo model. In particular, the study of the influence of the cortex curvature on spreading depression, the propagation of excitation waves in moving media, and the identification of chaotic elements as a source of certain diseases have been taken into account. Section 4 deals with the interacting neuronal populations, the appearance of chaos in neuronal networks and the interaction between synaptic inhibition and glial-potassium dynamics. Finally, the last Section 5 focuses on dynamic of cognition. In particular, the key role of metastable states in the execution of cognitive functions, the statistical description of neuronal ensembles in terms of a Fokker-Planck equation, and the unification of probabilistic inference and synaptic plasticity by using a neuronal network that implements the well-studied Helmholtz Machine are discussed.

Editor

ELEMENTS OF NEUROPHYSIOLOGY: AN INTRODUCTION

by Spezia Stefano

In this chapter, we introduce some basilar concepts about the neurophysiology of the neurons. In details, after a brief description of the morphology of both the neurons and the synapsis, we analyse the *action potentials*: the electrical signals representing the elementary unit of information exchanged between the neurons. Moreover, we describe the characteristics of a cell membrane, both in resting state and in that excited one, with particular attention to its electrical properties.

THE NEURONS

The neurons are the cells of the nervous system highly specialized and devoted to the transfer, the storage and the processing of information. Such processing occurs through electrical signals related to differences in electrical potential related to ionic-type electrical currents (i.e. the relevant ones are associated with the sodium, potassium, calcium and chlorine ions) which cross the cell membrane of the neuron. The neurons are usually very numerous and connected between them in a very complicated way (see Figure 1.1). Typically, even in the large variety of types of neuronal cells, in a neuron, we can ever identify three morphologically and functionally distinct parts: the body cell or soma, the dendrites and the axons.

The soma has a compact structure that appears approximately spherical whose size varies from about 5 micrometres to over 10 millimetres for some types of invertebrates, and it is that part of a neuron which processes the information.

The dendrites are the extensions of the soma and have a highly branched structure that extends for very long distances. Their role is to collect the signals coming from other neurons and to transmit them to the soma.

The axons are long protuberances of the soma which can reach over a metre in the motor neurons of the higher animals. Their function is to transmit the signals emitted by the soma to the dendrites of another neuron.

If we use the terminology used in signals and systems theory, we can say that the dendrites, the axons and the soma represent the input devices, the output devices and the information processing unity, respectively.

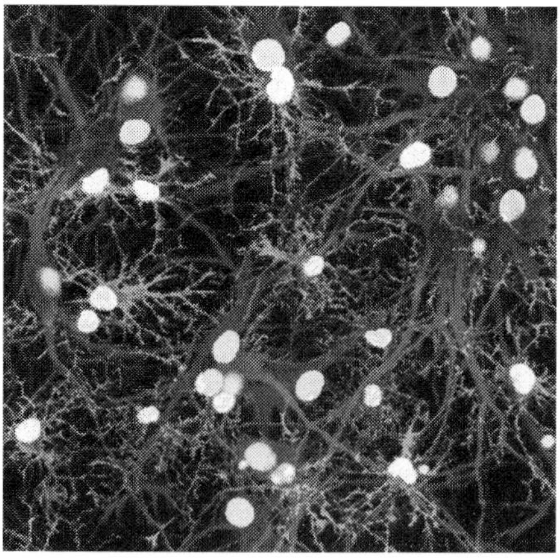

Figure 1.1: This image shows neurons (blue) and different types of glia (red and green) from the hippocampus of a rat. Published in https://www.flickr.com/photos/nichd/17104545478 by "NICHD/J. Cohen" under the terms of the Creative Commons Attribution License https://creativecommons.org/licenses/by/2.0/. No changes were made.

The morphology and the position of a particular kind of neuron in a neuronal network provide clues about its own role, e.g. the branching levels give an idea about the number of incoming signals that a neuron can receive, and how many own signals a neuron can emit to other neuronal cells (see Figure 1.2).

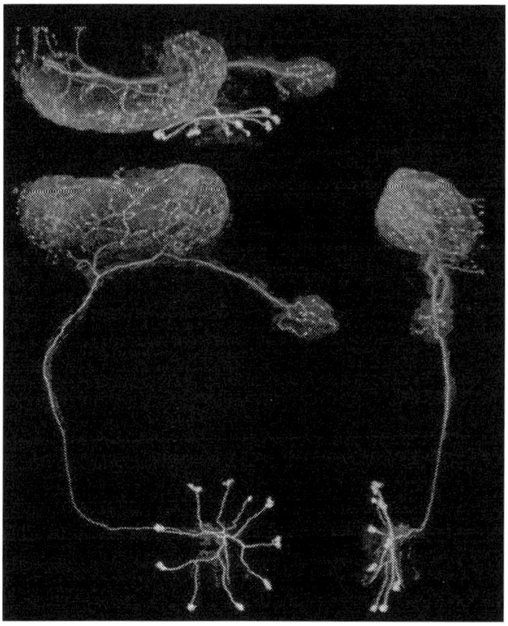

Figure 1.2: Three views of an olfactory projection neuron in the brain of an adult locust: Dorsal (top), anterior (bottom left), and lateral (bottom right). Published in https://www.flickr.com/photos/nichd/21076489782 by "NICHD/T. Miyazaki" under the terms of the Creative Commons Attribution L License https://creativecommons.org/licenses/by/2.0/. No changes were made.

THE NEURONAL SIGNALS AND THE ACTION POTENTIAL

In order to define what a neuronal signal is, we need to define the membrane potential. The latter is the difference of electrical potential measured between two electrodes, one located inside the neuronal cell and the other one placed within the surrounding extracellular liquid. When we talk about a neuronal signal, we refer to the temporal and spatial change of the membrane potential.

When a neuron is in resting state, i.e. it is not externally excited (and later we will clarify what it means), the membrane potential assumes a characteristic value called *resting potential.* Typically, this potential is approximately equal to -65 mV, namely the interior of the cell has an electrical potential lower than outside.

The action potentials are typical voltage impulses generated during neuronal dynamics, they have a stereotyped shape on time, and they are not affected by attenuation or distortion when they propagate along the axon.

In Figure 1.3 is shown the characteristic shape of an action potential, where it is possible to note that the voltage pulse has a duration of the order of 1 ms and an amplitude of about 100 - 120 mV measured between the maximum and the resting potential.

In detail, in the first phase of the pulse or *spike*, we can observe a very fast increase of the membrane potential until to reach a phase called *depolarization* where the membrane potential becomes positive, i.e. the interior of the cell achieves an electrical potential greater than to outside. In the next phase characterized by the descent, before that the neuron reaches the resting state, the pulse goes through a very slow phase called *hyperpolarization*, of typical duration of the order of 10 ms, in which the cell shows a membrane potential lower than the resting one.

The action potential, once generated in a neuronal cell and transmitted to other neurons through the axon, constitutes the elementary unit related to the neuronal signals transmission. Usually, when we refer to the signal emitted by a neuron, we mean the temporal sequence of these action potentials, also called *spike train*.

THE SYNAPSES

The synapse constitutes the junction between two neurons, that is the structure through which the information is transferred from a nervous cell to another one. In such contest, we define *presynaptic neuron* the transmitter neuron, i.e. the neuron that transmits the action potentials upstream of the synapse, and *postsynaptic neuron* the receiver neuron, i.e. the neuron that receives the spikes downstream of the synapse.

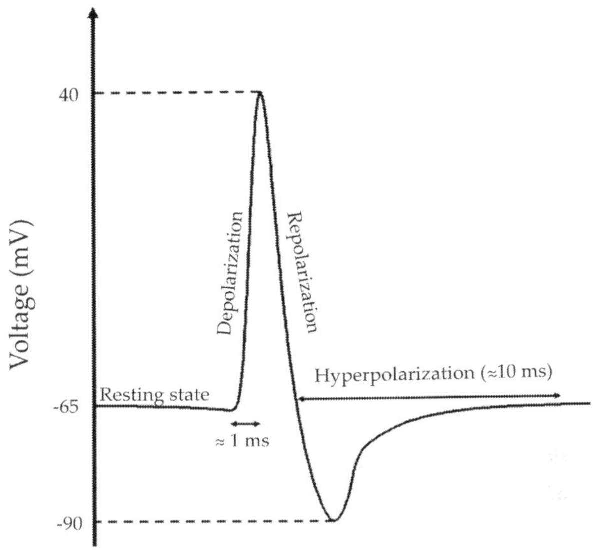

Figure 1.3: Typical shape of an action potential (or voltage pulse).

With this terminology, the synapse is therefore the region in which the presynaptic neuron interacts with that postsynaptic one. Moreover, it is possible to define the *postsynaptic potential* as the voltage response of the postsynaptic neuron resulting upon the arrival of the action potential coming from the presynaptic neuron.

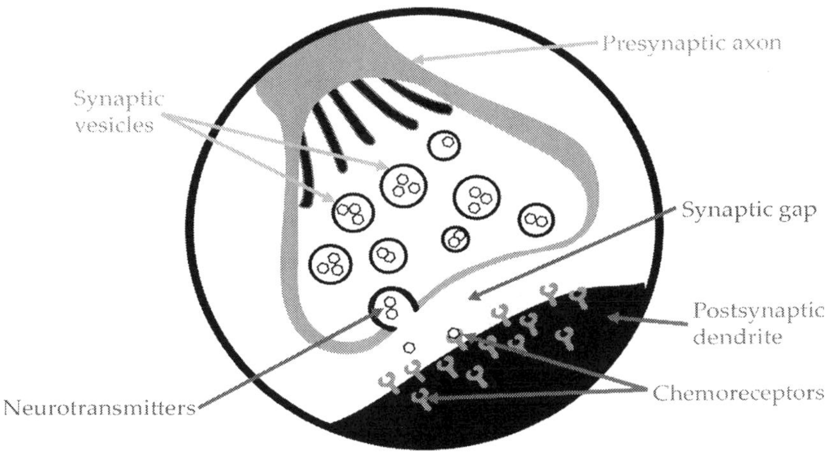

Figure 1.4: *Scheme of a chemical synapse: the presynaptic termination releases a chemical substance called neurotransmitter, in response to a depolarization.*

In literature, there are two types of synapses: the *chemical synapse* and the *electrical synapse* (differently called *gap-junction*). In Figure 1.4, it is shown a scheme of a chemical synapse, the most common type of synapse in vertebrates. The action potential generated by the presynaptic neuron, once reached the extremity of the axon, locally depolarizes the cell membrane causing the release within the synaptic gap of particular chemical substances called *neurotransmitters*. The above said the release is carried out by the *synaptic vesicles*: particular structures situated on the axon. The neurotransmitter, once it reaches the postsynaptic side, it is detected by specific molecules (*chemoreceptors*) placed on the postsynaptic membrane, whose role is to bring on the opening of channels through which an ion current can flow from the extracellular liquid to the cell. The input of these ions, in turn, leads to a change of the postsynaptic membrane potential. Therefore, in a chemical synapse, it firstly occurs the transduction on the presynaptic membrane of an electrical signal in a chemical one, and then the next transduction on the postsynaptic membrane of a chemical signal in an electrical one.

The electrical synapse, instead, realizes an electrical coupling between two neurons through highly specialized ion channels (called *gap-junctions*) which link the presynaptic membrane to the postsynaptic one, i.e. the electrical synapse directly allows a current flow between adjacent neurons.

THE CELL MEMBRANE OF A NEURON

The cell membrane of a neuron (see Figure 1.5) is made of lipid molecules and proteins. The first ones are placed in a bilayer with a thickness of 6 nm. In this lipid matrix, there are protein molecules which cross the entire thickness of the cell membrane, getting in touch with both the interior of the cell and the extracellular liquid. These particular proteins are called *protein channels*, or *membrane channels*, *aqueous channels* or *ion channels*.

The cell membrane can be traversed by the substances from outside to inside or vice versa through different mechanisms. For example, certain types of molecules, e.g. alcohols or glycerol, cross the membrane by melting in the lipid bilayer and emerging on the other side; in such case, the ease of penetration depends on the degree of solubility of liquids.

The inorganic ions of sodium, potassium, calcium and chlorine, form the ion currents on which is based the neuronal electrical activity and move through the membrane either by binding to specific molecules called *transport molecules* that carry them, either through the above-mentioned protein channels.

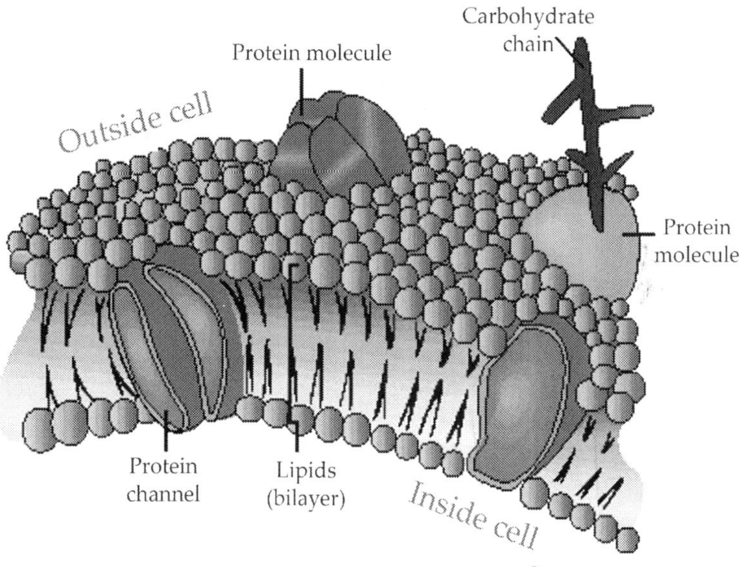

Figure 1.5: Structure of the cell membrane of a neuron. Photo credit: 140264jd *via* Foter.com / CC BY-SA.

Note, however, that the transport mechanism adopted during the generation of the action potential, is based on protein channels, inasmuch the involved flows of ions (of the order of 10^6 ions/s) are well beyond the capabilities of the transport molecules. In the structure of the protein channels (see Figure 1.5), we can recognize the following items:

- a central water-filled pore;
- a pore region which acts as selectivity filter regulating the transit of ions depending on both their size and chemical-physical characteristics;
- gates that open and close in a stochastic way, leading the channel state to oscillate between the open state and the closes state.

ION CHANNELS AND ION CURRENTS

When an ion channel is in an open state the channel is said *activated*, else when it is in a closed state, it is said *inactivated*. It is worth noting that the opening and the closure of channels are stochastic processes, and hence, the activation or the deactivation of a channel only means an increased or decreased probability of channel opening, and not an open state or closed state. There are various modes of activation, namely of the opening of a channel. In particular, if the opening of a channel can be regulated by the value of the membrane potential, the channels are said to be voltage-activated. About the ion selectivity or specificity, the channels can be distinguished in *cationic channels* and *anionic channels*, depending on their permeability to positive or negative ions. The membrane *permeability* p. related to an ionic species is an intrinsic characteristic of the membrane which measures the easiness wherewith the ions cross the membrane itself; it is defined in an empirical way as [1] [2]

$$J = -p\Delta[C],$$

where J. is the molar flux (measured in $mol/(cm^2 s)$.) and $\Delta[C]$ is the difference of ion concentration between the two sides of the membrane (measured in mol/cm^3). The permeability p is expressed as a speed and measured in cm/s, and it depends on both the type and the number of ion channels at the membrane. In particular, the cationic channel can be *specific* or *non-specific* if they are specialized or not to a particular ion species, e.g. there are specific channels for Na^+, K^+, Ca^{2+}). The anionic channels are essentially chloride channels (Cl^-): the chloride is the most permeating anion in biological solutions.

The conductance is instead a measure of the membrane capacity of transporting electric current, and it is usually measured in *Siemens*, with symbol *S*, where $1S = 1\Omega^{-1}$. Inasmuch as the electrical charge carriers are the ions, the membrane conductance does not only depend on the membrane characteristics, i.e. permeability, but also on the ion concentration inside and outside the cell expressed as the density of charge carriers. However, it is not possible to derive a general mathematical equation between the permeability and the conductance because such equation depends on the ways whereby the ions cross the channel, e.g. a simple diffusion through the water-filled pores or more complicate models which analyse the interaction between the channel and the permeating ion. In general, the current flowing through an ion channel depends on: [3] [4] [2]

- the channel conductance;
- the concentration gradient between the inside and the outside of the cell, which induces a flux from the area with the highest concentration to that at a lower concentration, in according to the Fick's law [2]

$$J_d = -D\frac{d[C]}{dx}$$

where J_d is the flux due to the diffusion measured in ion number/(cm^{2}s), D is the diffusion coefficient measured in cm^{2}s and $[C]$ is the ion concentration often expressed in ion number/cm^3;

- the difference in electrical potential applied to the membrane.

THE ORIGIN OF THE RESTING POTENTIAL

A cell contains ions of several types, i.e. Na^+, K^+, Cl^-, and other anions of big size and is surrounded by a solution of Na^+, K^+ and Cl^-. In particular, in the neuronal cells, there are other kinds of ions, as Ca^{2+} and Mg^{2+}, but their contribution to the resting potential is negligible. The cell membrane is completely impermeable to both Na^+ and other internal anions, but it is permeable to K^+, Cl^-. In such conditions, the potassium ion gets higher concentration inside the cell and tends to move outwards in according to its concentration gradient. On other hand, the inner surface of the membrane is negative than the outer one, and hence the potential gradient tends to attract K^+ ions inside the cell. When a cell is at rest, the concentration gradient and the electrical gradient are in equilibrium and have opposite directions. From these mechanisms, now we can better define the equilibrium potential of an ion species as the potential value of the membrane for which the net flow of that species is null [3].

The equilibrium potential of every ion species is related to the intracellular and extracellular concentrations through the so-called *Nernst's equation*

$$E_{ion} = kT \ln \frac{[n]_e}{[n]_i}$$

where E_{ion} is the equilibrium potential of the ion, $[n]_e$ and $[n]_e$ are respectively the extracellular and the intracellular concentrations, $k \cong 1.38 \cdot 10^{-23}\ J/K$ is the Boltzmann constant and T is the absolute temperature in Kelvin. It is possible to understand the Nernst's equation by considering the Boltzmann's statistical mechanics. In fact, for systems in thermal equilibrium, the probability $p(U' < U < U' + dU)$ that a molecule is at an energy state $U' < U < U' + dU$ is proportional to $e^{-U/kT}$ [5].

Now, let us consider ions with positive electrical charge q in a static electric field. Their energy in the point of coordinate x is $U(x) = qV(x)$, where $V(x)$ is the potential in that point of abscissa x. Hence, the probability density is proportional to $e^{-qV(x)/kT}$, and by reinterpreting the probability density as a proportional quantity to the ion density or concentration $[n(x)]$, one can write

$$\frac{p(U(x_1))}{p(U(x_2))} = \frac{[n(x_1)]}{[n(x_2)]} = e^{-\frac{q(V(x_1)-V(x_2))}{kT}}$$

and therefore at the thermodynamic equilibrium an electrostatic potential gradient $\Delta V = V(x_1) - V(x_2)$ generates an ion density gradient. Of course, this statement also applies in the other direction: a concentration gradient generates an electrostatic potential gradient. By solving the previous equation with respect to the potential gradient, one can find that at the

thermodynamic equilibrium $\Delta V = \frac{kT}{q} \ln \frac{[n]_2}{[n]_1}$, that is the Nernst's equation. The quantity kT/q is a difference of electrical potential and for a monovalent ionic species ($q = e$, where $e \cong 1.6 \cdot 10^{-19}$ C) is equal to 25 mV about at room temperature ($T \cong 300K$). For a cell model where the ratio between the concentrations of Cl^- and K^+ is equal to $1/30$, $E_K \cong E_{Cl} \cong -85$ mV. For such cell, where the unique charge carriers moving through the membrane are potassium and chloride ions characterized by equal electrical potential, the cell does not show any net flux of ions: the resting potential of such simplified model of cell coincides with the equilibrium potential of Cl^- and K^+.

THE EFFECT OF SODIUM AND ION PUMPS

The sodium ion (Na^+) has a concentration higher outside the cell than inside the cell, therefore the membrane must have a difference of electrical potential between inside and outside for opposing the entrance of Na^+ due to the concentration gradient. In fact, at $T \cong 300$ K the equilibrium potential of sodium is $E_{Na} \cong +34$ meV. In a neuronal cell, where the resting potential of the membrane is negative, both the concentration gradient and the membrane potential favour the incoming of sodium ions. However, the cell membrane is poorly permeable to Na^+ and the limited incoming of this ion species lightly depolarizes the membrane with respect to the equilibrium potential of the potassium ion, leading the latter one out of equilibrium, and consequently to flow outwards. To keep the equilibrium state, i.e. constant ion concentrations, there exist *ion transporters*, often called *ion pumps*, namely special transmembrane proteins moving ions across a cell membrane against the normal diffusion process due to their concentration gradient. In particular, such ion pumps transport Na^+ outside the cell and K^+ inside the cell to keep so as to maintain a steady state or a *dynamic equilibrium* state. The most important transport system is the sodium-potassium pump ($Na-K$ pump) which move three sodium ions outside the cell for every two potassium ions moved inside the cell: in such case, it is said that the *coupling ratio* $Na:K$ of the pump is $3/2$. In such way, the total flux is zero for each ion at equilibrium.

THE EFFECT OF THE IONIC PERMEABILITY AND OF THE TRANSPORT MECHANISMS

As we have seen, the resting potential of a membrane is mainly provided from the ratio of transmembrane concentrations of potassium ion. However, there is also depending on the permeability of the ion species. The equation that expresses such dependence is the Goldman–Hodgkin–Katz voltage equation, more commonly known as the Goldman equation, or GHK equation [6] [7]

$$V_m = \frac{kT}{e} \ln \frac{p_K [K^+]_e + p_{Na} [Na^+]_e + p_{Cl} [Cl^-]_i}{p_K [K^+]_i + p_{Na} [Na^+]_i + p_{Cl} [Cl^-]_e}$$

where V_m is the resting potential of the membrane, e indicates the charge of monovalent ions, the symbol $[I]_j$ indicates the indoor and outdoor concentration of the ion species ($I = K^+$, Na^+, Cl^- and $j = i, e$), and p_K, p_{Na}, p_{Cl} represent the ion permeability of each ion. This equation is based on the assumption that at a constant potential, the membrane charge does not change if the number of ions which cross the membrane is enough low. Consequently, all the ion currents through the membrane must give a zero-sum, else a charge accumulation or a charge loss different than zero brings the membrane potential to a continuous drift.

The GHK equation reminds the Nernst's equation, but it considers each ion species weighted with their permeability. Accordingly, the membrane potential tends to the equilibrium potential value of an ion species as its permeability is greater than those of other species. It is possible to show that there is not a significant dependence on the chloride ion of the resting potential, and therefore its contribute is often neglected; in such case, the GHK equation is rewritten as [3]

$$V_m = \frac{kT}{e} \ln \frac{[K^+]_e + b[Na^+]_e}{[K^+]_i + b[Na^+]_i}$$

where $b = p_{Na} / p_K$. A more accurate description is given by the *steady state equation* [8] which also considers the effects of the ion pump $Na - K$

$$V_m = \frac{kT}{e} \ln \frac{r[K^+]_e + b[Na^+]_e}{r[K^+]_i + b[Na^+]_i}$$

where r is the coupling ratio of the transport system ($r = 3/2$ for the ion pump $Na - K$). One can easily understand how the pump effects affect the membrane potential. In fact, the different flux of ions Na^+ and K^+ (three sodium ions moving outwards vs two potassium ions moving inwards) caused by the pump, generates a net ion current outwards. This flux of positive charges tends to hyperpolarise the membrane to a potential lightly more negative and closer to the equilibrium potential of K^+. At the aim to better comprehend the contribution of the ion pumps, one can compare the resting potential predicted by the GHK equation with that predicted by the steady state equation for the axon of squid in sea-water at room temperature of $T \cong 300$ K: by using the GHK equation one finds $V_m \cong -67$ mV, while by using the steady state equation one finds $V_m \cong -73$ mV. For the numerical evaluations, the following values have been used: $\left[K^+\right]_i = 400$ mM, $\left[K^+\right]_e = 10$ mM, $\left[Na^+\right]_i = 50$ mM, $\left[Na^+\right]_e = 460$ mM, $b = 0.04$ and $r = 3/2$ [7].

THE PASSIVE ELECTRICAL PROPERTIES OF THE MEMBRANE

From the electrical point of view, the simplest schematisation of a portion of cell membrane at rest makes use three circuital elements: a resistor of electrical resistance R, a capacitor of electrical capacity C, and a generator of constant voltage V equal to the resting potential (see Figure 1.6).

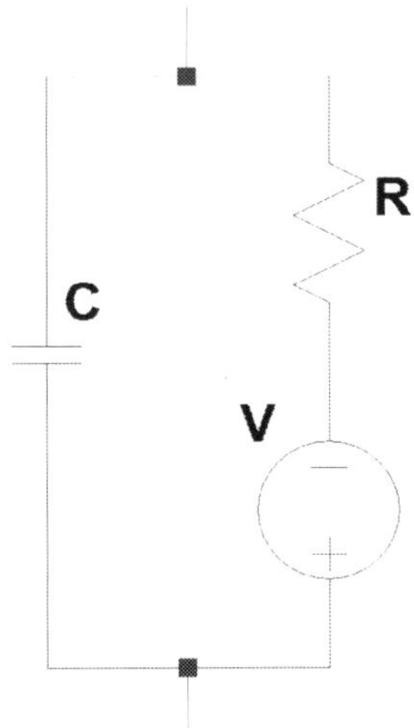

Figure 1.6: Electrical schematization of a portion of cell membrane of a neuron. V indicates the rest potential [1].

The electrical resistance R explains the existence of ion channels which realize the contact between inside and outside the cell. The membrane resistance is usually described as a specific membrane resistance, R_m, defined as a resistance for a unity of surface (measured in $\Omega \cdot cm^2$). R is hence obtained by dividing R_m for the area of the considered membrane. R_m is mainly determined by the permeability of potassium and chloride at rest. Typical values of R_m range from about $10^3 \Omega \cdot cm^2$ for membranes characterized by many ion channels to about $5 \cdot 10^4 \Omega \cdot cm^2$ for those characterized by few of them. In addition, allowing the passage of ion currents, the membrane accumulates electrical charges on the inner and outer surfaces, and it is precisely this separation that determines the membrane potential. The realization of this charge separation confers to the membrane the properties of a capacitor of specific membrane capacity, C_m, defined as a capacity for a unity of surface (measured in F / cm^2). C is hence obtained by multiplying C_m for the area of the considered membrane. C_m is typical the order of $1 \mu F/cm^2$. Through the equation $Q_m = C_m V$, where Q_m is the electrical charge for a unity of surface, and assuming $V \cong -65$ mV, one has that $Q_m \cong 6.5 \cdot 10^{-8} C / cm^2$, equivalent to about $4 \cdot 10^{11}$ monovalent ions per cm^2. It is important to stress that such schematisation only describes the passive behaviour of the membrane and does not take into account eventual nonlinear or active components, e.g. voltage-dependent conductance.

REFERENCES

1. C. Koch, Biophysics of computation, New York: Oxford University Press, 1999.
2. D. Johnston and S. Miao-Sin Wu, Foundations of Cellular Neurophysiology, Cambridge, Massachusetts: Massachusetts Institute of Technology, 1995.
3. J. G. Nicholss, R. A. Martin and B. G. Wallace, Dai neuroni al cervello, Bologna: Zanichelli, 1997.
4. E. R. Kandel, J. H. Schwartz and T. M. Jessell, Principles of neural science, McGraw-Hill, 2000.
5. K. Huang, Statistical mechanics, New York: John Wiley & Sons, 1987.
6. D. E. Goldman, "Potential, impedance and rectification in membranes," J. Gen. Physiol., vol. 342, pp. 37-60, 1943.

7. A. L. Hodgkin and B. Katz, "The effect of sodium ions on the electrical activity of the giant axon of the squid," J. Physiol., vol. 108, pp. 37-77, 1949.

8. L. J. Mullins and K. Noda, "The influence of sodium-free solutions on membrane potential of frog muscles fibers," J. Gen. Physiol., vol. 47, pp. 117-132, 1963.

SECTION I:

LEARNING THE HODGKIN-HUXLEY MODEL AND ITS PARAMETRIZATIONS

1 HODGKIN–HUXLEY REVISITED: REPARAMETRIZATION AND IDENTIFIABILITY ANALYSIS OF THE CLASSIC ACTION POTENTIAL MODEL WITH APPROXIMATE BAYESIAN METHODS

Aidan C. Daly[1], David J. Gavaghan[1], Chris Holmes[2] and Jonathan Cooper[1]

[1] Department of Computer Science
[2] Department of Statistics, University of Oxford, Oxford, UK

ABSTRACT

As cardiac cell models become increasingly complex, a correspondingly complex 'genealogy' of inherited parameter values has also emerged. The result has been the loss of a direct link between model parameters and experimental data, limiting both reproducibility and the ability to re-fit to new data. We examine the ability of Approximate Bayesian computation (ABC) to infer parameter distributions in the seminal action potential model of Hodgkin and Huxley, for which an immediate and documented connection to experimental results exists. The ability of ABC to produce tight posteriors around the reported values for the gating rates of sodium and potassium ion channels validates the precision of this early work, while the highly variable posteriors around certain voltage dependency parameters suggests that voltage clamp experiments alone are insufficient to constrain the full model. Despite this, Hodgkin and Huxley's estimates are shown to be competitive with those produced by ABC, and the variable behaviour of posterior parametrized models under complex voltage protocols suggests that with additional data the model could be fully constrained. This work will provide the starting point for a full identifiability analysis of commonly used cardiac models, as well as a template for informative, data-driven parametrization of newly proposed models.

INTRODUCTION

Cardiac cells have been one of the most popular targets of mathematical biological modelling since the field's inception in the late 1940s. This is partly owing to the difficulty of performing clinical studies on human hearts, as well as the high species dependence of electropysiological properties such as action potential (AP) shape, duration (APD) and restitution properties that limit the applicability of studies in model systems [**1**]. In the face of this paucity of data, simulations from these models are used as a primary level of inference for the impact of mutations or drugs on cellular- or organ-level properties [**2–4**], and can also inform the design of future experiments. Analysis of the ways in which simulations fail to reconstruct experimental recordings often leads to a better understanding of the data needed to correct the model in the next iteration [**5**].

One of the most influential and enduring cellular models is the Hodgkin–Huxley AP model [**6**]. Though the model was constructed for the squid giant axon, the experimental set-up and formulation of model components set a standard that was adopted by many cardiac models, and persists to this day. From their original voltage clamp experimental data [**7**], the authors determined that the ionic conductances were best explained by a system of nonlinear ordinary differential equations

(ODEs) which, unbeknown to them, can describe by direct biological analogy the opening and closing dynamics of membrane ion channels. These equations ((**2.6**)–(**2.16**) in §**2**) serve as the foundation for the so-called 'Hodgkin–Huxley style' formulation for ion channel modelling, where all channel transitions are modelled by ODEs. To this day, many cardiac models incorporate this formulation for modelling ion channels when performing large-scale simulations owing to its simplicity when compared to a more general Markov formulation [**8**].

At the time, parametrizing the Hodgkin–Huxley model required meticulously matching hand-drawn curves to experimental data. Despite the advances in model fitting techniques from these pen and paper methods, however, a commensurate increase in cardiac model complexity has made it difficult to obtain enough independent data to fully employ such methods. Advances in recording techniques such as the patch clamp experiment [**9**] have driven the inclusion of more ion channels and intracellular dynamics with the intention of forming a more complete representation of the true cellular environment, increasing the amount of experimentation required to fully observe a system (see [**5**,**10**] for a more complete discussion of the history of cardiac modelling). The modelling of ion channel kinetics in particular has posed a problem, as most cardiac data are macroscopic, and isolation/expression systems to study single channels can alter their native behaviour [**4**]. In the face of these difficulties in obtaining data for modelling experiments, many modellers borrow or adapt parameter values, or even entire model subunits, from previous models, which were often constructed for different experimental systems [**11**,**12**].

While much effort is taken to adjust model components for differences in species and/or experimental conditions, as well as to maintain certain macroscopic properties such as AP shape, comparative analyses have shown discrepancies in behaviour between sequentially fit models purporting to represent identical systems [**13**–**15**]. This indicates a fragility in these models that limits our confidence in their predictive power outside the range of their validated behaviour. While some of these discrepancies may be attributed to biological variability, poor documentation or justification of parameter inheritance has made it increasingly difficult to link parametrized components to original data that might capture or explain this variation [**14**,**16**], leading to a weakening link to the mechanistic underpinning of some of the more complex models being proposed today.

This devolving link between experimental data and parametrized models makes the application of modelling techniques which provide posterior distributions over parameters (rather than single optimal values) very difficult. Paradoxically, therefore, one of the most mature areas of systems biology has perhaps benefited least from the application of the advanced methods of parameter fitting and model selection that have been applied to more recently developed areas [**17**,**18**]. In light of this problem, the original Hodgkin–Huxley model becomes an ideal case study for illustrating and assessing the utility of the application of modern inference techniques to cardiac AP models based on the Hodgkin–Huxley formulation. All data used by Hodgkin and Huxley were originally and consistently produced by the authors, and therefore provide the ideal test-bed for exploring the relationship between parameter values and experimental data for these types of models.

Application of these new techniques also allows us to explore the issue of model identifiability for Hodgkin–Huxley type models, that is, the degree to which the 'optimal' parameters for the model are unique. Published models are underpinned by an assumption that the given parametrization is unique for the data that were used to fit them. This assumption is inherent in the use of classical fitting methods such as least-squares regression, which output only a single optimal point in parameter space. We know, however, that such a unique optimum is unlikely to exist in biological systems, as both inter- and intracellular variations may be reflected by variations in the biophysical parameters of the model. Therefore, by examining posterior distributions over model parameters, we can quantify and characterize model uncertainty, which can either inform us as to the ability of our data to constrain the model or the degree of biological variability in the system, as we outline below.

If a cellular model were to have a non-unique optimal parametrization, with either several or infinitely many equally likely possibilities, the model may be deemed 'unidentifiable'. Unidentifiability can be divided into two types: structural unidentifiability, where the model is overly complex to describe the system (this if often also called over-parametrization), and practical unidentifiability, where there are not enough data to fully constrain the model [**19**]. Distinguishing between the two types of unidentifiability can be important for experimental design. Pinpointing a structural unidentifiability, which is characterized by a functional relationship between several parameters (and thus infinitely many equally optimal parametrizations), may be a cause for concern, and prompt changes in model formulation before attempting further experimentation. However, if there is a high degree of prior faith in the model formulation, this may instead suggest a biologically important redundancy in the system characterized by the functional relationship between biophysical parameters. Practical unidentifiability, on the other hand, may inform how additional experimental data might be collected in an effort to further constrain the model [**19**,**20**], or may simply be a representation of inherent variability of the biophysical parameters in the experimental system.

While there is no universally accepted automated means to assess identifiability [**21**], many different methods, such as parameter sensitivity analysis [**2**,**22**,**23**] and analysis of curvature of an objective function [**24**,**25**], have been applied to cardiac cell models, and there now exist documented concerns for model identifiability in both widely used Hodgkin–Huxley and Markov style models under common experimental protocols [**21**,**24**,**26**].

When experimental data are available to us, however, an appropriately chosen Bayesian method for parameter fitting can also be used to assess the identifiability of a model. Bayesian inference involves calculating a full posterior probability distribution around the model parameters, the shape and width of which can inform the modeller as to the degree of variation about the optimal parameter value. A wide, flat posterior on a parameter, for example, indicates a large number of equally optimal values, which suggests that the parameter may be unidentifiable [**27**]. We expect variation arising from natural biological variation to manifest itself as a well-formed distribution, the statistical properties of which can be used to draw conclusion as to the nature or source of such variation, while variation arising from insufficient data would be more erratic, related to noise in the experimental data or random choices made by the algorithm when the observed data does not provide much information about a portion of the model dynamics. Because only further experimentation can firmly distinguish the two (inherent biological variation should remain unaffected by the inclusion of additional data), the automated design of 'optimal' experiments to attempt to reduce uncertainty is an area of current interest [**28**].

Because fully Bayesian inference involves potentially intractable integrals, much of the research in this area has been into methods to speed up or reasonably approximate calculation of the posterior [**29**,**17**]. Approximate Bayesian computation (ABC), first employed for parameter selection by Tavare et al. [**30**], gives a means to generate a population of solutions by repeatedly sampling from a prior distribution over model parameters and accepting draws based on an objective function evaluation. Bayesian inference also gives us the ability to include data from multiple sources in a principled manner, often weighted by the degree of noise present or prior knowledge as to its relative significance. This allows us to include data from multiple repetitions of an experiment at once, allowing both mean behaviour and variability of the experiment to inform posterior parameter distributions.

We aim to use ABC to investigate the identifiability of the Hodgkin and Huxley model by re-fitting both the basic (equations (**2.6**)–(**2.10**)) and expanded (equations (**2.11**)–(**2.16**)) form to the authors' original published data. We will assess the identifiability of the model by examining the width of the resulting posterior estimates for all parameters in each model and attempt to classify unidentifiabilities as structural or practical by examining model output fluctuations and parameter correlations across the posterior. Finally, we will examine the potential of using more complex voltage protocols proposed by the original authors to further constrain the voltage-dependent model. This analysis allows not only the assessment of the ability of the original manual parameter fitting methods to match modern automated techniques, but also serves as an example for the implementation of informative parameter estimation techniques in cardiac modelling.

We have released code implementing these techniques so that modellers can apply these Bayesian parameter fitting techniques to their own systems of study, providing an unambiguous link between components of published models and experimental data. This code was built on the functional curation extension [**31**,**32**] to the Chaste cardiac simulation library [**33**], which allows the specification of stand-alone simulation protocols that can be applied to a range of in silico models, in order to ease the mapping between simulated and real data. The adoption of such standards would allow modellers to quantify their certainty in a published model given the experimental data employed, and thus allow for the informed adoption of a model or its components by their peers.

In §**2**, we present both forms of the Hodgkin–Huxley model in full detail. In §**3**, we discuss the details of our implementation of ABC, as well as our use of functional curation. In §**4**, we present the results of our ABC re-fitting experiments on the two forms of the Hodgkin–Huxley model, and analyse the properties of the resulting posteriors. In §**5**, we discuss the implications of the identifiability analysis for potential improvements to the design of experiments used to fit the model, and finally conclude in §**6**.

THE HODGKIN–HUXLEY MODEL

Hodgkin and Huxley treated the squid axon as an electrical circuit, with current across the membrane being carried by a capacitor or by one of three ionic currents: I_K, the current carried by potassium ions, I_{Na}, the current carried by sodium ions, and I_l, a catch-all leakage current. Thus, the fundamental equations for simulating membrane potential changes were as follows:

$$I = C_M \frac{dV}{dt} + I_i, \quad (2.1)$$

$$I_i = I_{Na} + I_K + I_l, \quad (2.2)$$

$$I_{Na} = g_{Na}(V - V_{Na}), \quad (2.3)$$

$$I_K = g_K(V - V_K) \quad (2.4)$$

$$I_l = g_l(V - V_l), \quad (2.5)$$

where dV /dt is the rate of change in membrane potential, C_M is the membrane capacitance, and I_i is the sum of the three ionic currents. For each ionic current x, V x represents the reversal potential (the membrane potential at which there is no net flow of that ion) and g_x is the membrane conductance per unit area for that ion.

The ionic conductances, with the exception of the leakage current g_l which was assumed constant, were theorized to be explained by the following ODEs:

where $\bar{g}_K$ and $\bar{g}_{Na}$ describe the maximum ionic conductances. The dimensionless variables m, n and h describe the open probability of ion channels (and thus the expected fraction of the full ionic conductance across the membrane at a given time and voltage), and the α and β terms are rate constants.

conductance type	equations
potassium	$g_K = \bar{g}_K n^4,$ (2.6) $\frac{dn}{dt} = \alpha_n(1-n) - \beta_n n,$ (2.7)
sodium	$g_{Na} = \bar{g}_{Na} m^3 h,$ (2.8) $\frac{dm}{dt} = \alpha_m(1-m) - \beta_m m,$ (2.9) $\frac{dh}{dt} = \alpha_h(1-h) - \beta_h h,$ (2.10)

free parameters: $\alpha_n, \beta_n, \alpha_m, \beta_m, \alpha_h, \beta_h$

After choosing rate parameters α and β for a variety of clamped voltages, Hodgkin and Huxley empirically proposed functions that could explain the voltage dependency of these parameters, yielding the following equations:

conductance type	equations
potassium	$\hat{\alpha}_n(V) = \frac{0.01(V+10)}{\exp\left(\frac{V+10}{10}\right)-1},$ (2.11) $\hat{\beta}_n(V) = 0.125\exp(V/80),$ (2.12)
sodium	$\hat{\alpha}_m(V) = \frac{0.1(V+25)}{\exp\left(\frac{V+25}{10}\right)-1},$ (2.13) $\hat{\beta}_m(V) = 4\exp(V/18),$ (2.14) $\hat{\alpha}_h(V) = 0.07\exp(V/20),$ (2.15) $\hat{\beta}_h(V) = \frac{1}{\exp\left(\frac{V+30}{10}\right)+1}.$ (2.16)

METHODS

All Hodgkin–Huxley conductance data were obtained from figs **3** and **6** digitized from the original publication using Plot Digitizer (http://plotdigitizer.sourceforge.net). Plot L of the potassium conductance data (**figure 3**, depolarization of −6 mV) was eliminated owing to the lack of a reliable scale on the y-axis.

Model Simulations

All simulations for ABC were carried out using the current development version of the Python implementation of functional curation [**33**,**31**] (https://chaste.cs.ox.ac.uk/trac/browser/projects/FunctionalCuration).

Hodgkin–Huxley CellML Model

A CellML [**34**] model file for Hodgkin–Huxley was annotated with metadata tags to allow the adjustment of the α and β parameters (equations (**2.7**), (**2.9**) and (**2.10**)) by the fitting algorithm. Similarly, the constants in equations (**2.11**)–(**2.16**) were replaced with free, externally adjustable, variables as detailed in equations (**3.1**)–(**3.6**), giving five free parameters for the voltage-dependent potassium conductance g_K and nine free parameters for the voltage-dependent sodium conductance g_{Na}.

Voltage Clamp Protocol

A functional curation protocol, which details the external stimuli applied to a corresponding in silicomodel (parsed from a CellML file), was written to replicate the voltage clamp experiments in figs 3 and 6 of the original publication. Initial conditions for the ODEs and maximum conductance values were set as reported in **tables 1** and **2** of the same [**6**]. Each voltage clamp experiment was carried out by redefining the model's membrane voltage to a fixed value from within the protocol, then simulating the model over a 12 ms time course from the initial state, recording the two ionic conductance values every 0.1 ms.

Table 1. Specification of prior and kernel distributions employed by ABC for each of the parameters in the simplified (equations (**2.6**)–(**2.10**)) and expanded (equations (**3.1**)–(**3.6**)) Hodgkin–Huxley models.

variable(s)	prior distribution	kernel distribution
α_n, β_n	uniform(0, 1)	$\mathcal{N}(0, 0.1)$
$\alpha_m, \beta_m, \alpha_h, \beta_h$	uniform(0, 10)	$\mathcal{N}(0, 1)$
$k_{\alpha_n 1}, k_{\beta_n 1}, k_{\alpha_m 1}, k_{\alpha_h 1}$	uniform(0, 1)	$\mathcal{N}(0, 0.1)$
$k_{\beta_m 1}$	uniform(0, 10)	$\mathcal{N}(0, 1)$
$k_{\alpha_n 3}, k_{\beta_n 2}, k_{\alpha_m 3}, k_{\beta_m 2}, k_{\alpha_h 2}, k_{\beta_h 2}$	uniform(1, 100)	$\mathcal{N}(10)$
$k_{\alpha_n 2}, k_{\alpha_m 2}, k_{\beta_h 1}$	uniform(0, 100)	$\mathcal{N}(10)$

Table 2. Summary statistics of ABC posterior estimates for voltage dependency parameters of potassium conductance gating rates. (Parameter names in the left-most column refer to the corresponding labels in equations (**3.1**)–(3.2). 'Mean' and 'variance' columns indicate the summary statistics used to describe the final posterior estimate of ABC, while '5th, 95th percentiles' denote the points at which the posterior CDF records 5% and 95% density, respectively (analogous to 90% CI). The 'RMSE' row contains the value of the distance metric (equation (**3.2**)) attained by the reported parameters, as well as the minimum and maximum (bounded from above by the final ABC error constraint ϵT) values attained by particles in the final posterior estimate.)

		ABC posterior estimates		
parameter	reported value	mean	variance	5th, 95th percentiles
$k_{\alpha_n 1}$	0.01	0.0128	4.29×10^{-6}	0.0103, 0.0177
$k_{\alpha_n 2}$	10	44.8	162	27.1, 62.1
$k_{\alpha_n 3}$	10	30.7	48.2	18.7, 37.4
$k_{\beta_n 1}$	0.125	0.177	3.14×10^{-3}	0.0986, 0.253
$k_{\beta_n 2}$	80	65.1	218	48.9, 95.2
RMSE	0.642	min 0.559, max ($\leq\epsilon_T$) 0.794		

When fitting the limited six-parameter form of the model (equations (**2.7**), (**2.9**) and (**2.10**)), the rate parameters α_n, β_n, α_m, β_m, α_h and β_h are defined by the protocol to be constants, which are set by the fitting algorithm. When fitting the full 14-parameter voltage-dependent model, these six rate parameters are in turn parametrized according to equations (**3.1**)–(**3.6**), with the 14 free parameters set by the fitting algorithm:

conductance type	equations
potassium	$$\alpha_n(V) = \frac{k_{\alpha_n 1}(V + k_{\alpha_n 2})}{\exp\left(\frac{V + k_{\alpha_n 2}}{k_{\alpha_n 3}}\right) - 1}, \quad (3.1)$$ $$\beta_n(V) = k_{\beta_n 1}\exp(V/k_{\beta_n 2}), \quad (3.2)$$
sodium	$$\alpha_m(V) = \frac{k_{\alpha_m 1}(V + k_{\alpha_m 2})}{\exp\left(\frac{V + k_{\alpha_m 2}}{k_{\alpha_m 3}}\right) - 1}, \quad (3.3)$$ $$\beta_m(V) = k_{\beta_m 1}\exp(V/k_{\beta_m 2}), \quad (3.4)$$ $$\alpha_h(V) = k_{\alpha_h 1}\exp(V/k_{\alpha_h 2}), \quad (3.5)$$ $$\beta_h(V) = \frac{1}{\exp\left(\frac{V + k_{\beta_h 1}}{k_{\beta_h 2}}\right) + 1}. \quad (3.6)$$

free parameters: $k_{\alpha_n 1}, k_{\alpha_n 2}, k_{\alpha_n 3}, k_{\beta_n 1}, k_{\beta_n 2},$
$k_{\alpha_m 1}, k_{\alpha_m 2}, k_{\alpha_m 3}, k_{\beta_m 1}, k_{\beta_m 2},$
$k_{\alpha_h 1}, k_{\alpha_h 2}, k_{\beta_h 1}, k_{\beta_h 2}$

Other Functional Curation Protocols

Functional curation protocols were written for the four more complex voltage protocols described by Hodgkin and Huxley in figs 20–23 of their original publication [**6**]. Each of these protocols accepted the 14 voltage dependency parameters from equations (**3.1**)–(**3.6**) as input and set the corresponding model variables to provide input values at the start of the simulation. Each protocol subjected the model to an initial 1000 ms time course simulation to reach steady state for the new parametrization, after which all state variables n, m and h were shown to have reached constant values. Graphical representations of all protocols can be found in **figure 1**.

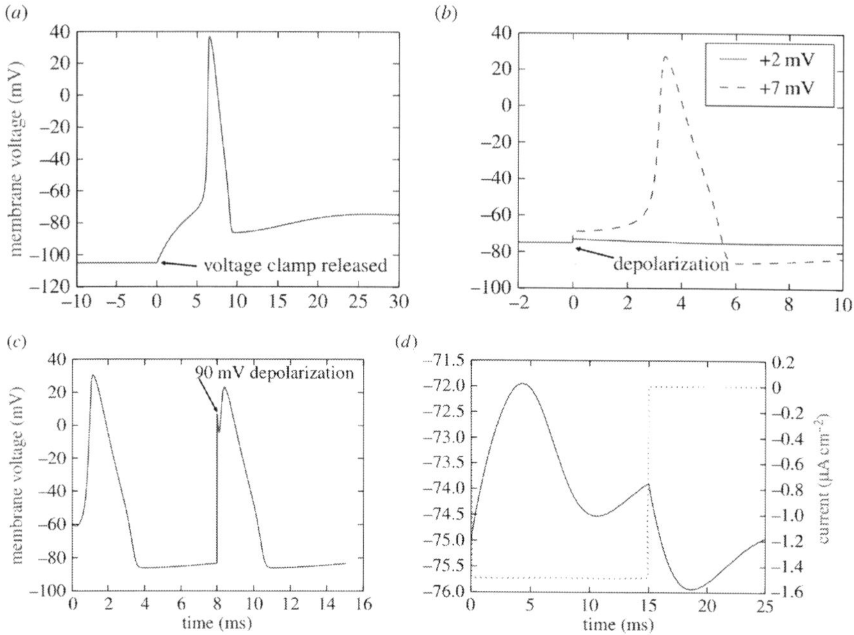

Figure 1. Stimulus-response representation of the four advanced voltage protocols proposed by Hodgkin and Huxley. Membrane voltage, the output quantity, is represented in blue, and in all figures is calculated with equations (**2.11**)–(**2.16**). In the anode break (a), threshold excitation (b) and positive phase depolarization (c) experiments, membrane voltage is manipulated directly as described in §**iii**. The graph representing the oscillation induction experiment (d) shows the induced stimulus current (dotted red) and the resulting fluctuation in membrane potential. In all graphs, the models have reached steady-state by time t=0 when stimulus is applied.

Anode break. After reaching steady state, the model was subjected to a 200 ms voltage clamp to 30 mV below resting potential. This clamp was then released and the membrane voltage recorded every 0.1 ms for the duration of a 30 ms time course simulation (**figure 1**a). The reversible voltage clamp was attained by redefining V in the model interface of the protocol to be constant between 1000 and 1200 ms and to follow the original model definition at all other time intervals.

Threshold excitation. After reaching steady state, the model was subjected to a membrane depolarization of either −10 mV, 2 mV, 5 mV 6 mV, or 7 mV, with only the latter being sufficient to trigger an AP under reported values for gating rate voltage dependency parameters (**figure**

1b). The membrane voltage was then recorded every 0.01 ms for the duration of a 10 ms time course simulation.

Positive phase depolarization. After reaching steady state, the model was subjected to a 15 mV membrane depolarization followed by a 5 ms, 6 ms or 8 ms time course simulation. After this, the model was subjected to an additional 90 mV depolarization and the remainder of a 15 ms total time course simulation (**figure 1**c). Membrane voltage was recorded every 0.01 ms for the total duration of the 15 ms time course following the initial depolarization.

Oscillation induction. After reaching steady state, the membrane current of the model was clamped to $-1.49\,\mathrm{mA\,cm^{-2}}$ for the duration of a 15 ms time course simulation. The clamp was then released and the model subjected to an additional 10 ms time course simulation, with membrane voltage being recorded every 0.1 ms for the entirety of the 25 ms time course (**figure 1**d). The reversible current clamp was attained by redefining the membrane stimulus current to be constant between 1000 and 1015 ms and to follow the original model definition at all other time intervals.

Approximate Bayesian Computation

Approximate Bayesian Computation General Settings and Adaptive Error Shrinking

The simplest form of ABC is known as the rejection sampler. In this scheme, parameters are continually sampled from a specified prior distribution and used to simulate model output. Parameter sets that generate simulated output close to the experimental data are accepted and are added as 'particles' of a population of solutions that estimate the true posterior. Thus, for the rejection sampler, all that is required is the specification of prior parameter functions, a distance function between simulated and experimental output, and an acceptance tolerance for this function.

When the prior and posterior distributions of parameters differ greatly, however, this method is impractical, as very few samples from the prior are expected to be accepted. More sophisticated variants of ABC create a series of posterior estimates, each one drawing samples from the one before it (rather than the prior). The error tolerance is initially relaxed, leading to a high number of acceptances, but is gradually tightened between rounds of sampling. Such a scheme will smooth the difference between the prior and the posterior, sequentially narrowing the range of accepted parameter values. We employed such a variation of ABC described by Toni et al. as 'sequential Monte Carlo' (ABC-SMC), which generates a series of posterior estimates of fixed size as follows:

- (i) draw N parameter vectors from the prior distribution $\pi(\boldsymbol{\theta})$ to form the initial posterior estimate $\boldsymbol{\theta}_0$. Each component particle $\boldsymbol{\theta}_0^{(i)}$ $(i \in [1, N])$ of this initial estimate will be assigned an initial, uniform weight of $w_0^{(i)} = 1/N$;
- (ii) for each subsequent posterior estimate $\boldsymbol{\theta}_t$ ($t \in [1,T]$), update each particle $\boldsymbol{\theta}_t^{(i)}$ $(i \in [1, N])$ in the following manner:
  - —draw a particle from the previous posterior estimate $\boldsymbol{\theta}_{t-1}$ and slightly perturb the parameter values according to a (stochastic) function K(θ) to obtain $\boldsymbol{\theta}^*$. Repeat this drawing and perturbation until $\boldsymbol{\theta}^*$ is legal (i.e. has non-zero likelihood under the original prior $\pi(\boldsymbol{\theta})$);
  - — simulate model output **y*** under parameters θ*. If the distance function between the simulated data and the original data $D(y*,y_{true})D(y*,y_{true})$ is less than $\epsilon_t < \epsilon_{t-1}$, ACCEPT $\theta_t^{(i)} = \theta^*$. Otherwise, REJECT and repeat from the previous step;
  - — set the weight $w_t^{(i)}$ of particle $\theta_t^{(i)}$ to $P(\theta_t^{(i)} | \theta_{t-1})$, the probability of obtaining the particle as a perturbed draw from the previous estimate. In order to calculate this probability, the kernel function K(θ) must either be reversible (P(K(θ)=θ*)=P(K(θ*)=θ)) or the so-called 'backward kernel' K^{-1} ($P(K(\theta)=\theta^*)=P(K^{-1}(\theta^*)=\theta)$) must be provided; and
- (iii) when all N particles are updated, normalize the weights $\mathbf{w}_t$ and begin the next iteration ($t \leftarrow t+1$).

The choice of the so-called kernel function K(θ) is important, as the perturbations it induces to the draws promotes exploration around previously accepted parameter estimates with the intention of generating better ones. Too much deviation from the original values, however, may lead to erratic behaviour in the draws and lower the acceptance rate.

Thus, a kernel function must balance between locality and exploration, and must take into account the scale of the model parameters in order to generate valid perturbations. The variance of the final posterior estimate can be used to quantify the identifiability of each parameter, or of the model as a whole.

The main drawback of ABC-SMC is the need to pick an appropriate 'cooling schedule' $[\epsilon_0,\ldots,\epsilon_T]$. If the reduction of error demanded between rounds is too great, the algorithm will behave like the rejection sampler. If it is too small, the algorithm may take an extremely long time to run. We propose a novel variant of ABC-SMC that adaptively sets ϵt at each round:

(i) initialize $\epsilon_t=0.5\epsilon_{t-1}$,

(ii) attempt to sample $\boldsymbol{\theta}_t$ from $\boldsymbol{\theta}_{t-1}$ with error bound ϵt,

(iii) if (ii) does not succeed within a given number of iterations, define $\Delta\epsilon=\epsilon_t-\epsilon_{t-1}$,

(iv) terminate if $\Delta\epsilon<\tau$. Otherwise, repeat (ii) with $\epsilon_t\leftarrow\epsilon_{t-1}+0.5\Delta\epsilon$.

A similar strategy for automated calculation of the cooling schedule is now implemented in the ABC-SysBio Python package [**35**]. While the SysBio implementation allows selection of an αparameter determining the quantile of the previous population to be used for the next error cut-off (arbitrarily initially set to 0.5 in our implementation), the implementation does not adaptively adjust α if the population is not filled in a certain number of iterations, unlike in our approach. Our adaptive error shrinking is thus relatively parameter-free, though a carefully chosen α (or heuristically determined cooling schedule) would be expected to show faster performance for a specific problem.

Our ABC algorithm maintained a posterior population of 100 particles and performed a maximum of 10 000 draws from the previous estimate before reattempting with a higher error threshold. A 'no improvement' threshold was set to cause the algorithm to terminate if successive rounds did not decrease the maximum error by more than 0.003. This cut-off was chosen based on the magnitude of the minimum RMSE attained by a simulated voltage clamp trace under reported parameters (equations (**2.11**)–(**2.16**)).

Approximate Bayesian computation settings for fitting of Hodgkin–Huxley parameters

Prior distributions over all parameters were assumed to be independently uniform with width roughly an order of magnitude greater than the reported value. A random walk kernel distributed as a zero-centred normal with variance roughly 10% of the width of the associated prior was applied to each draw. Exact specifications of prior and kernel distributions are detailed in **table 1**.

For each voltage clamp experiment, data were digitized from the original Hodgkin and Huxley publication in the form of a pair of vectors: time points, ***t***, and the ionic conductance (either sodium or potassium) at each time point, ***g***(t). When fitting the limited form of the Hodgkin–Huxley model (equations (**2.7**), (**2.9**) and (**2.10**)), the experimental data from each voltage clamp were used to fit the six parameters at the experimental depolarization. Thus, the distance function employed by ABC when fitting α and β was the squared distance between the experimental conductance and the simulated conductance $\hat{g(t)}$ at all reported time points t∈[0,T] of a given voltage clamp, which is represented visually in **figure 2**:

$$\mathcal{D} = \sqrt{\frac{1}{T}\sum_{t=0}^{T}(g(t) - \hat{g}(t))^2}. \tag{3.7}$$

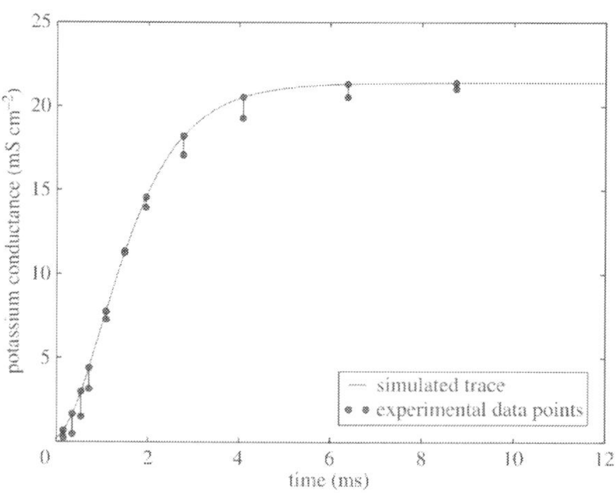

Figure 2. Representation of distance calculation for potassium conductance between experimental data points and data simulated with functional curation for

a single voltage clamp experiment (§3.1). Experimental data points are shown in red, while the simulated trace is shown in solid blue, with points corresponding to the time of experimental recordings marked with circles. Distances between these points and their experimental counterparts (shown in black lines) are used by equation (**3.1**) to calculate the objective function for ABC.

When fitting the full, voltage-dependent Hodgkin–Huxley model (equations (**3.1**)–(**3.6**)), all voltage clamp data were employed at once to capture time- and voltage-dependent effects. This is equivalent to including M equally weighted experimental repetitions. Thus, the distance function employed by ABC in this instance is simply the RMSE between the simulated data $\hat{g}_j(t)$ and the experimental data (defined in equation (**3.1**)) averaged over all M voltage clamp experimental traces $\boldsymbol{g}_j$ (reported for a single axon preparation in figs 3 and 6 of the original publication):

$$\mathcal{D} = \frac{1}{M}\sum_{j=1}^{M}\sqrt{\frac{1}{T}\sum_{t=0}^{T}(g_j(t) - \hat{g}_j(t))^2}. \qquad 3.8$$

RESULTS

Approximate Bayesian Computation Posteriors on Gating Rate Parameters α and β

We began with inference on the parameters of the simplified Hodgkin–Huxley model (equations (**2.6**)–(**2.10**)), as the data used by the authors to arrive at their reported parametrizations were visually reported in the original publication. Posteriors over the six gating rate parameters (α_n, β_n, α_m, β_m, α_h and β_n from equations (**2.7**), (**2.9**) and (**2.10**)) were inferred by ABC using the digitized voltage clamp data as described in §**ii**. Summaries of the final posterior estimates obtained by applying the adaptive error shrinking implementation of ABC (§**i**) are shown in **figure 3**a.

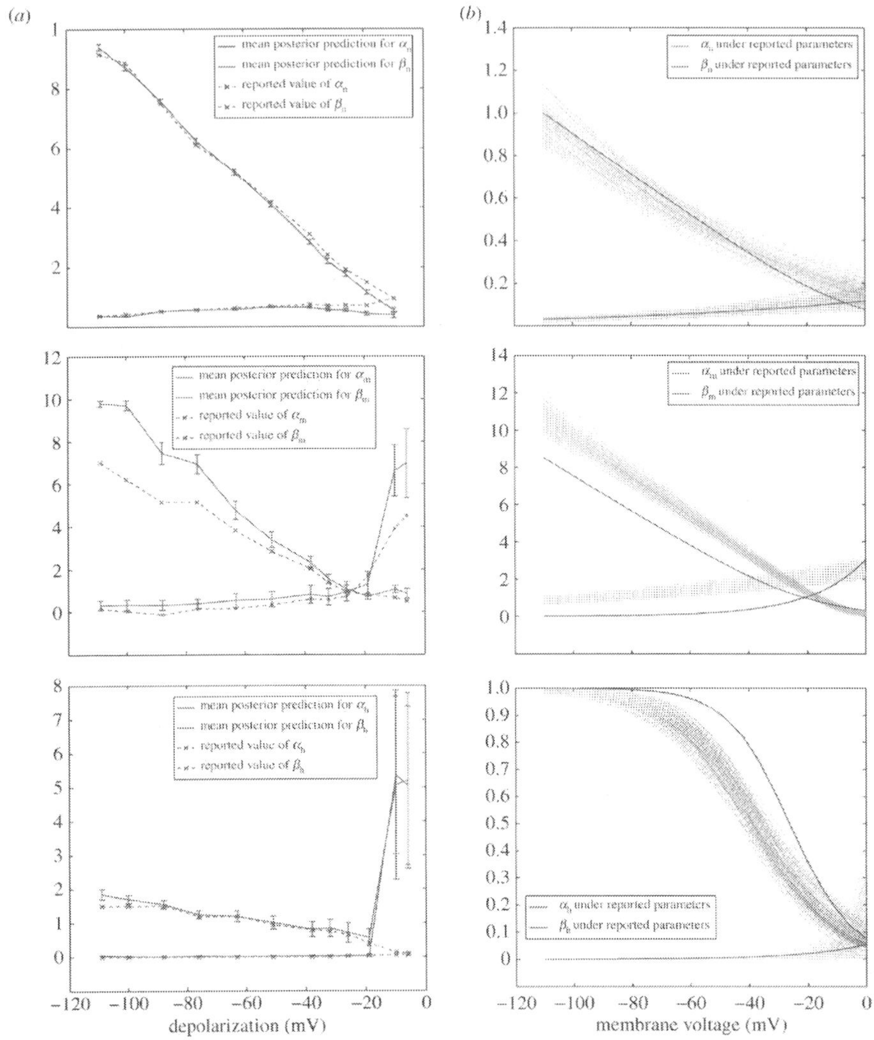

Figure 3. Visualizations of ABC posterior estimates for column (a): conductance gating rate parameters (equations **(2.7)**, **(2.9)** and **(2.10)**) and column (b): voltage dependency parameters (equations **(3.1)**–**(3.6)**). In (a), ABC mean estimates for α and β are depicted in solid lines, with error bars indicating 1 s.d. above and below these mean estimates. In (b), each particle in the ABC posterior estimate for the voltage dependency parameters is used to generate a trace of α, β according to equations **(3.1)**–**(3.6)** in a dotted line, while traces generated by reported parametrizations are represented by solid lines. Both (a, b) depict, from top to bottom, (α_n, β_n), (α_m, β_m) and (α_h, β_h).

In **figure 3**a, we see a high homology between the values of αn and βn reported for the data depicted in fig. **3** of the original publication and the mean values of the posterior estimates produced by ABC. We also see universally small standard deviations about the mean posterior estimates, indicating a high degree of confidence that they reflect a well-constrained optimal value. When the magnitude of the depolarization is smaller, we see a slight systematic deviation of the mean posterior estimates produced by ABC from the original reported values. We attribute this to the ability of our automated method to produce a better fit than the manual methods of Hodgkin and Huxley when fluctuations in the data ($\propto|\min(g_K)-\max(g_K)|$) are small. Indeed, we found the RMSE performances of all particles in these posterior estimates exceed that of the reported parametrization (not shown), indicating an improvement in fit to the experimental data by the ABC estimates.

Figure 3a shows a less perfect homology between ABC mean posterior values of $\alpha_m, \beta_m, \alpha_h, \beta_h$ and those reported by Hodgkin and Huxley, as well as generally larger standard deviations, which is not unexpected given the increased complexity of the function for sodium conductance (equation (**2.8**)). For the activation term m, which controls the initial increase of sodium conductance, we see that the 'spike' parameter α_m shows lower homology with reported values for high depolarizations (as well as looser bounds) yet traces from the final ABC estimates were found to achieve better RMSE performance than traces with reported values. The higher uncertainty at large depolarizations despite improvement of fit may be explained by the low frequency of sodium conductance sampling during the AP in **fig. 6** of [**6**], which results in very few data points falling on the spike itself. Poor definition of this portion of the curve would lead to multiple, equally fit choices for αm, and thus a larger variance in the ABC posterior. ABC mean values for the 'plateau' parameter βm are reasonably consistent with reported values with some exception towards low depolarizations, where the similar RMSE performance and loose bounds may indicate a non-unique parametrization for the complex function, resulting from low deviation in g_{Na} for $|\Delta V| < 10$ mV.

For the inactivation term h, ABC posterior estimates for α_h and β_h are both tightly bounded and show high mean homology with reported values (particularly αh, which is effectively 0 throughout) except for low-magnitude depolarizations. At $|\Delta V| < 10\,mV$, however, the sodium conductance inactivation gate h is effectively irrelevant, as the conductance never reaches a significant spike in activation. At these depolarization values, the shape of the sodium conductance curve begins to resemble that of potassium conductance, the model for which contains no decay term. This lack of impact of h on the observable quantity g_{Na} explains the large variations of the posteriors around α_h and β_h.

Approximate Bayesian Computation Posteriors on Voltage Dependency parameters *k*

After fitting the α and β parameters describing the gating rates for potassium and sodium current conductance, we attempted to fit the parameters of the full Hodgkin–Huxley model (equations (**3.1**)–(**3.6**)), which describe the voltage-dependency for the said rates. The forms of these functions were empirically chosen by the authors to fit the variation of their estimates of the α and β parameters of the simplified model over a range of depolarization values. These α and β values were in turn inferred from the shape of the ionic conductance data described in figs 3 and 6 of the original publication and represented as dashed lines in **figure 3**a. Rather than this indirect means of fitting—first deriving values of the rate constants and then parameters for the functions thought to describe them—we sought to fit the parameters of equations (**3.1**)–(**3.6**) directly to the experimental ionic conductance data (using the distance function described in equation (**3.2**)). This effectively expanded the parameter space for ABC inference from the six in the previous section to 14, but also expands the training dataset from one to 12 traces.

Summaries of the final ABC posterior estimates for the potassium conductance parameters are reported in **table 2**. We initially note that the mean posterior estimates of $k_{\alpha_n 2}$ and $k_{\alpha_n 3}$ demonstrate high deviation from the reported values. Additionally, we find the reported values of these parameters lie outside the 90-percentile range around this mean. The significance of this deviation, however, is negated by the large spread of posterior estimates for these parameters, quantified both by the variance about the mean and the width of the 90-percentile range, which suggests a lack of confidence in the mean posterior estimate. Similarly, while the reported value of $k_{\beta_n 2}$ lies within the 90-percentile range about the mean posterior estimate, the width of the distribution suggests the 'recovered' estimate is not unique. The width of the posterior estimates of these three parameters thus suggests an unidentifiability in equations (**3.1**) and (3.2).

As the three highly variable parameters $k_{\alpha_n 2}$, $k_{\alpha_n 3}$ and $k_{\beta_n 2}$ are all contained in the exponential portions of the equations, we sought to determine whether this could be a structural unidentifiability caused by the choice of an overly complex functional form. In **figure 3**b, we parametrized equations (**3.1**) and (3.2) according to each particle in the ABC posterior estimate and plotted the resulting values of α_n and β_n over a range of membrane voltages. In the case of a structural unidentifiability, we would expect low variation in the function output despite high variation in parameter space. Instead, we observe a relatively wide distribution around αn at all values of V , and a widening distribution around β_n at low values of V when k_{β_n} begins to dominate the exponential portion of the function. This suggests that fluctuations in parameter values lead to commensurate fluctuations in the observable output. Additionally, the biplot of the posterior estimate in **figure 4** fails to reveal any strong correlations between parameters that might be expected of a structural unidentifiability. The marginal distributions of the highly variable parameters $k_{\alpha_n 2}$, $k_{\alpha_n 3}$ and $k_{\beta_n 2}$ also show one or more distinct peaks, which indicate preference of the algorithm for certain values despite the overall uncertainty. This would not be expected if fluctuations in these parameters did not affect the observable quantity g_K. Together, this suggests that the full Hodgkin–Huxley model for potassium conductance exhibits a practical, rather than structural, unidentifiability.

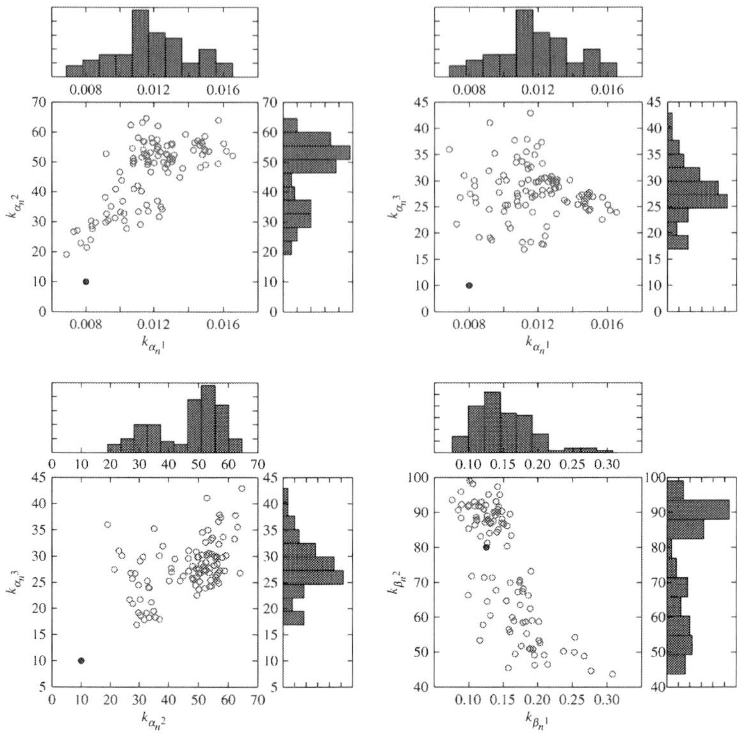

Figure 4. Biplot visualization of relationships between voltage dependency parameters for potassium conductance activation (equations (**3.1**) and (3.2)) in an ABC posterior estimate. In each plot, red circles indicate a particle from the ABC posterior estimate, while the solid blue dot indicates the reported value's location in parameter space. Histograms along each axis indicate the marginal frequencies for each value of the parameter.

ABC posterior estimates for sodium conductance voltage dependency parameters are comparatively well constrained. While **table 3** shows several parameters with high mean deviation from reported values ($k_{\alpha_m}2$, $k_{\alpha_m}3$, $k_{\beta_m}2$ and $k_{\alpha_h}2$), posterior estimates for all parameters show low relative variance and 90-percentile spread with the exception of $k_{\alpha_h}2$. This, coupled with the substantial decrease in RMSE of all particles in the posterior when compared to the model under reported parametrization (indicated in the bottom row of **table 3**), suggests that ABC has arrived at a relatively tightly bounded optimum exceeding that of the manual methods of the original paper. **Figure 3**b seems to support this, as posterior traces around α_m, β_m and β_h notably deviate from the reported traces yet maintain a constrained form across all values of V.

Table 3. Summary statistics of ABC posterior estimates for voltage dependency parameters of sodium conductance gating rates. (Parameter names in the left-most column refer to the corresponding labels in equations (**3.3**)–(**3.6**). 'Mean' and 'variance' columns indicate the summary statistics used to describe the final posterior estimate of ABC, while '5th, 95th percentiles' denote the points at which the posterior CDF records 5% and 95% density, respectively (analogous to 90% CI). The 'RMSE' row contains the value of the distance metric (equation (**3.2**)) attained by the reported parameters, as well as the minimum and maximum (bounded from above by the final ABC error constraint ϵT) values attained by particles in the final posterior estimate.)

		ABC posterior estimates		
parameter	reported value	mean	variance	5th, 95th percentiles
k_{α_m1}	0.1	0.110	3.69×10^{-5}	0.102, 0.121
k_{α_m2}	25	13.3	3.43	10.8, 16.1
k_{α_m3}	10	3.73	1.36	1.68, 5.66
k_{β_m1}	4	2.59	0.0915	2.17, 3.12
k_{β_m2}	18	95.9	9.87	91.1, 99.6
k_{α_h1}	0.07	0.414	0.0978	0.0315, 0.917
k_{α_h2}	20	5.73	17.3	1.08, 13.0
k_{β_h1}	30	40.6	8.74	36.8, 45.6
k_{β_h2}	10	13.6	3.01	11.5, 16.6
RMSE	1.93		min 1.20, max $(\leq \epsilon_T)$ 1.22	

In **figure** 3b, α_h shows high input sensitivity to variation in the unconstrained parameter k_{α_h2} at low values of V , where k_{α_h2} dominates the exponential portion of equation (**3.5**). This is not suggestive of structural unidentifiability, as fluctuation in the posterior distribution is manifest in the observable quantity αh. A biplot visualization of the ABC posterior (**figure 5**) fails to show any correlation between the voltage dependency parameters of h, and the marginal distribution of k_{α_h2} has clear skew, indicating the ability of the algorithm to discern and penalize fluctuations in its value. This supports classification of k_{α_h2} as a practically, rather than structurally, unidentifiable parameter.

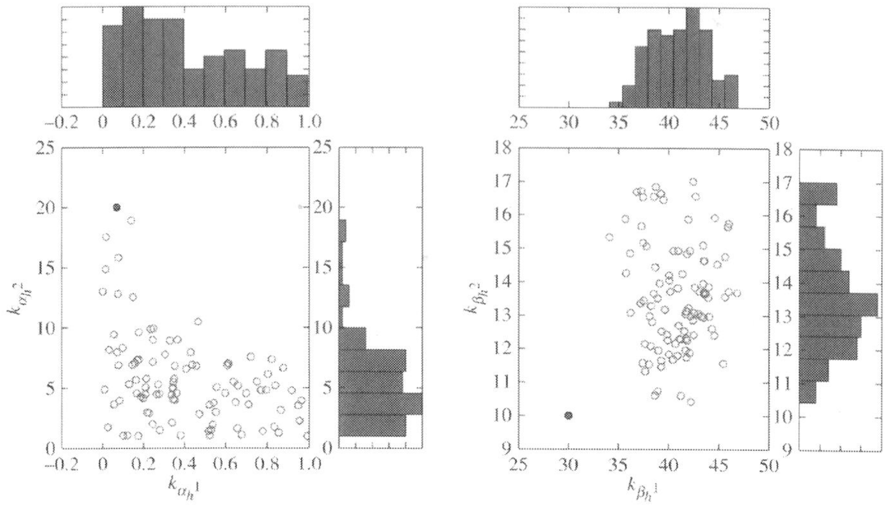

Figure 5. Biplot visualization of relationships between voltage dependency parameters for sodium conductance inactivation (equations (**3.1**) and (3.2)) in an ABC posterior estimate. In each plot, red circles indicate a particle from the ABC posterior estimate, while the solid blue dot indicates the reported value's location in parameter space. Histograms along each axis indicate the marginal frequencies for each value of the parameter.

Posterior Performance on Voltage Protocols

To investigate experimental designs that could be employed to further constrain the ambiguous parameters returned from our ABC analysis (or, failing that, support classification of said variation as inherently biological), we investigated the emergent behaviour of the parametrized models when subjected to the four voltage protocols proposed by Hodgkin and Huxley in figs 20–23 of their original publication. Under these protocols, Hodgkin and Huxley's numerical solutions to equations (**2.11**)–(**2.16**) showed general qualitative agreement with experimental behaviour [**6**]. Two of these protocols were designed to probe the ability of the cell to respond to membrane potential injections of variable magnitude or timing. The 'positive phase depolarization' protocol assessed the degree to which a voltage stimulus during the recovery phase of the AP could trigger a secondary activation, while the 'sub-threshold depolarization' experiment sought to assess behaviour of the membrane potential under depolarizations insufficient to trigger a full AP. The other two protocols, 'anode break excitation' and 'oscillation induction', were designed to probe the behaviour of the membrane under new clamping conditions. The anode break protocol introduced an anodal polarization, lowering potassium conductance activation and sodium conductance inactivation, reversing the membrane current at resting potential and causing full excitation after release of the clamp. The oscillation induction experiment probed the observed oscillation of the membrane potential in response to small, sustained current injections.

For each experiment, either the sodium or potassium conductance model was allowed to vary according to the ABC posterior estimate (described in §4.2), while the other was held to default values. Simulating the membrane voltage under these conditions allowed us to attribute differences from reported behaviour to variation in the behaviour of a single channel, and thus assess the ability of the protocol to further constrain the conductance models.

Variable Behaviour of Potassium Posterior

Under the anode break excitation experimental protocol, the model's membrane voltage showed variable behaviour over the potassium posterior ranging from total inactivation to full excitation nearly matching the experimental trace in both magnitude and temporal placement (**figure 6**a). This suggests that the experiment might be informative for further constraining one or more parameters that showed high posterior variation under voltage clamp data alone, which is not entirely surprising given that the hyperpolarizing current alters the potassium conductance at resting potential, potentially revealing ion channel kinetics that were absent under the simpler protocols.

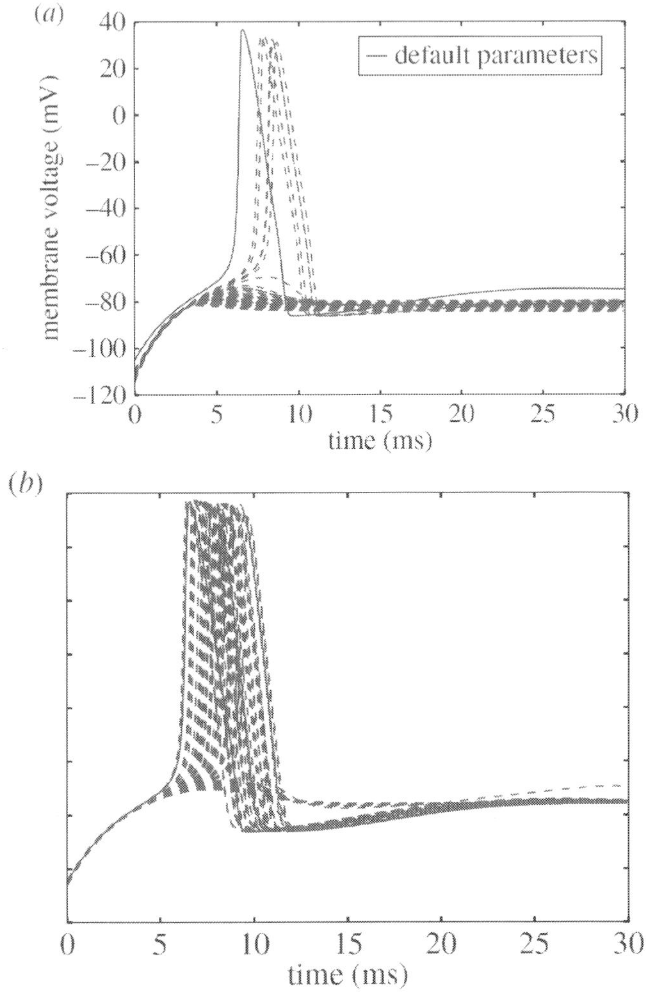

Figure 6. Membrane voltage response to the anode break excitation protocol. Models parametrized according to the posterior estimates for potassium (a) or sodium (b) conductance parameters of equations (**3.1**)–(**3.6**) are shown in red, whereas response under default parameters is shown in blue.

Both the positive phase depolarization experiment and the 7 mV (**figure** 7a) threshold excitation experiment reveal the inability of the posterior parametrized models to trigger an AP in response to the same membrane stimulus sufficient for the reported parametrization. The positive phase depolarization experiments (**figure 8**a–c) also show a more exaggerated second activation response to a stimulation during restitution across all posterior parametrized models. At low depolarization (**figure** 7b) or slight polarization (**figure** 7c) of the membrane, as well as the oscillation induction experiment (**figure** 9a), the posterior parametrized models showed the same qualitative behaviour, yet at a notable downward shift in voltage.

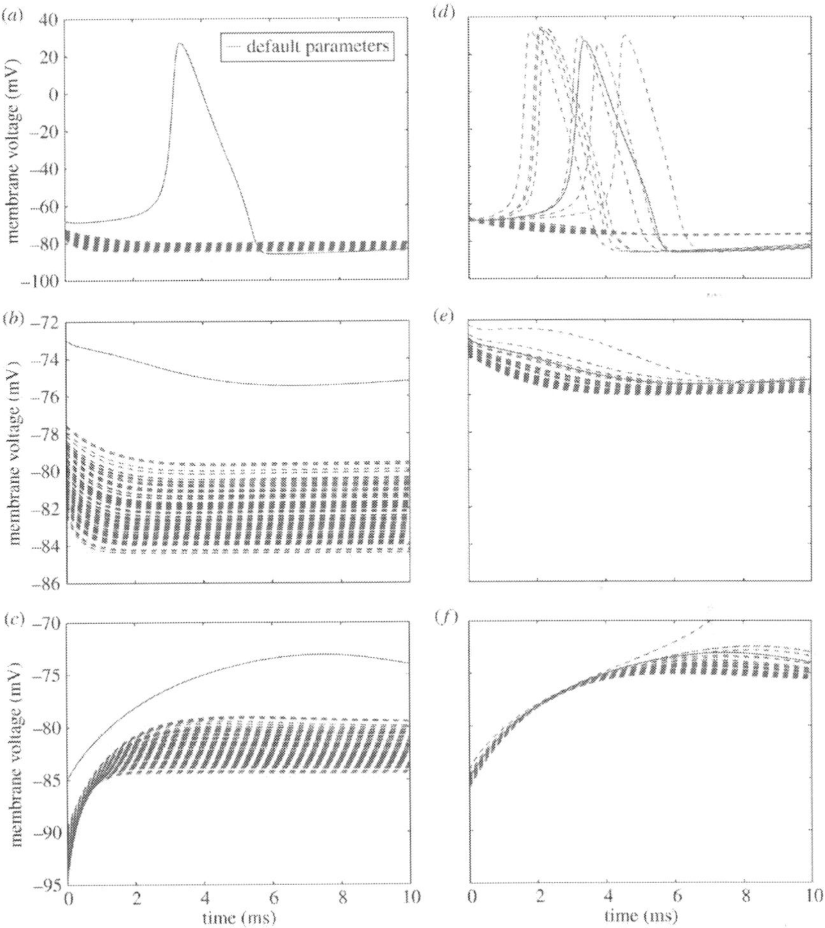

Figure 7. Membrane voltage response to the threshold depolarization protocol for (from top to bottom) 7 mV, 2 mV, −10 mV. Models parametrized according to the posterior estimates for either potassium (a–c) or sodium (d–f) conductance parameters of equations (**3.1**)–(**3.6**) are shown in red, whereas response under default parameters is shown in blue.

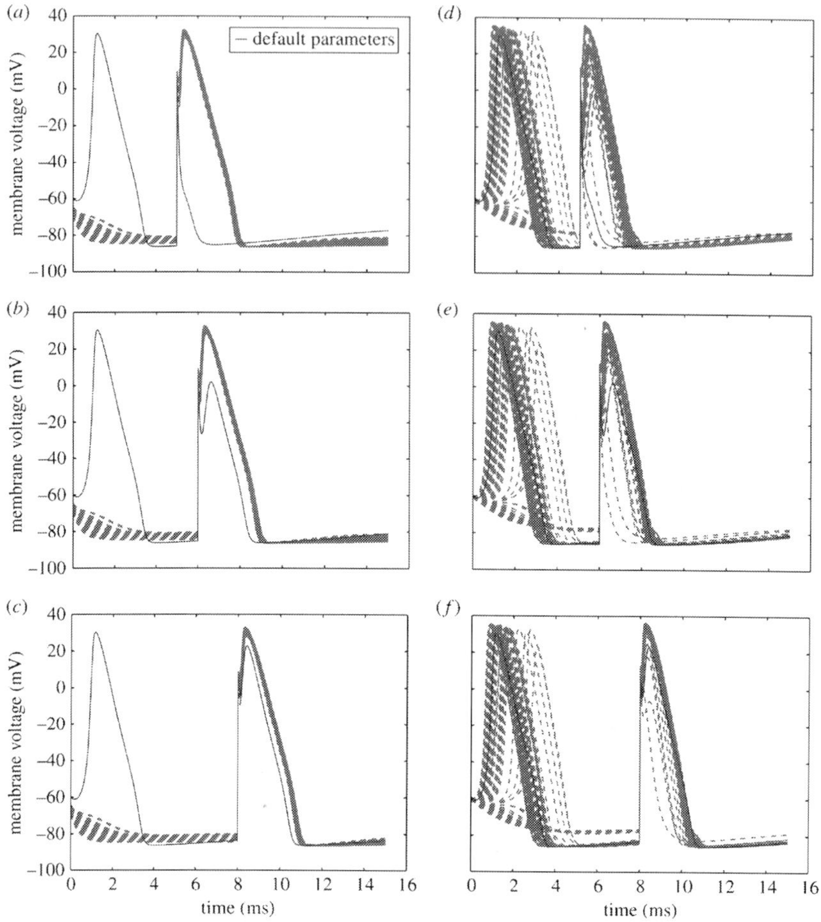

Figure 8. Membrane voltage response to the positive phase depolarization protocol at (from top to bottom) 5 ms, 6 ms, 8 ms. Models parametrized according to the posterior estimates for either potassium (a–c) or sodium (d–f) conductance parameters of equations (**3.1**)–(3.2) are shown in red, whereas response under default parameters is shown in blue.

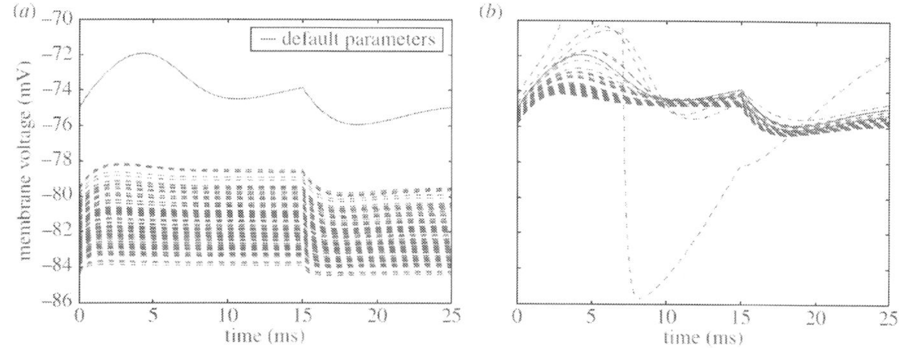

Figure 9. Membrane voltage response to the oscillation induction protocol. Models parametrized according to the posterior estimates for potassium (a) or sodium (b) conductance parameters of equations **(3.1)**–**(3.6)** are shown in red, whereas response under default parameters is shown in blue.

This inability to recreate the reported behaviour may be owing to the fact that the leakage current component of the model (I_l in equation **(2.2)**) was set by the authors after fitting the sodium and potassium conductances as a 'catch-all' to ensure matching to experimental behaviour. As such, this value is probably not a biological truth, and employing it without a similar adjustment for our parametrizations may lead to this failure to capture the same behaviour. Regardless, it appears from these results that separately fitting the potassium conductance parameters to the voltage clamp experiments with ABC failed to produce a parametrized model capable of exhibiting all of the behaviour reported by the authors.

Variable Behaviour of Sodium Posterior

Under the anode break excitation experimental protocol, the qualitative behaviour under-reported values is captured nearly universally across all posterior parametrized sodium conductance models, with variation largely constrained to the temporal placement of the excitation spike (**figure 6**b). This suggests that information may be gained from this protocol as to the unidentifiable parameter $k_{\alpha_h 2}$, which as in the case of the potassium n gate parameters is unsurprising, given that the hyperpolarization of the membrane also decreases sodium current inactivation (controlled by h), leading to the reversal of membrane current and potentially exposing new ion channel kinetics.

Unlike the behaviour of the models parametrized under the potassium posterior, at least some models parametrized under the sodium posterior capture the membrane potential behaviour after nearly any depolarization (**figure** 7d–f). We additionally see a range of behaviour, including the reported behaviour, when examining AP activation during both the first and second depolarizations of the positive phase depolarization experiment (**figure** 8d–f). Interestingly, and again unlike the potassium posterior, we see that a particle in the posterior has a large outlier with regard to response to the current clamp experiment, and thus even the well-constrained model is sensitive to fluctuations within its estimates (**figure** 9b). We may conclude that some further information on the parameters of the model could be gleaned by examining model response to any of these protocols as well as that of the anode break experiment.

DISCUSSION

ABC posterior estimates for the parameters of the simplified Hodgkin–Huxley model were tightly constrained across most magnitudes of depolarization, indicating that the model was identifiable from the data under most experimental conditions. Not only this, but the mean estimates showed high homology with the values reported by Hodgkin and Huxley in their original publication. Such a well-constrained recovery of the original parameters highlights the remarkable degree of accuracy achieved by the authors without the aid of modern computational tools. Even when the posterior estimates of the ABC approach widened at low magnitude depolarizations, particularly around the parameters controlling the sodium inactivation gate h, simulations employing the estimates of the original authors were able to reproduce the experimental data.

Performing ABC parameter inference on the full Hodgkin–Huxley model highlighted the value of the method's ability to quantify model uncertainty. The wide posteriors around certain parameters in the potassium ($k_{\alpha_n 2}, k_{\alpha_n 3}$ and $k_{\beta_n 2}$) and sodium ($k_{\alpha_h 2}$) conductance components suggested an inability to fully fit the model under the provided experimental data. Because the output of ABC is the full population estimating the posterior over the parameters, we were able to take advantage of additional information, such as pairwise correlations and marginal distributions over the parameters, to support classification of the unidentifiabilities as practical rather than structural. This analysis suggests that additional data may be useful in further constraining the model, although without further experimentation we cannot rule out the unidentifiability being attributed to natural biological variability. In either case, this study highlights the usefulness of ABC in reporting all relevant statistics about the posterior within the final estimating population.

The classification of the model unidentifiability as practical rather than structural suggests that Hodgkin and Huxley arrived at a model of the appropriate degree of complexity to describe their system. This also suggests that the authors could not have found a better parametrization of their model without the inclusion of additional data that could further inform the ion channel dynamics within the model. Despite this, we do observe instances where the ABC posterior estimates show notable performance gains over the reported model parametrization (**table 3**). This is probably a result of our fitting to data from a single recording of a single axon, whereas the authors employed unreported data from several different axons, which would be expected to exhibit variation in their conductance recordings. Using only the single-recording data available to us, it would be unwise to conclude anything significant from the deviations in mean posterior estimates and resulting gains in RMSE performance for particles in the posterior. Instead, the value of our approach lies in the quantification of confidence around the mean parameter estimates given just a sample of the data that was available to the original authors.

To assess what the original authors might have been able to accomplish with the experimental data for the protocols they proposed at the end of their paper (and armed with modern computational tools), we examined the differences in response to several complex protocols over the particles in the posterior estimates for both potassium and sodium conductance. We noted that the particles of the sodium conductance posterior showed greater consistency in the response to these protocols, both internally and when compared to the response of the reported parametrization, than the particles of the potassium posterior. This is probably owing to the presence of higher posterior variability in the potassium conductance model. The failure of all particles in the posterior to produce the same behaviour as the reported parametrization could be indicative of dependencies between the parameters of the two conductance models not captured by the separate fitting approach, or simply another constraint on the model not captured by the voltage clamp protocol.

While there appears to be additional information that could be leveraged from these experiments to further constrain both conductance models, the leveraging of this information to potentially decrease model uncertainty would be non-trivial. Each of these experimental protocols, including the standard voltage clamp, will only constrain a subset of the model parameters. Thus, fully constraining, the model would require leveraging a weighted combination of the deviation from experimental data under each protocol in our distance function. Such a weighted combination of data introduce so-called 'hyperparameters' into the ABC algorithm—parameters that control how much importance the algorithm gives to each source of data. These parameters are difficult to assign, especially without knowledge of the precision of the measurements being included, and might require an additional level of parameter fitting to set. Given the sparsity of the experimental data, it is unlikely that these hyperparameters could be reliably constrained. Even using the posteriors from the voltage clamp ABC as priors for a new round of ABC employing more complex voltage protocols would require a weight to be assigned for the trade-off between prior likelihood and experimental performance. This amounts to attempting to maintain the behaviour of the model under the voltage clamp protocol while simultaneously seeking to improve performance under the new protocol. In light of these limitations, we believe full integration of simulated data from these protocols to be beyond the scope of the paper. We can only theorize as to the degree of constraint they could have produced on the model, or the support they would lend to classifying the uncertainty as inherent biological variation by failing to further constrain it.

Arriving at an optimal means to combine data from multiple protocols, or simply determining a protocol that provides the most information on a given model, are problems of 'experimental design', and will be a focus of future work with ABC parameter fitting techniques, owing to its close relationship to model identifiability.

CONCLUSION

ABC parameter fitting of the Hodgkin and Huxley model to the voltage clamp data reported in their original paper was able to recover precisely the rate parameters controlling sodium and potassium channel gating, but was less able to recover the parameters controlling the voltage dependency of said rates. The width of the distributions around these unidentifiable parameters, as well as their lack of correlation with each other and the relative sensitivity of the output to their fluctuations, suggested that these were practically unidentifiable, potentially requiring more data to constrain. Investigation of differential behaviour under more complex voltage protocols thought to probe the conductance dynamics more fully revealed several promising sources of further model constraint.

We hope that this study serves as both a strong affirmation of the quality of the original work done by Hodgkin and Huxley to fit their model, which appears to be near-optimal under the data available to them, as well as a template for similar identifiability analyses in current cardiac models. While we doubt that many present-day models will be able to be constrained as effectively as the Hodgkin–Huxley model, given the increased complexity and the myriad sources of data, we believe the uncertainty quantification provided by these analyses can pinpoint directions for future experimental work, or provide insight into the degree and distribution of biological variability in the system. The adoption of these Bayesian parameter fitting methods would foster a documented link between experiment and data that has slowly been lost since the original work done by the founders of the field of computational biology, Hodgkin and Huxley.

Data accessibility

A repository containing the digitized Hodgkin–Huxley data, CellML files, ABC implementation, functional curation protocols and Python scripts used to generate the results presented in this paper are freely available to the public in the Paper Tutorials section of the Chaste wiki (https://chaste.cs.ox.ac.uk/trac/wiki/PaperTutorials/HodgkinHuxleyABC), along with instructions on how to compile and execute the code, or by browsing the source code directly (https://chaste.cs.ox.ac.uk/trac/browser/projects/HodgkinHuxleyABC).

Authors' contributions

A.C.D. wrote all code used in the study, performed all statistical analyses, participated in the design of the study and drafted the manuscript. D.J.G. conceived of the study of the Hodgkin–Huxley model and coordinated the effort. C.H. conceived of the use of ABC and participated in the design of the study. J.C. participated in the design and coordination of the study, adapted the functional curation library to meet the needs of the study and aided in drafting the manuscript. All authors participated in revising the manuscript and gave final approval for publication.

Competing interests

We declare we have no competing interests.

REFERENCES

1. ten Tusscher K, Noble D, Noble P, Panafilov A. 2003 A model for human ventricular tissue. Am. J. Physiol. Heart Circ. Physiol.286, H1573–H1589. (doi:10.1152/ ajpheart.00794.2003)
2. Romero L, Pueyo E, Fink M, Rodriguez B. 2009 Impact of ionic current variability on human ventricular electrophysiology. Am. J. Physiol. Heart Circ. Physiol.297, H1436–H1445. (doi:10.1152/ ajpheart.00263.2009)
3. Clancy C, Rudy Y. 2001 Cellular consequences of HERG mutations in the long QT syndrome: precursors to sudden cardiac death. Cardiovasc. Res. 50, 301–313. (doi:10.1016/S0008-6363(00)00293-5)
4. O'Hara T, Virag L, Varro A, Rudy Y. 2011 Simulation of the undiseased human cardiac ventricular action potential: model formulation and experimental validation.PLoS Comput. Biol.7, e1002061. (doi:10.1371/journal.pcbi.1002061)
5. Noble D, Rudy Y. 2001 Models of cardiac ventricular action potentials: iterative interactions between experiment and simulation. Phil. Trans. R. Soc. Lond. A 359, 1127–1142. (doi:10.1098/ rsta.2001.0820)
6. Hodgkin A, Huxley A. 1952 A quantitative description of membrane current and its application to conduction and excitation in nerve. J.

Physiol. 117, 500–544. (doi:10.1113/jphysiol.1952.sp004764)

7. Halliwell JV, Plant TD, Robbins J, Standen NB. 1994 Voltage clamp techniques. In Microelectrode techniques: the Plymouth workshop handbook (ed. D Ogden), pp. 17–35. Cambridge, UK: The Company of Biologists Limited.
8. Lee J, Smaill B, Smith N. 2006 Hodgkin-Huxley type ion channel characterization: an improved method of voltage clamp parameter estimation. J. Theor. Biol.242, 123–134. (doi:10.1016/j.jtbi.2006.02.006)
9. Neher E, Sakmann B. 1992 The patch clamp technique.Sci. Am.266, 44–51. (doi:10.1038/scientificamerican0392-44)
10. Rudy Y, Silva J. 2006 Computational biology in the study of cardiac ion channels and cell electrophysiology. Q. Rev. Biophys.39, 57–116. (doi:10.1017/S0033583506004227)
11. Cannon R, D'Alessandro G. 2006 The ion channel inverse problem: neuroinformatics meets biophysics.PLoS Comput. Biol.2, 862–869. (doi:10.1371/journal.pcbi.0020091)
12. Burger M. 2011 Inverse problems in ion channel modeling. Inverse Probl.27, 1–34. (doi:10.1088/0266-5611/27/8/083001)
13. Cherry E, Fenton F. 2006 A tale of two dogs: analyzing two models of canine ventricular electrophysiology. Am. J. Physiol. Heart Circ. Physiol.292, H43–H55. (doi:10.1152/ajpheart. 00955.2006)
14. Niederer S, Fink M, Noble D, Smith N. 2009 A meta-analysis of cardiac electrophysiology models. Exp. Physiol. 64, 486–495. (doi:10.1113/expphysiol. 2008.044610)
15. ten Tusscher K, Bernus O, Hren R, Panfilov A. 2006 Comparison of electrophysiological models for human ventricular cells and tissues. Prog. Biophys. Mol. Biol. 90, 326–345. (doi:10.1016/j.pbiomolbio. 2005.05.015)
16. Niederer S, Smith S. 2012 At the heart of computational modeling. J. Physiol.590, 1331–1338. (doi:10.1113/jphysiol.2011.225045)
17. Toni T, Welch D, Strelkowa N, Ipsen A, Stumpf M. 2009 Approximate Bayesian computation scheme for parameter inference and model selection in dynamical systems. J. R. Soc. Interface 6, 187–202. (doi:10.1098/rsif.2008.0172)

18. Wilkinson D. 2012Stochastic modelling for systems biology, 2nd edn. London, UK: Chapman and Hall/CRC.
19. Raue A, Kreutz C, Maiwald T, Klingmuller U, Timmer J. 2011 Addressing parameter identifiability by model-based experimentation. IET Syst. Biol. 4, 120–130. (doi:10.1049/iet-syb.2010.0061)
20. Csercsik D, Hangos K, Szederkenyi G. 2012 Identifiability analysis and parameter estimation of a single Hodgkin–Huxley type voltage dependent ion channel under voltage step measurement conditions. Neurocomputing77, 178–188. (doi:10.1016/j.neucom.2011.09.006)
21. Roper R, Saccomani M, Vicini P. 2010 Cellular signaling identifiability analysis: a case study. J. Theor. Biol.264, 528–537. (doi:10.1016/j.jtbi.2010. 02.029)
22. Sobie E. 2009 Parameter sensitivity analysis in electrophysiological models using multivariable regression. Biophys. J. 96, 1264–1274. (doi:10.1016/j. bpj.2008.10.056)
23. Sarkar A, Sobie E. 2010 Regression analysis for constraining free parameters in electrophysiological models of cardiac cells.PLoS Comput. Biol. 6, e1000914. (doi:10.1371/journal.pcbi.1000914)
24. Fink M, Noble D. 2009 Markov models for ion channels: versatility versus identifiability and speed.Phil. Trans. R. Soc. A 367, 2161–2179. (doi:10.1098/rsta.2008.0301)
25. Brown KS, Sethna JP. 2003 Statistical mechanical approaches to models with many poorly known parameters.Phys. Rev. E. Stat. Nonlin. Soft Matter Phys. 68, 021904. (doi:10.1103/PhysRevE.68. 021904)
26. Milescu L, Akk G, Sachs F. 2005 Maximum likelihood estimation of ion channel kinetics from macroscopic currents. Biophys. J. 88, 2494–2515. (doi:10.1529/ biophysj.104.053256)
27. Siekmann I, Sneyd J, Crampin E. 2012 MCMC can detect nonidentifiable models. Biophys. J.103, 2275–2286. (doi:10.1016/j. bpj.2012.10.024)
28. Pronzato L. 2008 Optimal experimental design and some related control problems. Automatica 44, 303–325. (doi:10.1016/j. automatica.2007. 05.016)

29. Calderhead B, Girolami M, Lawrence N. 2009 Accelerating Bayesian inference over nonlinear differential equations with Gaussian processes. In Advances in neural information processing systems, vol. 21 (eds D Koller, D Schuurmans, Y Bengio, L Bottu), pp. 217–224. Red Hook, NY: Curran Associates, Inc.
30. Tavare S, Balding D, Griffiths R, Donnelly P. 1997 Inferring coalescence times from DNA sequence data. Genetics145, 505–518.
31. Cooper J, Mirams G, Niederer S. 2011 High-throughput functional curation of cellular electrophysiology models.Prog. Biophys. Mol. Bio.107, 11–20. (doi:10.1016/j.pbiomolbio.2011. 06.003)
32. Cooper J, Vik J, Waltemath D. 2015 A call for virtual experiments: accelerating the scientific process. Prog. Biophys. Mol. Bio.117, 99–106. (doi:10.1016/j. pbiomolbio.2014.10.001)
33. Mirams G et al.2013 Chaste: an open-source C++ library for computational physiology and biology. PLoS Comput. Biol. 9, e1002970. (doi:10.1371/ journal.pcbi.1002970)
34. Lloyd C, Lawson J, Hunter P, Nielsen P. 2008 The CellML model repository. Bioinformatics 24, 2122–2123. (doi:10.1093/ bioinformatics/ btn390)
35. Liepe J, Barnes C, Cule E, Erguler K, Kirk P, Toni T, Stumpf M. 2010 ABC-SysBio: approximate Bayesian computation in Python with GPU support. Bioinformatics26, 1797–1799. (doi:10.1093/ bioinformatics/btq278)

2 CAN THE THERMODYNAMIC HODGKIN-HUXLEY MODEL OF VOLTAGE-DEPENDENT CONDUCTANCE EXTRAPOLATE FOR TEMPERATURE?

Michael D. Forrest [1,2]

[1] Department of Computer Science, University of Warwick, Coventry, CV4 7AL, UK

[2] Institute for Adaptive and Neural Computation, School of Informatics, University of Edinburgh, Edinburgh, EH1 2QL, UK

ABSTRACT

Hodgkin and Huxley (H-H) fitted their model of voltage-dependent conductances to experimental data using empirical functions of voltage. The thermodynamic H-H model of voltage dependent conductances is more physically plausible, as it constrains and parameterises its empirical fit by assuming that ion channel transition rates depend exponentially on a free energy barrier that in turn, linearly or non-linearly, depends on voltage. The original H-H model contains no explicit temperature terms and requires Q_{10} factors to describe data at different temperatures. The thermodynamic H-H model does have explicit terms for temperature. Do these endow the model with extrapolation for temperature? We utilised voltage clamp data for a voltage-gated K^+ current, recorded at three different temperatures. The thermodynamic H-H model's free parameters were fitted (Marquardt-Levenberg algorithm) to a data set recorded at one (or more) temperature(s). Then we assessed whether it could describe another data set, recorded at a different temperature, with these same free parameter values and its temperature terms set to the new temperature. We found that it could not.

Keywords: thermodynamic; Hodgkin-Huxley; model; voltage; temperature; computational neuroscience; action potential; $Q10$; transition state

INTRODUCTION

Hodgkin and Huxley (H-H) quantitatively characterised the voltage-dependence of membrane currents in the giant squid axon, and showed how they can generate and propagate action potentials [1]. Their model can describe ion channels in the membrane, opening and closing in a voltage-dependent manner [2]. However, membrane currents are not just voltage-dependent, but temperature dependant also, and the H-H description is without any explicit temperature term. To describe data over a range of temperatures, the H-H model requires an empirical multiplicative constant, Q_{10}, to be applied to a number of its parameters [1].

Hodgkin and Huxley fitted their model of voltage-dependent conductances to experimental data using empirical functions of voltage. Thermodynamic Hodgkin-Huxley models describe voltage-dependence empirically as well, but can be construed to be more physically plausible, as they constrain and parameterise their fit with thermodynamic principles of transition state theory [3,4,5,6,7]. They consider that the rate of transition between channel states depends exponentially on the free energy barrier that separates them. The free energy barrier, in turn, is voltage dependent. It can vary linearly or non-linearly with voltage: linear and non-linear thermodynamic H-H models respectively [6,7].

In contrast to the original H-H model, thermodynamic H-H models do have explicit terms for temperature, on account of their incorporation of thermodynamic theory. We hypothesise that these temperature terms permit these models to describe voltage-gated currents at different temperatures *i.e.*,describe how currents change with voltage *and* temperature.

Voltage clamp data for a K^+ current was taken from [8]. The data was grouped at three temperatures: T15 (15 °C), T25 (25 °C) and T35 (35 °C). The thermodynamic H-H model (linear and non-linear) was fitted to each temperature set separately; the Marquardt-Levenberg algorithm [9] optimised the model's free parameters to find a best fit in each case. This was a success, with close fits being produced. We then investigated if the model, with its free parameters fitted to one temperature data set, could replicate the data of a different temperature set, when the temperature terms of the model were changed to the new temperature. We found that it could not.

We repeated this process but, in the fitting phase, we fitted the thermodynamic H-H model (linear and non-linear) to the combined data sets of two different temperatures: (T15 + T25) or (T15 + T35) or (T25 + T35). We then investigated if the model, with its free parameters fitted to two temperature data sets, could replicate the data of a third temperature set, when the temperature terms of the model were changed to the new temperature. We found that it could not.

We find that the thermodynamic H-H model (linear or non-linear) cannot accurately predict data at temperatures that it is not specifically fitted to *i.e.*, it cannot extrapolate for temperature.

EXPERIMENTAL SECTION

K^+ *Current Data*

Tiwari and Sikdar studied non-inactivating K^+ currents, in a gonadotroph cell line, across a temperature range [8]. They assembled whole cell, voltage clamp data at a number of depolarising potentials, grouped at three temperatures: T15 (15 °C, 288 K), T25 (25 °C, 298 K), T35 (35 °C, 308 K). We obtained this data via personal communication and modified it, as shown in Figure 1.

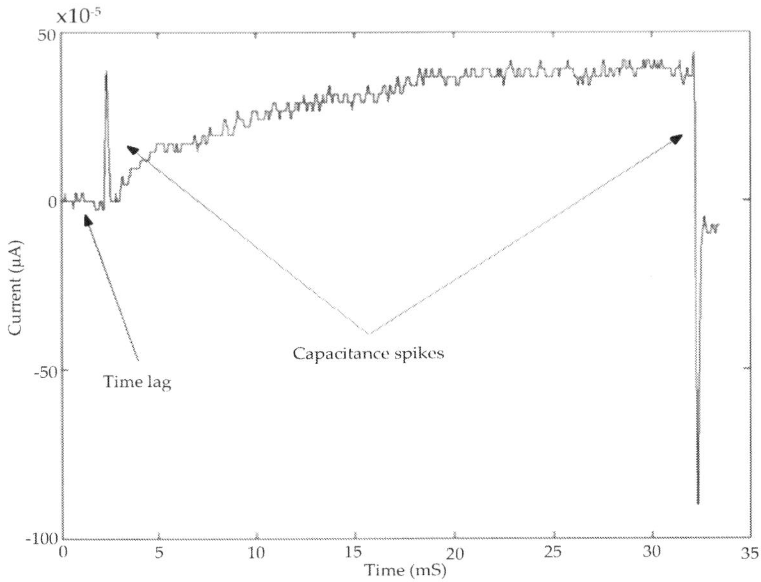

Figure 1. K^+ current in a gonadotroph cell, when the holding voltage is stepped from −10 mV to +116 mV using a voltage clamp. We modified this data before we used it ourselves, removing the labelled capacitance spikes and time lag. We repeated this action for all other current data used. The *x*-axis presents time (mS); the *y*-axis presents K^+ current.

Thermodynamic H-H Models

α and β are forward and backward rate constants between the Closed (*C*) and Open (*O*) channel states.

$$C \underset{\beta}{\overset{\alpha}{\rightleftarrows}} O \tag{1}$$

When the channel switches between these states, it must pass through a high energy intermediate: the transition state (not shown). α and β are voltage (*V*) and temperature (*T*) dependent:

Linear Variant

$$\alpha(V) = \alpha_0 e^{-(a_1 + b_1 V)/RT} \tag{2}$$

$$\beta(V) = \beta_0 e^{-(a_2 + b_2 V)/RT} \tag{3}$$

R is the gas constant, V is the membrane voltage. α_0, a_1, b_1 are parameters describing the free energy barrier between the Closed and Transition state. β_0, a_2, b_2 are parameters describing the free energy barrier between the Open and Transition state. T is temperature; so this model has terms for temperature, unlike the original Hodgkin-Huxley formulation [1].

Non-Linear Variant (Quadratic)

$$\alpha(V) = \alpha_0 e^{-(a_1 + b_1 V + c_1 V^2)/RT} \tag{4}$$

$$\beta(V) = \beta_0 e^{-(a_2 + b_2 V + c_2 V^2)/RT} \tag{5}$$

R is the gas constant, V is the membrane voltage. α_0, a_1, b_1, c_1 are parameters describing the free energy barrier between the Closed and Transition state. β_0, a_2, b_2, c_2 are parameters describing the free energy barrier between the Open and Transition state. T is temperature; so this model has terms for temperature, unlike the original Hodgkin-Huxley formulation [1].

Thermodynamic H-H Model of a Non-Inactivating K⁺ Current (I_K), Recorded in a Gonadotroph Cell Line

$$I_K = g_K(V - E_K) \tag{6}$$

$$g_K = \bar{g}_K \cdot n \tag{7}$$

$$\frac{dn}{dt} = \frac{n_\infty(V) - n}{\tau_n(V)} \tag{8}$$

$$n_\infty = \frac{\alpha_n}{\alpha_n + \beta_n} \tag{9}$$

$$\tau_n = \frac{1}{\alpha_n + \beta_n} \tag{10}$$

I_K is the K^+ current, g_K is the K^+ conductance, $\bar{g}_K$ is the maximal K^+ conductance, V is the membrane potential, t is time and E_K is the reversal/Nernst potential for K^+ (set to −9.8 mV; [8]). To produce a linear thermodynamic H-H model, α_n and β_n are set by Equations (2) and (3) (respectively). To produce a non-linear thermodynamic H-H model, α_n and β_n are instead set by Equations (4) and (5) (respectively).

The maximal conductance ($\bar{g}_K$) for the T15, T25 and T35 membrane patches was calculated by deriving a maximal current ($\bar{I}_K$) value for the largest depolarising potential explored at each temperature (Figure 1D; [8]). Then:

$$\bar{g}_K = \frac{\bar{I}_K}{(V - E_K)} \tag{11}$$

Fitting the Thermodynamic H-H Model to the K$^+$ *Current* (I_K) Data

The free parameters in the thermodynamic H-H model, adjusted in order to fit the model to the data, are α_0, β_0, a_1, b_1, a_2, b_2 for the linear variant (refer to Equations (2) and (3)); with the addition of c_1 and c_2 for the non-linear variant (quadratic) (refer to Equations (4) and (5)). The fitting was performed using the criterion of least squares minimisation, implemented by the Marquardt-Levenberg algorithm [9] in Matlab (MathWorks Inc., Natick, MA, USA). Initial free parameter values were chosen arbitrarily. Repeated runs with different initial free parameter values checked that our fits were globally, and not just locally, optimal. No constraints were set for free parameter values.

Temperature Extrapolation

The thermodynamic H-H model (linear and non-linear) was fitted to each temperature set separately: T15, T25 and T35. We then investigated if the model, with its free parameters fitted to one temperature data set, could replicate the data of a different temperature set, when the temperature terms of the model (in Equations (2)–(5)) were changed to the new temperature.

We repeated this process but, in the fitting phase, we fitted the thermodynamic H-H model (linear and non-linear) to the combined data sets of two different temperatures: (T15 + T25), (T25 + T35), (T15 + T35). At each temperature, the temperature setting in the thermodynamic model equations was set appropriately and the model's free parameters were optimised with experimental data spanning more than one temperature. We then investigated if the model, with its free parameters fitted to two temperature data sets, could replicate the data of a third temperature set, when the temperature terms of the model were changed to the new temperature.

So, to elaborate further on what was done for both the linear and non-linear models: where *p* denotes parameters and *d* is data: (T15p on T25d), (T15p on T35d), (T25p on T15d), (T25p on T35d), (T35p on T15d) and (T35p on T25d). Combined data sets were also used to "train" the free parameter values: (T15+T25p on T35d), (T15+T35p on T25d), (T25+T35p on T15d).

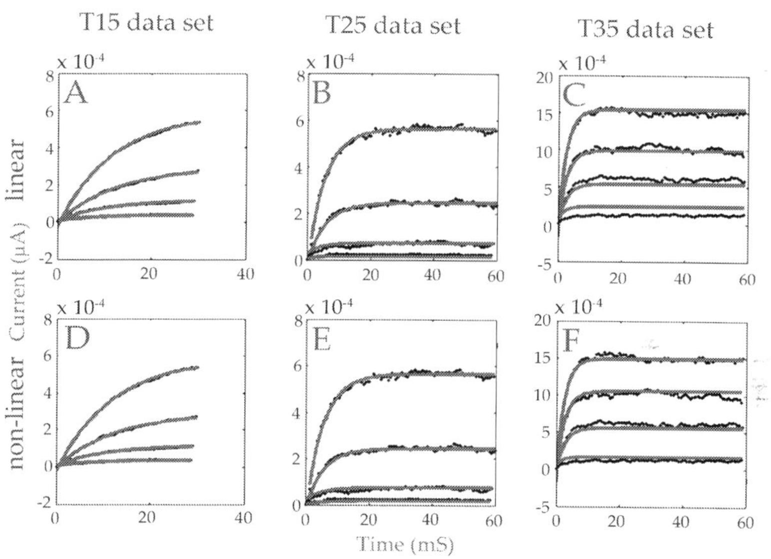

Figure 2. Thermodynamic Hodgkin-Huxley models, linear and non-linear, (red lines) fitted to K^+ current data (black circles). (**A**) Linear thermodynamic H-H model fitted to the T15 data set; (**B**) Linear thermodynamic H-H model fitted to the T25 data set; (**C**) Linear thermodynamic H-H model fitted to the T35 data set; (**D**) Non-linear thermodynamic H-H model fitted to the T15 data set; (**E**) Non-linear thermodynamic H-H model fitted to the T25 data set; (**F**) Non-linear thermodynamic H-H model fitted to the T35 data set. For all panels: the *x*-axis presents time (ms), the *y*-axis presents K^+ current (μA).

RESULTS AND DISCUSSION

Curve Fitting

The thermodynamic H-H model (linear and non-linear) was fitted to each temperature data set separately. Figure 2presents the linear and non-linear models fitted separately to the T15, T25 and T35 data sets. Each panel shows K^+ current in a gonadotroph cell (black circles; every tenth data point plotted) when the holding voltage is stepped from −10 mV to +57 mV, from −10 mV to +77 mV, from −10 mV to +96 mV and from −10 mV to +116 mV [8]. So, there are 4 data plots at each temperature. Greater depolarisation produces greater K^+ current. Red lines show the thermodynamic Hodgkin-Huxley (H-H) model—linear or non-linear—fitted to the data; its free parameters optimised to the data by the Marquardt-Levenberg algorithm [9]. This was a success, with close fits being produced. For example, the non-linear model fit to the T15 data set has a logged reduced chi-square metric: $\log_{10}(x^2_{red}) = -0.05$. The worst fit was still good in absolute terms: the linear model fit to the T35 data set has a logged reduced chi-square metric: $\log_{10}(x^2_{red}) = 1.69$. Temperature and goodness of fit were inversely related, possibly due to data variance being a function of temperature. The non-linear model provided a slightly better fit than the linear model.

The thermodynamic H-H model (linear and non-linear) was then fitted to the combined data sets of two different temperatures: (T15 + T25), (T15 + T35), (T25 + T35). At each temperature, the temperature setting in the thermodynamic model equations was set appropriately and the model's free parameters were optimised with experimental data spanning more than one temperature. Figure 3 presents the linear and non-linear models fitted separately to the (T15 + T25), (T15 + T35) and (T25 + T35) data sets. This was a reasonable success, with reasonably close fits being produced. But, the fit to an amalgamation data set (recorded at 2 temperatures) is inferior to the fit of any of its individual component data

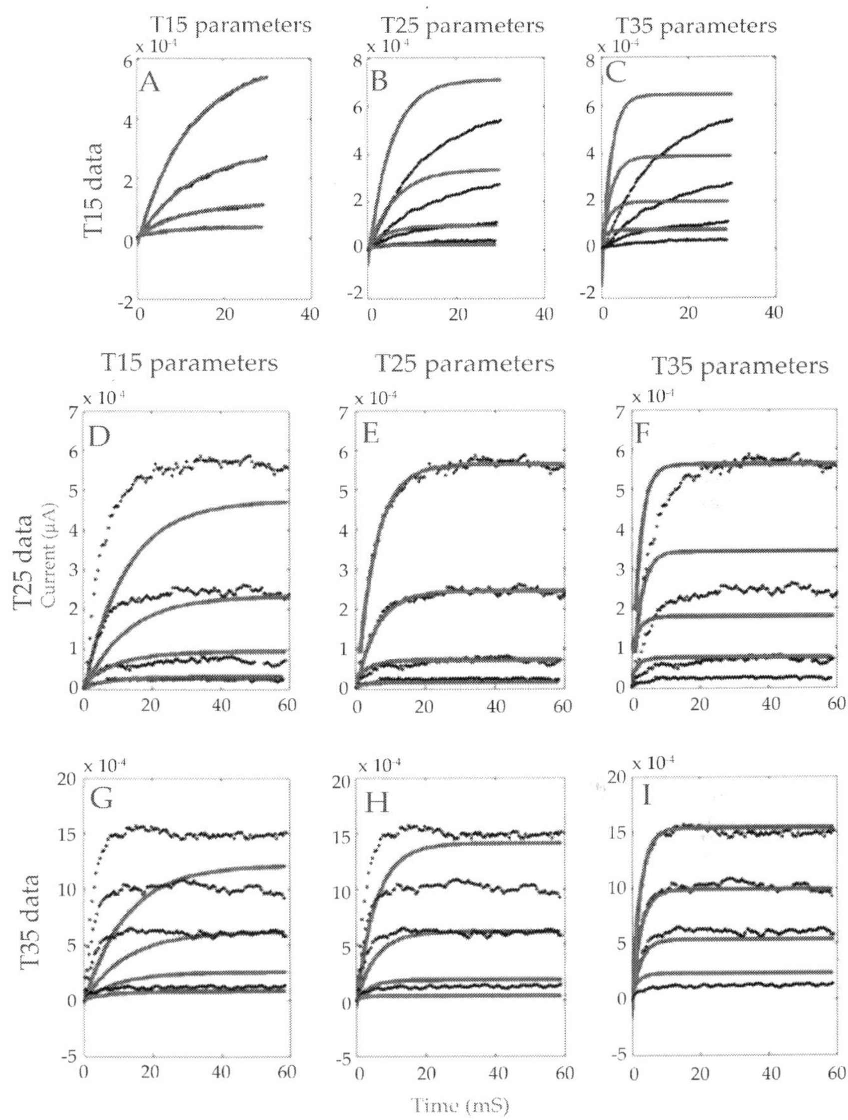

Figure 5. The NON-LINEAR thermodynamic Hodgkin-Huxley (H-H) model, with its free parameters fitted to one temperature data set, cannot replicate the data of a different temperature set, when its temperature terms are changed to the new temperature. So, the model's temperature terms do not permit it to replicate data at temperatures that it has not been fitted to. (**A**) Experimental K^+ current data recorded at 15 °C (black circles). Plots of the non-linear thermodynamic H-H model (red lines), with its free parameters having been previously fitted to

data recorded at 15 °C, and its temperature terms set to 15 °C. The model describes the data well; (**B**) Experimental K^+ current data recorded at 15 °C (black circles). Plots of the non-linear thermodynamic H-H model (red lines), with its free parameters having been previously fitted to data recorded at 25 °C, and its temperature terms set to 15 °C. The model does not describe the data well; (**C**) Experimental K^+ current data recorded at 15 °C (black circles). Plots of the non-linear thermodynamic H-H model (red lines), with its free parameters having been previously fitted to data recorded at 35 °C, and its temperature terms set to 15 °C. The model does not describe the data well; (**D**) Experimental K^+ current data recorded at 25 °C (black). Plots of the non-linear thermodynamic H-H model (red lines), with its free parameters having been previously fitted to data recorded at 15 °C, and its temperature terms set to 25 °C. The model does not describe the data well; (**E**) Experimental K^+ current data recorded at 25 °C (black circles). Plots of the non-linear thermodynamic H-H model (red lines), with its free parameters having been previously fitted to data recorded at 25 °C, and its temperature terms set to 25 °C. The model describes the data well; (**F**) Experimental K^+ current data recorded at 25 °C (black circles). Plots of the non-linear thermodynamic H-H model (red lines), with its free parameters having been previously fitted to data recorded at 35 °C, and its temperature terms set to 25 °C. The model does not describe the data well; (**G**) Experimental K^+ current data recorded at 35 °C (black circles). Plots of the non-linear thermodynamic H-H model (red lines), with its free parameters having been previously fitted to data recorded at 15 °C, and its temperature terms set to 35 °C. The model does not describe the data well; (**H**) Experimental K^+ current data recorded at 35 °C (black circles). Plots of the non-linear thermodynamic H-H model (red lines), with its free parameters having been previously fitted to data recorded at 25 °C, and its temperature terms set to 35 °C. The model does not describe the data well; (**I**) Experimental K^+ current data recorded at 35 °C (black circles). Plots of the non-linear thermodynamic H-H model (red lines), with its free parameters having been previously fitted to data recorded at 35 °C, and its temperature terms set to 35 °C. The model describes the data well. For all panels: the *x*-axis presents time (mS), the *y*-axis presents K^+ current (μA).

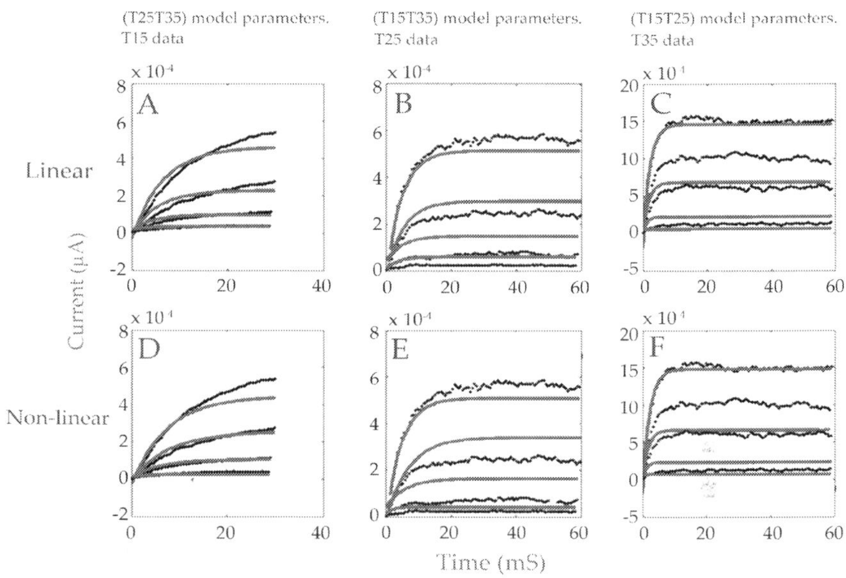

Figure 6. The thermodynamic Hodgkin-Huxley (H-H) model (linear and non-linear), with its free parameters fitted to data recorded at two different temperatures, does not replicate data recorded at a third temperature, when its temperature terms are changed to this third temperature. So, the model's temperature terms do not permit it to replicate data at temperatures that it has not been fitted to. (**A**) Experimental K^+current data recorded at 15 °C (black circles). Plots of the linear thermodynamic H-H model (red lines), with its free parameters having been previously fitted to data recorded at 15 °C and 25 °C, and its temperature terms set to 15 °C. The model does not describe the data well; (**B**) Experimental K^+ current data recorded at 25 °C (black circles). Plots of the linear thermodynamic H-H model (red lines), with its free parameters having been previously fitted to data recorded at 15 °C and 35 °C, and its temperature terms set to 25 °C. The model does not describe the data well; (**C**) Experimental K^+ current data recorded at 35 °C (black circles). Plots of the linear thermodynamic H-H model (red lines), with its free parameters having been previously fitted to data recorded at 15 °C and 25 °C, and its temperature terms set to 35 °C. The model does not describe the data well; (**D**) Experimental K^+ current data recorded at 15 °C (black). Plots of the non-linear thermodynamic H-H model (red lines), with its free parameters having been previously fitted to data recorded at 25 °C and 35 °C, and its temperature terms set to 15 °C. The model does not describe the data well; (**E**) Experimental K^+ current data recorded at 25 °C (black circles). Plots of the non-linear thermodynamic H-H

model (red lines), with its free parameters having been previously fitted to data recorded at 15 °C and 25 °C, and its temperature terms set to 25 °C. The model does not describe the data well; (**F**) Experimental K^+ current data recorded at 35 °C (black circles). Plots of the non-linear thermodynamic H-H model (red lines), with its free parameters having been previously fitted to data recorded at 15 °C and 25 °C, and its temperature terms set to 35 °C. The model does not describe the data well. For all panels: the *x*-axis presents time (mS), the *y*-axis presents K^+ current (μA).

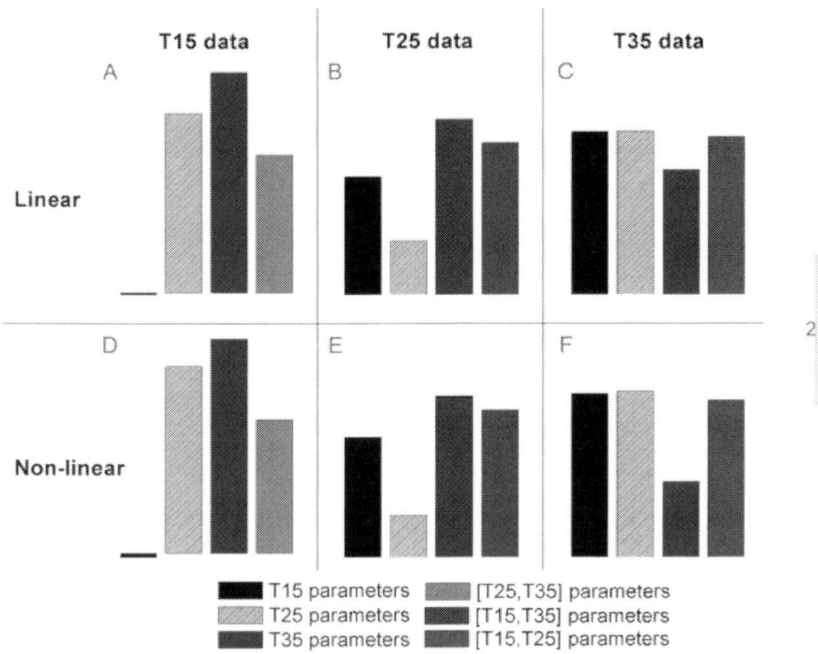

Figure 7. Logged ($\log_{10}$), reduced chi-square values indicating if the linear and non-linear thermodynamic H-H models—with their parameters optimised to data recorded at one (or more) temperature(s)—can describe different data recorded at a different temperature, if their temperature terms are set to this new temperature. (**A**) Bars show how well the linear variant describes K^+ current data recorded at 15 °C, if its free parameters have been fitted to this data (black bar) or if they have been fitted to different data recorded at 25 °C (turquoise bar), or 35 °C (blue bar) or (25 °C and 35 °C) (olive bar); (**B**) Bars show how well the linear variant describes K^+ current data recorded at 25 °C, if its free parameters have been fitted to this data (turquoise bar) or if they have been fitted to different data recorded at 15 °C (black bar), or 35 °C (blue bar) or (15 °C and 35 °C) (wine bar); (**C**) Bars show how well the linear variant describes K^+ current data recorded at 35 °C, if its free parameters have been fitted to this

data (blue bar) or if they have been fitted to different data recorded at 15 °C (black bar), or 25 °C (turquoise bar) or (15 °C and 25 °C) (purple bar); (**D**) Bars show how well the non-linear variant describes K^+ current data recorded at 15 °C, if its free parameters have been fitted to this data (black bar) or if they have been fitted to different data recorded at 25 °C (turquoise bar), or 35 °C (blue bar) or (25 °C and 35 °C) (olive bar); (**E**) Bars show how well the non-linear variant describes K^+ current data recorded at 25 °C, if its free parameters have been fitted to this data (turquoise bar) or if they have been fitted to different data recorded at 15 °C (black bar), or 35 °C (blue bar) or (15 °C and 35 °C) (wine bar); (**F**) Bars show how well the non-linear variant describes K^+ current data recorded at 35 °C, if its free parameters have been fitted to this data (blue bar) or if they have been fitted to different data recorded at 15 °C (black bar), or 25 °C (turquoise bar) or (15 °C and 25 °C) (purple bar). All bars are scaled by the presented *y*-scale bar.

CONCLUSIONS

The Hodgkin-Huxley (H-H) model of voltage-dependent conductances, for which Hogkin and Huxley were awarded the Nobel Prize in 1963, was and is a triumph. However, their model cannot intrinsically account for temperature. A modified variant of their model has been proposed, the thermodynamic H-H model, which suggests that channel transition rates depend on a free energy barrier by analogy with reaction rates. In this study, we investigate if this variant *can* intrinsically account for temperature unlike the original H-H description, which can't. We hypothesise that it might because this variant—unlike the original H-H form—has temperature terms in its equations.

We conclude that the temperature terms of the thermodynamic H-H model (linear or non-linear) do not permit it to describe voltage-gated K^+ current data, over a range of temperatures, with a single set of free parameter values. The model cannot extrapolate for temperature. Why this failure?

Eyring rate theory, which underpins the thermodynamic H-H formalisms, assumes that temperature has little or no effect on the relative energies of the states (Closed, Transition, Open). Its tenet is that temperature changes the probability of moving between states, it does not significantly change the states themselves [4]. However, an ion channel is likely to breach this assumption. Protein structure is temperature dependent [10,11,12] and membrane lipid structure is temperature dependent. The latter is relevant because it can change ion channel structure [13,14] and also because voltage-gated channels are gated by "voltage sensor paddles", which float and move in the lipid medium [15].

Our study only investigates one K^+ current data set and so the conclusions we draw are somewhat provisional. We hope that others will build on this work and follow our methodology with other, further data sets; because our hypothesis is an important issue to address.

The Hodgkin-Huxley model, in all its forms, is an abstraction. The reality is better approximated by a Markov model [2]. It can represent different voltage-gated channel states, and how these states change with voltage and time. They can be accurate and powerful but are computationally expensive to simulate (e.g., [16,17]). So, neuron modelling studies typically use the H-H model as opposed to Markov descriptions (e.g., [18,19]). Hence the H-H model is still relevant and important [20]. Presently, thermodynamic H-H variants are not typically to be found in neuron modelling studies. However, if they can be found to provide intrinsic temperature tenability, then their incorporation will increase the predictive power of neuron models immensely.

Acknowledgments

The author conducted this work in 2003, as a component of his Master degree at the University of Edinburgh [21]. He was funded on this course by the Medical Research Council (MRC) of the UK. The funders had no role in study design, data collection and analysis, decision to publish, or preparation of the manuscript. This work was reviewed by the author, and written up for publication, in 2012 (with no funding appropriated for this task). The author would like to thank David C. Sterratt and Andrew Gillies at the University of Edinburgh who conceived this investigation and who provided much help and guidance; they were the author's project supervisors.

Conflicts of Interest

The authors declare no conflict of interest.

REFERENCES

1. Hodgkin, A.L.; Huxley, A.F. A quantitative description of membrane current and its application to conduction and excitation in nerve. *J. Physiol. (Lond.)* **1952**, *117*, 500–544.
2. Hille, B. *Ion Channels of Excitable Membranes*, 3rd ed.; Sinauer Associates: Sunderland, MA, USA, 2001.
3. Eyring, H. The activated complex in chemical reactions. *J. Chem. Phys.* **1935**, *3*, 107–115.
4. Tsien, R.W.; Noble, D. A transition state theory approach to kinetics of conductance changes in excitable membranes. *J. Membr. Biol.* **1969**, *1*, 248–273.
5. Hill, T.L.; Chen, Y.-D. On the theory of ion transport across the nerve membrane. VI. Free energy and activation free energies of conformational change. *Proc. Natl. Acad. Sci. USA* **1972**, *69*, 1723–1726.
6. Destexhe, A.; Huguenard, J.R. Which formalism to use for modelling voltage-dependent conductances? In *Computational Neuroscience: Realistic modelling for experimentalists*; CRC Press: Boca Raton, FL, USA, 2000.
7. Destexhe, A.; Huguenard, J.R. Nonlinear thermodynamic models of voltage dependent currents. *J. Comput. Neurosci.* **2000**, *9*, 259–270.
8. Tiwari, J.K.; Sikdar, S.K. Temperature dependent conformational changes in a voltage-gated potassium channel. *Eur. Biophys. J.* **1999**, *28*, 338–345.
9. Press, W.H.; Flannery, B.P.; Teukolsky, S.A.; Vetterling, W.T. *Numerical Recipes in C: The Art of Scientific Computing*; Cambridge University Press: Cambridge, UK, 1992.
10. Creighton, T.E. *Proteins*; WH Freeman and Company: New York, NY, USA, 1984.
11. Alberts, B.; Bray, D.; Lewis, J.; Raff, M.; Roberts, K.; Watson, J.D. *Molecular Biology of the Cell*; Garland Science: New York, NY, USA, 1994.
12. Stryer, L. *Biochemistry*; WH Freeman and Company: New York, NY, USA, 1995.

13. Chapman, D. Phase transitions and fluidity characteristics of lipids and cell membranes. *Q. Rev. Biophys.* **1975**, *8*, 185–235.
14. Rosen, A.D. Nonlinear temperature modulation of sodium channel kinetics in GH_3 cells. *Biochim. Biophys. Acta* **2001**, *1511*, 391–396.
15. Jiang, Y.; Ruta, V.; Chen, J.; Lee, A.; MacKinnon, R. The principle of gating charge movement in a voltage-dependent K^+ channel. *Nature* **2003**, *423*, 42–48.
16. Vandenberg, C.A.; Bezanilla, F. A model of sodium channel gating based on single channel, macroscopic ionic, and gating currents in the squid giant axon. *Biophys. J.* **1991**, *60*, 1511–1533.
17. Perozo, E.; Bezanilla, F. Phosphorylation affects voltage gating of the delayed rectifier K^+ channel by electrostatic interactions. *Neuron* **1990**, *5*, 685–690.
18. Forrest, M.D.; Wall, M.J.; Press, D.A.; Feng, J. The Sodium-Potassium Pump Controls the Intrinsic Firing of the Cerebellar Purkinje Neuron. *PLoS One* **2012**, *7*, e51169.
19. Forrest, M.D. Mathematical Model of Bursting in Dissociated Purkinje Neurons. *PLoS One* **2013**, *8*, e68765.
20. Vandenberg, J.I.; Waxman, S.G. Hodgkin and Huxley and the basis for electrical signalling: A remarkable legacy still going strong. *J. Physiol.* **2012**, *590*, 2569–2570.
21. Forrest, M.D. Can Thermodynamic Models of Voltage-Dependent Conductances Extrapolate for Temperature? Master's Thesis, September 2003.

SECTION II: INTEGRATE-AND-FIRE NEURON MODELS: EXTENSIONS, CIRCUITS AND STOCHASTIC ANALYSIS

3 EXTENDING INTEGRATE-AND-FIRE MODEL NEURONS TO ACCOUNT FOR THE EFFECTS OF WEAK ELECTRIC FIELDS AND INPUT FILTERING MEDIATED BY THE DENDRITE

Florian Aspart[1,2], Josef Ladenbauer[1,2], Klaus Obermayer[1,2]

[1] Department of Software Engineering and Theoretical Computer Science, Technische Universität Berlin, Berlin, Germany

[2] Bernstein Center for Computational Neuroscience Berlin, Berlin, Germany

ABSTRACT

Transcranial brain stimulation and evidence of ephaptic coupling have recently sparked strong interests in understanding the effects of weak electric fields on the dynamics of brain networks and of coupled populations of neurons. The collective dynamics of large neuronal populations can be efficiently studied using single-compartment (point) model neurons of the integrate-and-fire (IF) type as their elements. These models, however, lack the dendritic morphology required to biophysically describe the effect of an extracellular electric field on the neuronal membrane voltage. Here, we extend the IF point neuron models to accurately reflect morphology dependent electric field effects extracted from a canonical spatial "ball-and-stick" (BS) neuron model. Even in the absence of an extracellular field, neuronal morphology by itself strongly affects the cellular response properties. We, therefore, derive additional components for leaky and nonlinear IF neuron models to reproduce the subthreshold voltage and spiking dynamics of the BS model exposed to both fluctuating somatic and dendritic inputs and an extracellular electric field. We show that an oscillatory electric field causes spike rate resonance, or equivalently, pronounced spike to field coherence. Its resonance frequency depends on the location of the synaptic background inputs. For somatic inputs the resonance appears in the beta and gamma frequency range, whereas for distal dendritic inputs it is shifted to even higher frequencies. Irrespective of an external electric field, the presence of a dendritic cable attenuates the subthreshold response at the soma to slowly-varying somatic inputs while implementing a low-pass filter for distal dendritic inputs. Our point neuron model extension is straightforward to implement and is computationally much more efficient compared to the original BS model. It is well suited for studying the dynamics of large populations of neurons with heterogeneous dendritic morphology with (and without) the influence of weak external electric fields.

INTRODUCTION

Extracellular electric fields in the brain and their impact on neural activity have gained a considerable amount of attention in neuroscience over the past decade. These electric fields can be generated endogenously [1–3] or through transcranial (alternating) current stimulation [4–6], and can modify the activity of neuronal populations in various ways [1, 7–9]. Although the fields generated by this type of noninvasive brain stimulation are rather weak ($\leq$1 V/m [4, 5]) and do not directly elicit spikes, they can modulate spiking activity and lead to changes in cognitive processing, offering a range of possible clinical interventions [10–12]. How external fields lead to changes of the membrane voltage in single cells has been studied in detail [13–15]. However, their effects on population spike rate and the underlying mechanisms are largely unexplored.

Computational models of neurons exposed to electric fields offer a useful tool to gain a better understanding of these mechanisms. Multi-compartment models of neurons are well suited for corresponding investigations at the level of single cells and small circuits [16] but are too complex for a purposeful application in large populations. Single-compartment (point) neuron models of the integrate-and-fire (IF) type are well applicable to study the dynamics of large neuronal populations, due to their computational efficiency and analytical tractability [17]. However, typical IF model neurons lack the dendritic morphology required for a biophysical description of electric field effects. Furthermore, even in the absence of an extracellular field, the dendritic morphology strongly shapes neuronal response properties [18].

In this contribution, we extend the popular class of IF point neuron models to quantitatively account for morphology dependent modulations of neural activity due to: (i) dendritic influences on the integration of synaptic inputs and (ii) the effects of extracellular electric fields. Furthermore, we describe how oscillatory electric fields affect neuronal subthreshold and spiking activity and identify field-induced spike rate resonance. Specifically, we considered a canonical spatial pyramidal neuron model which consists of a somatic compartment and one (apical main) passive dendritic cable, and which is exposed to in-vivo like fluctuating synaptic input and an electric field. Based on that model we analytically derived an extension for the classical leaky and the refined exponential, [19], IF point neuron models in order to exactly reproduce the subthreshold dynamics of the spatial model for arbitrary parametrizations. We then evaluated the extended IF models by quantitatively comparing their spiking activity with the spiking activity of the corresponding spatial model. Finally, we used these models to study the effects of an oscillating electric field (due to the presence of the dendritic cable) on the spike rate dynamics.

RESULTS

Our derivation of the extended point neuron model consists of two steps. We first calculate the somatic membrane voltage of a ball-and-stick (BS) model in response to subthreshold synaptic inputs at the soma and the distal dendrite and to a time-varying, spatially homogeneous, extracellular electric field. This involves solving a generalized cable equation [20]. Second, we seek to exactly reproduce this voltage response in the point neuron model by deriving additional model components (see Fig 1): two linear temporal filters, one for each input location, to be applied to the "raw" synaptic input and one additional input current to describe the field effect. The model components are given in analytical form and depend on the parameters of the BS model and the electric field. We refer to the model equipped with the new components as the extended point (eP) neuron model. We first derive this extension for the well-known leaky IF (LIF) neuron model, and present the extension adapted for the exponential IF (EIF) neuron model in a separate section.

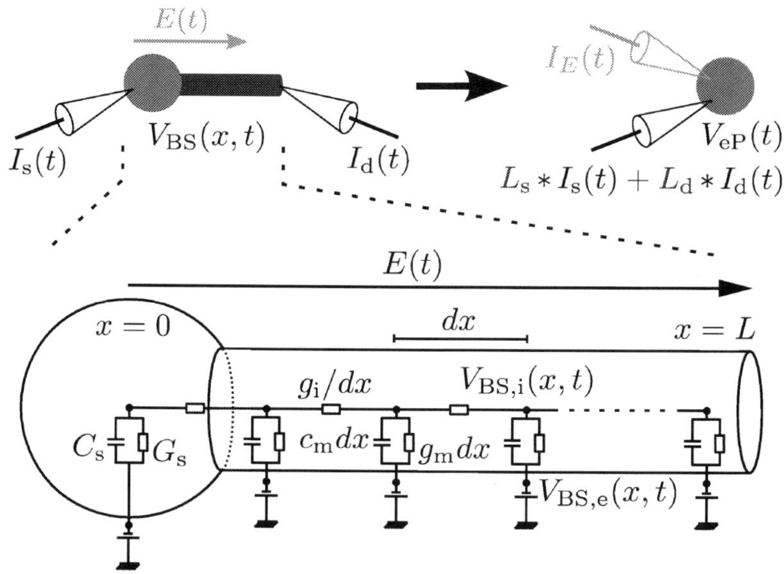

Figure 1. Diagram of the extended point neuron model. Top: Visualization of the ball-and-stick, BS, (left) and the extended point, eP, (right) neuron models. Both models receive synaptic input currents at the soma and the distal dendrite, $I_s(t)$ and $I_d(t)$, and are exposed to an external electric field E(t). $L_s(t)$ and $L_d(t)$ denote the additional input filters describing the dendritic effects. $I_E(t)$ denotes the additional input current describing the field effect. $V_{eP}(t)$ and $V_{BS}(0, t)$ denote the corresponding membrane voltages (at the soma). Bottom: Electrical circuit diagram for the subthreshold dynamics of the BS model. For a description of the parameters and their values see Table 1.

Models

The BS neuron model consists of a lumped soma attached to a passive dendritic cable of length L. The dynamics of its membrane voltage, when receiving synaptic inputs at the soma, $I_s(t)$, and the distal dendrite, $I_d(t)$, and when exposed to a spatially homogeneous external electric field, E(t), are governed by the cable equation:

$$c_m \frac{\partial V_{BS}}{\partial t} - g_i \frac{\partial^2 V_{BS}}{\partial x^2} + g_m V_{BS} = 0, \qquad 0 < x < L, \tag{1}$$

subject to the boundary conditions:

$$C_s \frac{\partial V_{BS}}{\partial t} - g_i \frac{\partial V_{BS}}{\partial x} + G_s V_{BS} - G_s \Delta_T e^{\frac{V_{BS} - V_T}{\Delta_T}} = I_s(t) - g_i E(t), \qquad x = 0, \tag{2}$$

$$\frac{\partial V_{BS}}{\partial x} = \frac{I_d(t)}{g_i} + E(t), \qquad x = L, \tag{3}$$

at the soma (x = 0) and the end of the dendrite (x = L). V_{BS} denotes the deviation of the membrane voltage from rest, V_{rest}, $V_{BS}(x, t) := V_{BS,i}(x, t) - V_{BS,e}(x, t) - V_{rest}$, where $V_{BS,i}$ and $V_{BS,e}$ are the intra- and extracellular potentials. The effects of a spike are described by the IF-type reset condition for the soma:

$$\text{if } V_{BS}(0, t) \geq V_s \text{ then } V_{BS}(0, t) := V_r \tag{4}$$

and by a short refractory period of length T_{ref} during which $V_{BS}(0, t)$ is clamped at the reset value V_r. Spike times are defined by the times at which the somatic membrane voltage $V_{BS}(0, t)$ crosses the spike voltage value V_s from below. c_m denotes the membrane capacitance, g_m the membrane conductance, and g_i the internal (axial) conductance of a dendritic cable segment of unit length. C_s and G_s are the somatic membrane capacitance and leak conductance. The exponential term with threshold slope factor Δ_T and effective threshold voltage V_T approximates the somatic sodium current at spike initiation [19]. For details see Methods.

In the proposed IF point neuron extension, that is, the eP model, the deviation of the membrane voltage, V_{eP}, from rest is governed by

$$C_{eP} \frac{dV_{eP}}{dt} + G_{eP} V_{eP} - \alpha G_{eP} \Delta_T e^{\frac{V_{eP} - V_T}{\Delta_T}} = [L_s * I_s](t) + [L_d * I_d](t) + I_E(t), \tag{5}$$

and by the reset condition:

$$\text{if } V_{eP} \geq V_s \text{ then } V_{eP} := V_r', \tag{6}$$

where V_{eP} is clamped to V_r' for the duration of the refractory period T_{ref} after every spike. C_{eP} and G_{eP} are the membrane capacitance and leak conductance. The scaling factor α ensures an equal membrane voltage response to the depolarizing current described by the exponential terms in both models (BS and eP). We consider two versions of these models separately. First, we treat the LIF versions in detail, where we omit the exponential terms in Eqs 2 and 5; specifically, by taking the limit $\Delta_T \rightarrow 0$ (and setting $V_s = V_T$). In the subsequent part we then consider the (full) EIF versions of the BS and eP models. Below we explain in detail how the components of the point model extension are derived: the linear input filters $L_s(t)$, $L_d(t)$, the additional input current equivalent to the field effect, $I_E(t)$, and, in case of the (full) EIF type models, the scaling factor α. The analytical expressions of these model components are given in Eqs 10, 13, 20 and 21 (for the LIF case), and in Eqs 22–26 (for the EIF case). To mimic the remaining depolarization along the dendritic cable after each spike, we choose an elevated reset voltage for all point neuron models: $V_r' = (V_r + V_T)/2$.

For comparison we also use a point neuron model (of LIF and EIF type, respectively) without the extension, that is, $L_s(t) = L_d(t) = \delta(t)$ and $\alpha = 1$, and we fit the parameters of that model to best reproduce the activity of the BS model for equal synaptic inputs (details see below). We refer to this model as the P model.

The Somatic Input Filter for the LIF Model

We first consider the BS and eP model neurons of the LIF type (i.e, $\Delta_T \rightarrow 0$, $V_s = V_T$) receiving subthreshold synaptic input at the soma in the absence of an electric field ($E(t) = 0$, $I_E(t) = 0$, $I_d(t) = 0$). To avoid ambiguity we use the superscript I_s for the membrane voltage variables in this case. The somatic membrane voltage response of the BS model (Eqs 1–3) can be calculated as (see Methods)

$$\hat{V}_{BS}^{I_s}(0,\omega) = \frac{\hat{I}_s(\omega)}{C_s i\omega + G_s + z(\omega) g_i \tanh(z(\omega)L)}, \tag{7}$$

$$z(\omega) = \sqrt{\frac{g_m + \sqrt{g_m^2 + \omega^2 c_m^2}}{2g_i}} + \operatorname{sgn}(\omega) i \sqrt{\frac{-g_m + \sqrt{g_m^2 + \omega^2 c_m^2}}{2g_i}}, \tag{8}$$

Fig 2B–2D show the amplitudes and phases of the input filter $\hat{L}_s(\omega)$ for various sets of parameters for the BS morphology. $\hat{L}_s(\omega)$ is always a high-pass filter, which attenuates the somatic inputs at lower and amplifies them at higher frequencies. This effect is more pronounced for a larger dendritic and a smaller somatic compartment. It becomes stronger with increasing ratio of dendritic over somatic size. Nevertheless, the filter does not differ qualitatively for changes in neuron morphology.

We next compare how well the point neuron models eP and P reproduce the spiking activity of the BS model neuron. For this purpose we consider an in vivo-like fluctuating synaptic input current $I_s(t)$ described by an Ornstein-Uhlenbeck process (see Methods). The model outputs are compared over a range of values for the input mean I_s^{0} and standard deviation σ_s. The parameter values of the P model were adjusted to best reproduce the spike train of the BS model (see Methods for details). Fig 3A displays the time series of the somatic membrane voltage of the three models in response to the same input currents—a weak (small I_s^{0}, σ_s) and a strong current (large I_s^{0}, σ_s). For both input currents, the eP model well reproduces the somatic voltage dynamics of the BS model. Consequently, the spike times are also well reproduced. There is, however, a mismatch between the voltage traces during short periods (of less than approximately 10 ms duration) after spikes have occurred. This discrepancy is a result of the remaining dendritic depolarization after a spike has occurred in the BS model, which is only approximated by the elevated reset voltage V_r' (see section Models above) in the point neuron models. In comparison, the P model performs worse in reproducing the BS membrane voltage dynamics, particularly the fast fluctuations are poorly recovered. This is expected from the mismatch in the impedance for high frequencies (cf. Fig 2A).

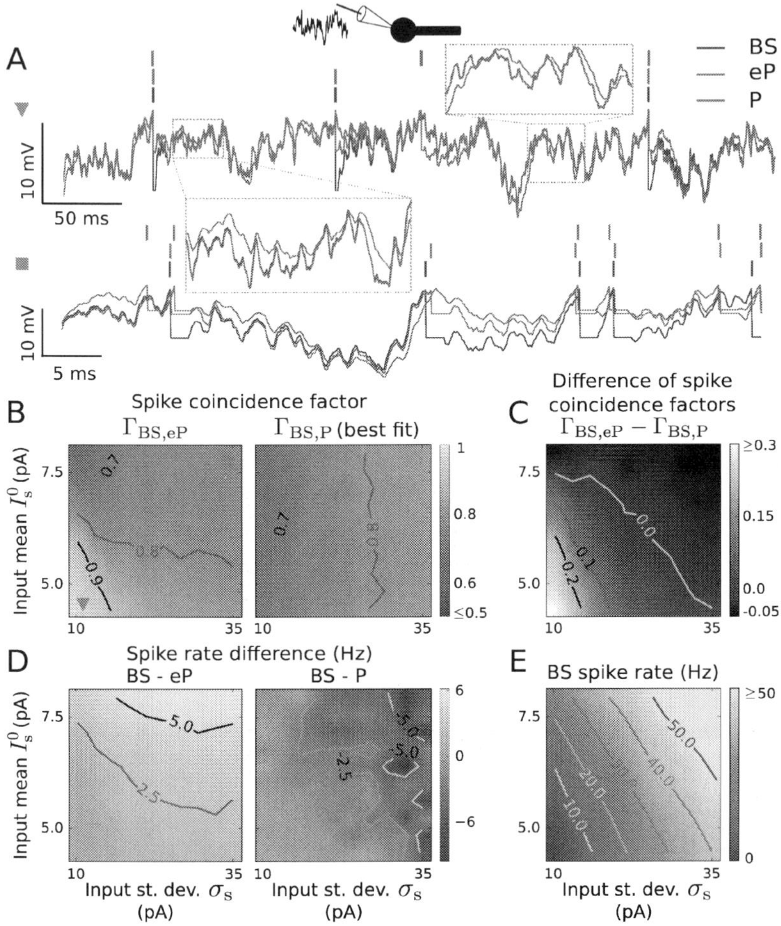

Figure 3. Reproduction of spiking activity for somatic inputs using LIF type models. A: Membrane voltage traces of the BS (blue), eP (green) and P (red) neuron models in response to a weak ($I_s^0 = 4.68$ pA, σ_s = 11.94 pA, top) and a strong input current ($I_s^0 = 7.69$ pA, σ_s = 33.34 pA, bottom). The parameter values of the P model were tuned independently to maximize the coincidence factor $\Gamma_{BS,P}$ for each set of input parameters. B: Coincidence factor for the BS and eP model spike trains, $\Gamma_{BS,eP}$ (left), and for the BS and P model spike trains, $\Gamma_{BS,P}$ (right) as a function of input mean I_s^0 and standard deviation σ_s. C: Difference $\Gamma_{BS,eP} - \Gamma_{BS,P}$ between the coincidence factors shown in B. D: Spike rate difference of the BS and eP models (left) and of the BS and P models (right) as a function of I_s^0 and σ_s. E: Spike rate of the BS neuron model. The input parameters used in A are indicated in B. Results presented in B-E are averages over 6 noise realizations. The parameter values of the BS model are listed in Table 1.

In Fig 3B–3E we compare spiking activity in terms of spike coincidences and spike rates for a wide range of input parameters. We used the spike coincidence measure Γ which quantifies the similarity between two spike trains for a given precision of 3 ms (see Methods). The maximum value of 1 indicates an optimal match, i.e., spike times always coincide, a value of 0 corresponds to pure chance, i.e., the degree of coincidences for two Poisson spike trains with equal rates. The P model was fitted to the BS model for each input (in terms of I_s^0, σ_s) separately. The parameters of the eP model, on the other hand, are constant and do not depend on the input at all. The eP model very accurately reproduces the BS spike times for small spike rates ($\Gamma \geq 0.9$ for small I_s^0 and σ_s), see Fig 3B and 3E. This performance decreases only slightly for increasing σ_s (noise dominated input) and somewhat stronger for increasing I_s^0 (mean dominated input). Generally, Γ decreases with increasing spike rates. This can be attributed to the transient periods after spikes during which the dendritic cable is still loaded and the membrane voltages of both neuron models deviate. Those periods do not depend on the spike rate and therefore have a stronger deteriorating effect when the interspike intervals are smaller. In addition, when σ_s is small the model neurons spike repetitively in a rather clock-like manner, with comparable rate but most likely out of phase due to mismatches caused by the membrane voltage resets. This helps understand the rather low values of Γ for mean dominated inputs. The spike rate of the BS model is also reproduced quite well by the eP model, which underestimates it only slightly (Fig 3D). Spike coincidence and spiking rate reproduction of the eP model can be improved even further by additionally tuning the reset voltage V_r' using Γ or the spike rate distance as a cost function. The P model, in comparison, is substantially worse in reproducing the spike times at small spike rates and only slightly better than the eP model for large spike rates (Fig 3B and 3C). The spike rate of the BS model is slightly overestimated by the P model (Fig 3D). Even though the parameters of the P model were optimized in an input-dependent manner the eP model leads to an improved reproduction of the BS spiking activity overall.

In summary, the dendritic cable implements a high pass filter for inputs at the soma. Due to the derived filter for somatic inputs, the eP model—without having fitted any of its parameters—well reproduces the BS model dynamics for subthreshold and suprathreshold inputs. Notably, the computation time required for the BS model was at least 25 times that of the eP model, using measurements on a single core of a desktop computer.

The Distal Input Filter for the LIF Model

We next consider subthreshold synaptic input at the distal dendrite instead of somatic input, but otherwise the same setup as in the previous section. Here we use superscipt I_d for the membrane voltage variables to better distinguish from the previous scenario. The somatic membrane voltage response of the BS model can be expressed as (see Methods)

$$\hat{V}^{I_d}_{BS}(0,\omega) = \frac{\hat{I}_d(\omega)\,\mathrm{sech}(z(\omega)L)}{C_s i\omega + G_s + z(\omega)\,g_i\,\tanh(z(\omega)L)}, \tag{12}$$

where z(ω) is given by Eq 8. In order to reproduce that voltage response using the eP model, for which $\hat{V}^{I_d}_{eP}(\omega) = \hat{L}_d(\omega)\hat{I}_d(\omega)/(C_{eP}i\omega + G_{eP})$ (cf. Eq 9), we obtain

$$\hat{L}_d(\omega) = \frac{(C_{eP}i\omega + G_{eP})\,\mathrm{sech}(z(\omega)L)}{C_s i\omega + G_s + z(\omega)\,g_i\,\tanh(z(\omega)L)}. \tag{13}$$

As in the previous section, we choose $C_{eP} = C_s$, $G_{eP} = G_s$. In contrast to the somatic input filter L_s the filter L_d for distal inputs exhibits low pass properties for various BS morphologies, see Fig 4A. The shape of this filter is largely independent of the soma size. Compared to the attenuation of low frequency in case of somatic input, the filter gain for high frequency dendritic input is much lower. This results in a stronger filtering effect for dendritic inputs than for somatic inputs.

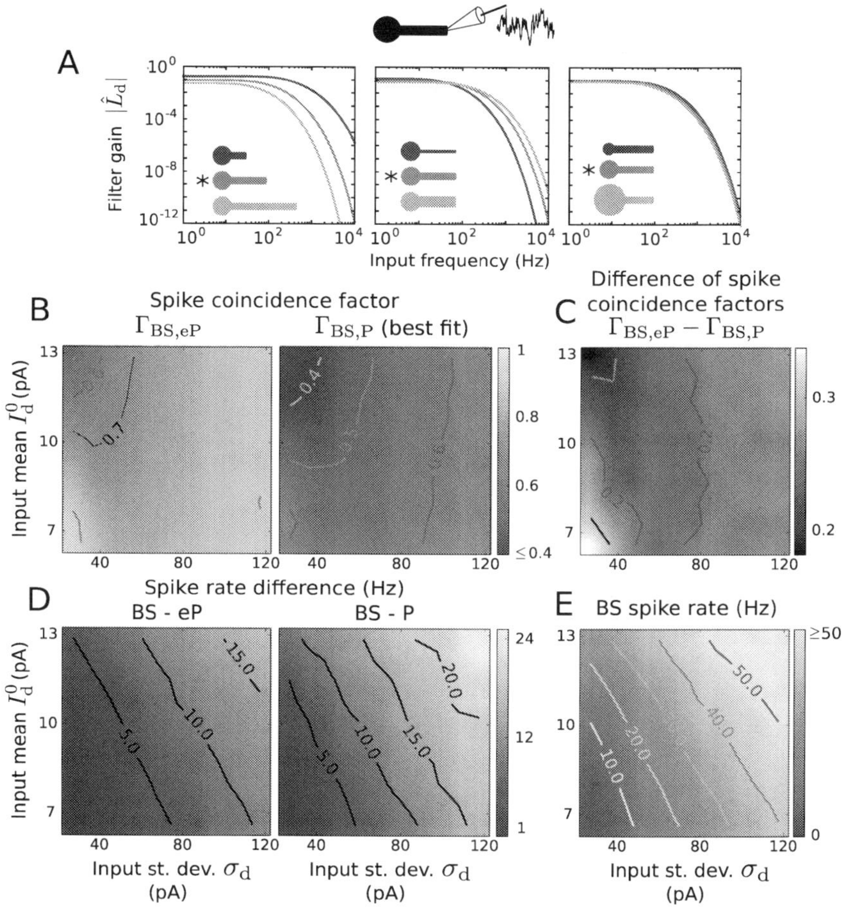

Figure 4. Distal input filter and reproduction of spiking activity using LIF type models. A: Gain and phase of the input filter $\hat{L}_d$ as a function of frequency. The neuronal morphology varied as indicated, in terms of dendritic cable length (350 µm, 700 µm, 1050 µm), cable diameter (0.6 µm, 1.2 µm, 1.8 µm) and soma diameter (5 µm, 10 µm, 15 µm). * indicates the default parameter values. B: Coincidence factor for the BS and eP model spike trains, $\Gamma_{BS,eP}$ (left), and for the BS and P model spike trains, $\Gamma_{BS,P}$ (right) as a function of input mean I_d^0 and standard deviation σ_d. C: Difference $\Gamma_{BS,eP} - \Gamma_{BS,P}$ between the coincidence factors shown in B. D: Spike rate difference of the BS and eP models (left) and of the BS and P models (right) as a function of I_d^0 and σ_d. E: Spike rate of the BS neuron model. Results presented in B-E are averages over 6 noise realizations. The default parameters values of the BS model are listed in Table 1.

An evaluation of the distal input filter in terms of reproduction of BS spiking activity (Γ and rates) is shown in Fig 4B–4E for a range of input mean I_d^0 and standard deviation σ_d values. For comparison we used the P model (without filter) whose parameters were tuned to best reproduce the spike train of the BS model for each input (i.e., (I_d^0, σ_d)-pair) separately. The eP model very accurately reproduces the BS spike times for small spike rates ($\Gamma \geq 0.9$ for small I_d^0 and σ_d). The accuracy drops somewhat as I_d^0 increases, which can be explained as in the previous section. Interestingly, the performance does not deteriorate with increasing spike rate in general; it remains high if the noise intensity σ_d is sufficiently strong ($\Gamma \geq 0.8$ for $\sigma_d \geq 80$ pA, independent of I_d^0 in the considered range). The spike rate of the BS model is somewhat underestimated by the eP model (Fig 4D). It should be noted that the spike rate reproduction could be substantially improved by an increased reset voltage value V_r', as the remaining dendritic depolarization after spikes is more pronounced in case of distal input compared to somatic input. The computational speed-up of the eP model here is the same as in the previous section. The P model, in comparison, is less accurate across all inputs (Fig 4B–4D), even though its parameters depend on the input.

In summary, the dendritic cable implements a low pass filter for inputs at the distal dendrite, and due to the corresponding derived filter the eP model reproduces the BS model dynamics for subthreshold and suprathreshold inputs much better than the P model.

Effect of an Extracellular Electric Field on the Neuronal Dynamics

We now consider an extracellular electric field—in addition to the synaptic inputs—to which the neuron is exposed to. We characterize the effects of that field on the subthreshold somatic membrane voltage and spiking dynamics of the BS neuron and we determine an explicit expression for the additional input current of the extended point neuron model to reproduce these effects. The electric fields we are interested in are oscillatory, spatially uniform on the neuronal scale and weak such as induced by transcranial brain stimulation [6]. In the following, we consider a field with amplitude E_1 and angular frequency φ,

$$E(t) = -\frac{\partial V_{\mathrm{BS,e}}}{\partial x}(t) = E_1 \sin(\varphi t). \tag{14}$$

Recall that $V_{\mathrm{BS,e}}(x, t)$ is the extracellular potential. The BS subthreshold somatic membrane voltage response to this field, $V_{\mathrm{BS}}^{E}(0, t)$, is determined by Eqs 1–3. Using the temporal Fourier transform the solution can be expressed analytically as

$$\hat{V}_{\mathrm{BS}}^{E}(0, \omega) = \frac{\hat{E}(\omega) g_{\mathrm{i}} \left[\operatorname{sech}(z(\omega)L) - 1\right]}{C_{\mathrm{s}} \mathrm{i}\omega + G_{\mathrm{s}} + z(\omega)\, g_{\mathrm{i}} \tanh(z(\omega)L)}, \tag{15}$$

where z(ω) is given by Eq 8 (see Methods). Note, that we again neglect the exponential current in this section (LIF case, but see next section for the EIF case). In the time domain this yields

$$V_{\mathrm{BS}}^{E}(0, t) = |A(\varphi)| \sin(\varphi t + \arg(A(\varphi))), \tag{16}$$

$$A(\varphi) = \frac{E_1 g_{\mathrm{i}} \left[\operatorname{sech}(z(\varphi)L) - 1\right]}{C_{\mathrm{s}} \mathrm{i}\varphi + G_{\mathrm{s}} + z(\varphi)\, g_{\mathrm{i}} \tanh(z(\varphi)L)}, \tag{17}$$

where arg(x) denotes the argument of the complex number x. The overall subthreshold response in presence of the electric field and synaptic input can be decomposed as

$$\hat{V}_{\mathrm{BS}}(0, \omega) = \hat{V}_{\mathrm{BS}}^{I_{\mathrm{s}}}(0, \omega) + \hat{V}_{\mathrm{BS}}^{I_{\mathrm{d}}}(0, \omega) + \hat{V}_{\mathrm{BS}}^{E}(0, \omega), \tag{18}$$

with $\hat{V}_{\mathrm{BS}}^{I_{\mathrm{s}}}(0, \omega)$, $\hat{V}_{\mathrm{BS}}^{I_{\mathrm{d}}}(0, \omega)$ and $\hat{V}_{\mathrm{BS}}^{E}(0, \omega)$ given by Eqs 7, 12 and 15. For the eP model, on the other hand, we have

$$\hat{V}_{\mathrm{eP}}(\omega) = \frac{\hat{L}_{\mathrm{s}}(\omega)\hat{I}_{\mathrm{s}}(\omega) + \hat{L}_{\mathrm{d}}(\omega)\hat{I}_{\mathrm{d}}(\omega) + \hat{I}_{E}(\omega)}{C_{\mathrm{eP}} \mathrm{i}\omega + G_{\mathrm{eP}}}. \tag{19}$$

To guarantee an equal subthreshold response in both models, i.e., $\hat{V}_{\mathrm{eP}}(\omega) = \hat{V}_{\mathrm{BS}}(0, \omega)$, we obtain the following expression for the additional input current,

$$I_E(t) = |B(\varphi)| \sin(\varphi t + \arg(B(\varphi))), \tag{20}$$

$$B(\varphi) = \frac{E_1 g_{\mathrm{i}} (C_{\mathrm{eP}} \mathrm{i}\varphi + G_{\mathrm{eP}})\left[\operatorname{sech}(z(\varphi)L) - 1\right]}{C_{\mathrm{s}} \mathrm{i}\varphi + G_{\mathrm{s}} + z(\varphi)\, g_{\mathrm{i}} \tanh(z(\varphi)L)}, \tag{21}$$

where we set $C_{\mathrm{eP}} = C_{\mathrm{s}}$ and $G_{\mathrm{eP}} = G_{\mathrm{s}}$ (as in the previous sections). It should be noted that these results are not restricted to sinusoidal field variations, as considered here, and can be easily adjusted for any time-varying or constant description of the electric field using its Fourier transform.

The equivalent input current $I_E(t)$ as well as the somatic subthreshold sensitivity to the field, $|A(\varphi)|/E_1$ and the phase shift between oscillating membrane voltage and field, $\arg(A(\varphi))$, with $A(\varphi)$ from Eq 17, are shown in Fig 5. Interestingly, the amplitude of $I_E(t)$ increases with increasing field frequency (Fig 5A), while the sensitivity decreases (Fig 5B). The sensitivity curve changes quantitatively, but not qualitatively, with varying neuronal morphology (Fig 5B). Specifically, its dependence on the field frequency becomes more pronounced with increasing ratio of dendritic size over somatic one. The cable length has the strongest impact in this respect. Notably, the morphology parameters can be adjusted such that the sensitivity curve well matches with empirical results obtained from rat hippocampal pyramidal cells in vitro. The phase shift between the somatic membrane voltage and field oscillations also depends on the field frequency. It exhibits an anti-phase relation for slow oscillations, and decreases with increasing frequency (Fig 5B).

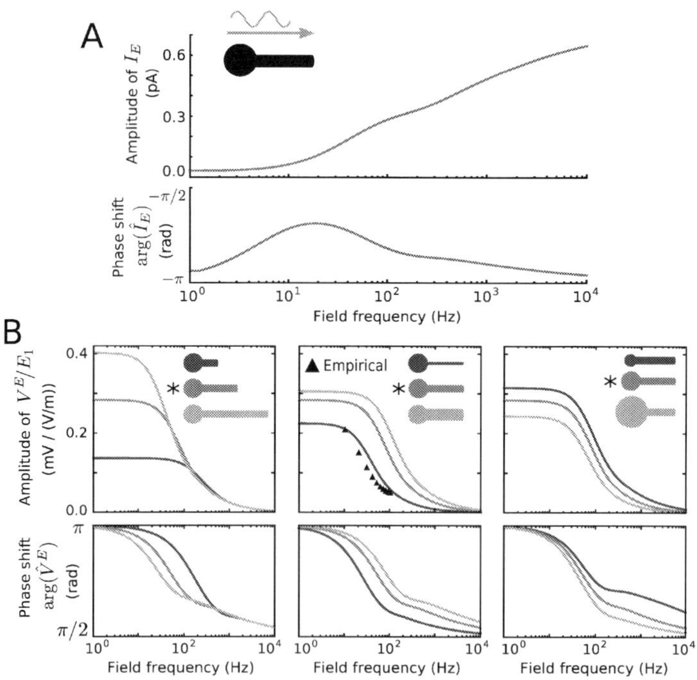

Figure 5. Input current equivalent to the field effect and somatic sensitivity. A: Input current I_E to reproduce the effect of a 1 V/m field in the eP model. Its

amplitude and phase shift relative to the field as a function of field frequency. B: Neuron sensitivity to the field, i.e., the ratio between its somatic membrane voltage amplitude and the field amplitude, and phase shift between the oscillatory membrane voltage and the field. The neuronal morphology varied as indicated, in terms of dendritic cable length (350 μm, 700 μm, 1050 μm), diameter (0.6 μm, 1.2 μm, 1.8 μm) and soma diameter (5 μm, 10 μm, 15 μm). ▲ indicate values obtained from electrophysiological recordings of rat hippocampal pyramidal cells [15]. * indicates the default parameter values. For all other parameter values used see Table 1.

We next assess how the electric field affects spiking activity for a range of field frequencies using the BS and eP models. For that purpose, we simulated both model neurons subject to the field and noisy synaptic input at the soma or at the distal dendrite. The synaptic drive alone is strong enough to cause stochastic spiking with rate r_0. The oscillatory field leads to an oscillatory spike rate modulation quantified as $r_1(\varphi)\sin(\varphi t + \psi(\varphi))$ around the constant baseline spike rate r_0 (see Methods for details). Note that this spike rate modulation measure is related to the frequently used spike field coherence measure.

The amplitude r_1 and phase shift ψ of the spike rate modulation for various somatic inputs (in terms of I_s^0 and σ_s), a range of field oscillation frequencies and two field strengths are shown in Fig 6. The eP model well reproduces the spike rate dynamics of the BS model exposed to the field for all considered field and input parameter values. The amplitude r_1 increases linearly with increasing field magnitude E_1. In contrast to the subthreshold sensitivity to the field (cf. Fig 5B), the spike rate modulation exhibits a clear resonance in the beta and gamma frequency bands across the different inputs. In other words, the spike rate oscillations are strongest for field oscillations of that frequency range. The amplitude peak is more pronounced for stronger inputs and most prominent when the input is dominated by its mean (large I_s^0, small σ_s). This resonance amplitude rapidly increases with increasing baseline spike rate—by increasing both, mean and standard deviation of the background input from small values—and saturates at about $r_0 = 30$ Hz (Fig 6, center). The resonance frequency shifts rather gradually from the beta to the gamma range as the baseline spike rate increases from a few spikes per second to about 60 Hz. The phase shift ψ varies around π, depending on the input and field frequency. Note that $\psi = \pi$ implies that the probability of spiking is largest at the trough of the field oscillation. This results from the orientation of the field, which, in case of $E(t) = E_0 > 0$, induces a (hyper-)polarized somatic membrane voltage.

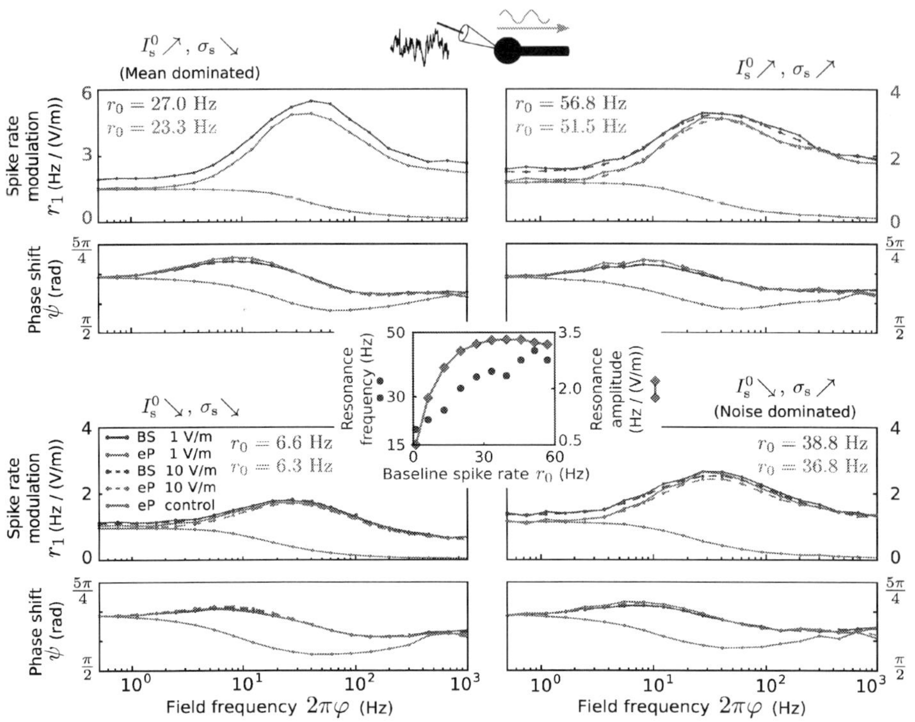

Figure 6. Spike rate modulation due to an electric field for somatic inputs. Top/bottom, left/right: Spike rate modulation of the BS (blue) and the eP (green) models due to an oscillating electric field as a function of its frequency, for different field amplitudes (E_1 = 1 V/m, solid lines; E_1 = 10 V/m, dashed lines) and somatic inputs: $I_s^0 = 7.69$ pA, σ_s = 11.94 pA (top left), $I_s^0 = 7.69$ pA, σ_s = 33.34 pA (top right), $I_s^0 = 4.68$ pA, σ_s = 11.94 pA (bottom left), and $I_s^0 = 4.68$ pA, σ_s = 33.34 pA (bottom right). Magenta lines show the spike rate modulation of the eP model for which I_E was given by $I_E(t) = I_1 \sin(\varphi t + \phi)$ with constant amplitude $I_1 = |B(0.5/(2\pi))|$, phase shift $\phi = \arg(B(0.5/(2\pi)))$, B from Eq 21 and E_1 = 10 V/m. Note the different amplitude scales in the top panel. Results for larger field amplitude (E_1 = 10 V/m) are not displayed for the mean driven regime (top right), because spike rate modulation amplitudes exceeded the baseline rate in that case, which impedes the modulation quantification procedure (see Methods). Center: Resonance frequency $\text{argmax}(r_1)$ and amplitude $\max(r_1)$ of the spike rate modulation of the eP model as a function of baseline spike rate r_0,

which was changed by simultaneously increasing (I_s^0, σ_s) from (4.25 pA, 8.89 pA) to (8.12 pA, 36.40 pA).

To examine the importance of the specific shape of $I_E(t)$, we also considered an alternative sinusoidal input current $I_E(t) = I_1 \sin(\varphi t + \phi)$ for the eP model. Note that the amplitude and phase shift of that current are constant across different field frequencies. Using that current, the typical resonance of the spike rate modulation due to the field cannot even roughly be reproduced (Fig 6).

Let us now inspect spike rate modulation due to the field in presence of distal dendritic inputs instead of somatic ones. In Fig 7 the results are shown for various distal inputs (in terms of I_d^0 and σ_d). Interestingly, for all considered distal dendritic inputs, spike rate modulation amplitudes increase monotonically with the field frequency over the whole considered range (up to 1 kHz, see Discussion for an explanation). Similarly as for somatic inputs, modulation is strongest for mean dominated (large I_d^0, small σ_d) distal inputs, and the phase shift ψ varies around π. Overall, the eP model well reproduces the modulation observed in the BS model.

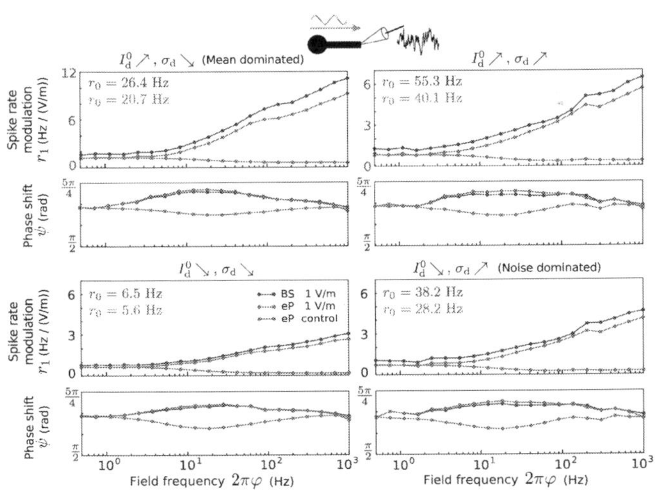

Figure 7. Spike rate modulation due to an electric field for distal dendritic inputs. Spike rate modulation of the BS (blue) and the eP (green) models due to an oscillating electric field as a function of its frequency, for different distal dendritic inputs: $I_d^0 = 12.44$ pA, $\sigma_d = 33.04$ pA (top left), $I_d^0 = 12.44$ pA, $\sigma_d = 111.2$ pA (top right), $I_d^0 = 7.03$ pA, $\sigma_d = 33.04$ pA (bottom left), and $I_d^0 = 7.03$ pA, $\sigma_d = 111.2$ pA

(bottom right). Magenta lines show the spike rate modulation of the eP model for which I_E was given by $I_E(t) = I_1 \sin(\varphi t + \phi)$ with constant amplitude $I_1 = |B(0.5/(2\pi))|$ and phase shift $\phi = \arg(B(0.5/(2\pi)))$ with B from Eq 21 and $E_1 = 10$ V/m. Note the different amplitude scales in the upper panel.

Extension for EIF Model Neurons

In the previous sections, we considered only capacitive and leak currents through the neuronal membrane; the model extension presented there applies to the LIF type model neurons. Here, we consider the BS and eP models described by Eqs 1–3, 5 without neglecting the exponential term, that approximates the voltage dependent sodium current at spike initiation. That is, we derive and evaluate the model extension for model neurons of the EIF type.

To derive the required model components $L_s(t)$, $L_d(t)$, α and $I_E(t)$ we linearize the exponential terms in Eqs 2 and 5 around a baseline voltage value V_0 and then proceed similarly as above. Specifically, we calculate the subthreshold somatic membrane voltage response of the BS model, using the (temporal) Fourier transform, and obtain four response components: $\hat{V}_{\mathrm{BS}}(0,\omega) = \hat{V}_{\mathrm{BS}}^{I_s}(0,\omega) + \hat{V}_{\mathrm{BS}}^{I_d}(0,\omega) + \hat{V}_{\mathrm{BS}}^{\Delta_T}(0,\omega) + \hat{V}_{\mathrm{BS}}^{E}(0,\omega)$, where $V_{\mathrm{BS}}^{I_s}$, $V_{\mathrm{BS}}^{I_d}$ and V_{BS}^{E} denote the voltage response components to I_s, I_d and E, respectively, and the additional term $V_{\mathrm{BS}}^{\Delta_T}$ is due to the (linearized) exponential term. These four voltage response components are given by the explicit expressions Eqs 41–43 in the Methods section. For the eP model, on the other hand, we can also calculate the subthreshold membrane voltage response in the Fourier domain, $\hat{V}_{\mathrm{eP}}(\omega)$, given by Eq 48. By requiring equal subthreshold responses, $\hat{V}_{\mathrm{eP}}(\omega) = \hat{V}_{\mathrm{BS}}(0,\omega)$, we obtain the following explicit expressions for the components L_s, L_d, $\alpha$and I_E, considering the electric field defined in Eq 14:

$$\hat{L}_s(\omega) = \frac{C_{\mathrm{eP}} i\omega + G_{\mathrm{eP}}\left(1 - \alpha e^{\frac{V_0 - V_T}{\Delta_T}}\right)}{C_s i\omega + G_s\left(1 - e^{\frac{V_0 - V_T}{\Delta_T}}\right) + z(\omega)\, g_i \tanh(z(\omega) L)}, \tag{22}$$

$$\hat{L}_{\mathrm{d}}(\omega) = \frac{\left[C_{\mathrm{eP}}\mathrm{i}\omega + G_{\mathrm{eP}}\left(1 - \alpha e^{\frac{V_0 - V_{\mathrm{T}}}{\Delta_{\mathrm{T}}}}\right)\right]\operatorname{sech}(z(\omega)L)}{C_{\mathrm{s}}\mathrm{i}\omega + G_{\mathrm{s}}\left(1 - e^{\frac{V_0 - V_{\mathrm{T}}}{\Delta_{\mathrm{T}}}}\right) + z(\omega)\,g_{\mathrm{i}}\tanh(z(\omega)L)}, \tag{23}$$

$$\alpha = \frac{G_{\mathrm{s}}}{G_{\mathrm{s}} + \tanh(L/\lambda)\,g_{\mathrm{i}}/\lambda}, \tag{24}$$

$$I_E(t) = |B(\varphi)|\,\sin(\varphi t + \arg(B(\varphi))), \tag{25}$$

$$B(\varphi) = \frac{E_{\mathrm{i}}g_{\mathrm{i}}\left[C_{\mathrm{eP}}\mathrm{i}\varphi + G_{\mathrm{eP}}\left(1 - \alpha e^{\frac{V_0 - V_{\mathrm{T}}}{\Delta_{\mathrm{T}}}}\right)\right][\operatorname{sech}(z(\varphi)L) - 1]}{C_{\mathrm{s}}\mathrm{i}\varphi + G_{\mathrm{s}}\left(1 - e^{\frac{V_0 - V_{\mathrm{T}}}{\Delta_{\mathrm{T}}}}\right) + z(\varphi)\,g_{\mathrm{i}}\tanh(z(\varphi)L)}, \tag{26}$$

where z(ω) is given by Eq 8. The scaling factor α guarantees that the voltage response component caused by the exponential term, $V_{\mathrm{BS}}^{\Delta_{\mathrm{T}}}$, is reproduced. In other words, α ensures that the spike initiation current, described by the exponential term, leads to an equal steady state in both models. Note that the two filters for EIF neurons and those for LIF neurons depend on input frequency in qualitatively the same way (by comparing Eqs 22 and 23 with Eqs 10 and 13).

We assessed the reproduction of BS spiking activity by the extended EIF model for somatic inputs using the spike coincidence factor Γ and estimated spike rates (Fig 8). Here again the parameter values of the P model were adjusted to maximize $\Gamma_{\mathrm{BS,P}}$ for each input separately. The range of input parameter values was chosen to obtain similar spike rates as in Fig 3. Despite the linearization in the derivation, the eP model achieves a correct reproduction of the BS spike trains ($\Gamma \geq 0.7$ for a wide range of input parameters). In particular, $\Gamma_{\mathrm{BS,eP}}$ is large for small spike rates (small I_{s}^{0} and σ_{s}) and decreases for increasing I_{s}^{0} (towards mean dominated input), see Fig 8A and 8D. The eP model tends to underestimate the firing rate of the BS model (Fig 8C). This discrepancy in the rate could be reduced by optimizing the point model reset voltage, V_{r}', to better account for the remaining dendritic cable depolarization in the BS model. Similarly, an improved performance of the eP model in terms of spike train reproduction could be achieved by tuning this reset voltage. The P model, on the other hand, rather poorly reproduces the BS spiking dynamics for small input noise intensity ($\Gamma \leq 0.6$ for $\sigma_{\mathrm{s}} \leq 30$

pA, see Fig 8A). Overall, also in presence of the exponential term the eP model clearly outperforms the simpler P model for small spike rates ($\Gamma_{BS,eP} - \Gamma_{BS,P} \geq 0.3$ for small I_s^0 and σ_s) and achieves similar performance for high spiking activity (Fig 8B).

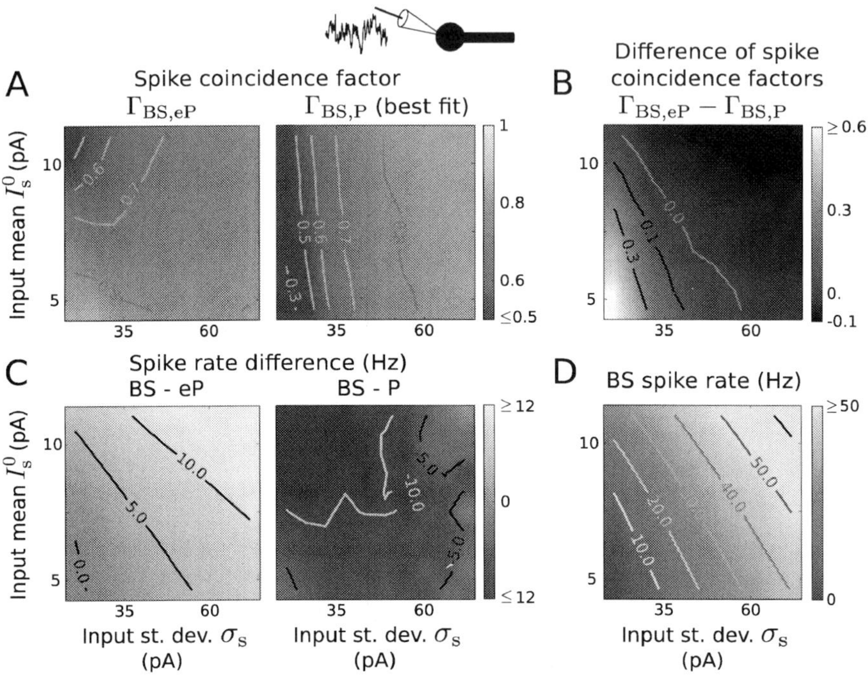

Figure 8. Reproduction of spiking activity for somatic inputs using EIF type models. A: Coincidence factor for the BS and eP model spike trains, $\Gamma_{BS,eP}$ (left), and for the BS and P model spike trains, $\Gamma_{BS,P}$ (right) as a function of input mean I_s^0 and standard deviation σ_s. The parameter values of the P model were optimized to maximize $\Gamma_{BS,P}$ for each input (i.e., (I_s^0, σ_s)-pair) independently. B: Difference $\Gamma_{BS,eP} - \Gamma_{BS,P}$ between the coincidence factors shown in B. C: Spike rate difference of the BS and eP models (left) and of the BS and P models (right) as a function of I_s^0 and σ_s. D: Spike rate of the BS neuron model. Results presented in A-D are averages over 6 noise realizations. The parameter values of the BS model are listed in Table 1.

The reproduction of spiking activity of the BS model was also assessed for distal dendritic inputs. The range of input parameters (I_d^0 and σ_d) was adjusted to obtain similar BS spike rates as for the LIF case. The eP model performs well, in particular for small spike rates or sufficiently strong noise intensity; its performance decreases in the mean driven regime (Fig 9A). On the contrary the P model fails to reproduce the BS spiking activitiy (see Fig A in S1 Text for more details).

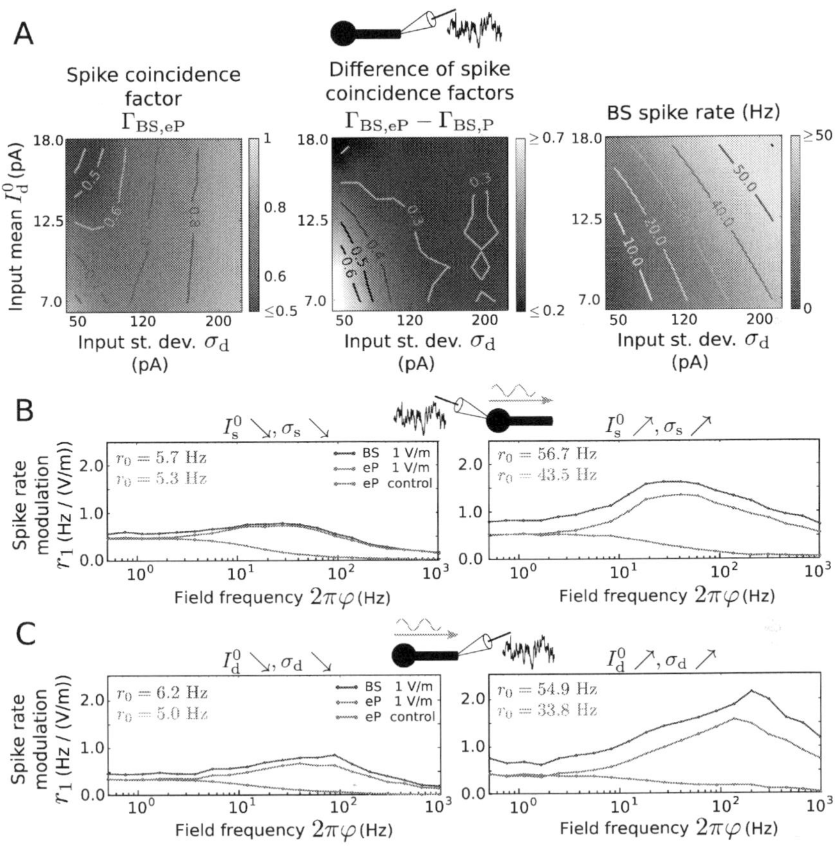

Figure 9. Reproduction of spiking activity for dendritic inputs and spike rate modulation due to an electric field using EIF type models. A: Coincidence factor for the BS and eP model spike trains, $\Gamma_{BS,eP}$ (left) as a function of input mean I_s^0 and standard deviation σ_s. Difference $\Gamma_{BS,eP} - \Gamma_{BS,P}$ between the coincidence factors obtained with the eP and the P models (center). The parameter values of the P model were optimized to maximize $\Gamma_{BS,P}$ for each input (i.e., (I_s^0, σ_s)-pair) independently. Spike rate of the BS neuron model (right). Results are averages over 6 noise realizations. The parameter values of the BS model are listed in Table 1. B: Spike rate modulation of the BS (blue)

and the eP (green) models due to an oscillating electric field as a function of its frequency, for different distal somatic inputs: $I_s^0 = 5.05$ pA, $\sigma_s = 24.08$ pA (left), $I_s^0 = 10.69$ pA, $\sigma_s = 68.21$ pA (right). Magenta lines show the spike rate modulation of the eP model for which I_E was given by $I_E(t) = I_1 \sin(\varphi t + \phi)$ with constant amplitude $I_1 = |B(0.5/(2\pi))|$ and phase shift $\phi = \arg(B(0.5/(2\pi)))$ with B from Eq 21 and $E_1 = 10$ V/m. C: Same as B for dendritic synaptic input instead of somatic one: $I_d^0 = 7.56$ pA, $\sigma_d = 57.73$ pA (left), $I_d^0 = 16.73$ pA, $\sigma_d = 203.41$ pA (right).

In summary, the somatic and distal dendritic input filters obtained for EIF neurons are qualitatively similar to the ones obtained for LIF neurons. The eP model, in contrast to the P model, well reproduces the BS model dynamics for subthreshold and suprathreshold inputs—also for the EIF case.

Spike rate modulations due to an oscillatory electric field using EIF type model neurons for synaptic background input at the soma or distal dendrite are displayed in Fig 9 (see also Fig B and Fig C in S1 Text for additional parameter values of the background input). Similarly to the LIF case, spike rate modulation amplitudes do not decrease monotonically with the field frequency. For somatic background input, we find spike rate resonance in the beta and gamma frequency range, similarly as shown by LIF type models. However, in case of distal dendritic input, EIF neurons exhibit resonance peaks in the high gamma frequency band, in contrast to LIF neurons, whose resonance frequency is substantially larger (see Discussion for an explanation). For both input locations the spike rate modulations shown by the BS model are well reproduced by the eP model and resonance amplitudes are stronger for large spike rates (i.e., large I_s^0, σ_s and large I_d^0, σ_d, respectively).

DISCUSSION

In this contribution we presented an extension for IF point model neurons to accurately reflect the filtering of synaptic inputs caused by the presence of a dendrite and the effects of weak, oscillatory electric fields on neuronal activity. Based on a canonical BS neuron model, we analytically derived additional components for LIF and EIF point neuron models to exactly reproduce the subthreshold voltage dynamics of the spatially extended BS neuron.

These new components consist of (i) two linear filters applied to synaptic inputs depending on their location (soma or distal dendrite) and (ii) an additional input current quantifying the field effect on the membrane voltage. The EIF point model requires an additional scaling parameter to accurately match the BS voltage dynamics. Exhaustive evaluations for suprathreshold in-vivo like fluctuating inputs demonstrated that the BS spiking activity is well reproduced by the extended point neuron model in both cases (LIF and EIF). Optimizing the parameters of the standard LIF and EIF models without the derived extension components, however, does not suffice to adequately reproduce the BS model dynamics.

Due to their computational efficiency the extensions of the point neuron models are well suited for application in large networks to investigate, for example, the effects of neuronal morphology and electrical fields on neuronal spiking activity at the population level. Additionally, our methodological results serve as a building block to derive mean-field descriptions for the collective (spike rate) dynamics of large coupled populations [17, 21, 22, 23, chapter 4.2]. An implementation of the presented models using Python (for the eP model) and NEURON (for the BS model) is freely available at https://github.com/nigroup/IF_extension.

Below, we summarize our results on the obtained input filters and the field effects on neuronal dynamics.

Synaptic input Filtering due to the Dendrite

We have demonstrated that synaptic input is integrated at the soma in distinct ways due to the presence of the dendrite, depending on the input site. Distal dendritic input is low-pass filtered (cf. Fig 4A), in accordance with previous results [24], whereas somatic input is high-pass filtered (cf. Fig 3B). The latter effect is consistent with recent measurements from Purkinje cells and with theoretical results [18] which show a similar change in somatic impedance due to the presence of a dendritic tree (Fig. 4 in [18], in comparison with Fig 2A here). Consequently, the presence of a dendrite can lead to an enhanced neuronal spiking response to high-frequency somatic inputs [18], which may be further amplified by the dendritic effect on the sharpness of spikes at the axon initial segment [25]. The derived IF model extension enables efficient analyses of the BS spike rate response to modulations of the input current—which are, however, not within the scope of this paper.

There are two different strategies for taking into account complex neuron morphologies in models while keeping numerical simulation computationally efficient. One option is to reduce the number of compartments while retaining important properties of the dendritic tree [26]. Alternatively, one can extend point neuron models with temporal kernels which are calibrated to reproduce the somatic membrane voltage response to synaptic inputs as observed in complex morphological cells [27, 28]. Our approach is of the latter type, with the advantage that the temporal kernels (filters) are analytically derived from the underlying morphological BS model.

A similar extension for point model neurons to reproduce dendritic input integration of model cells with complex morphology has been recently proposed in [29]. Using the Green's function formalism a synapse model was developed, whose computational complexity practically allows for only a small number of synaptic input locations. Based on the BS model we were able to derive input filters for point model neurons using only the Fourier transform (without having to rely on the Green's function) and these filters are simple to implement.

We have demonstrated that our extended model outperforms the simpler point neuron model in terms of spike train reproduction. Overall, it performs well for suprathreshold inputs, particularly in case of distal inputs and for somatic inputs that are not too strong. That performance could be further enhanced by optimizing the reset voltage to better reflect the remaining dendritic membrane depolarization in the BS model after each spike, as was mentioned previously.

In our study we have considered passive dendrites. Nonlinear (spike-generating) currents along the dendrite, which cause nonlinear synaptic input integration [30–32], could be incorporated using our approach in a "quasi-active" framework [24]. This would involve solving the cable equation with linearized nonlinear components, similarly as for the exponential terms used here (EIF case).

Effects of Weak Electric Fields on Neuronal Activity

We investigated in detail the effects of a spatially homogeneous, oscillating, weak electric field, as induced by transcranial electrical stimulation, on the activity of the BS neuron. Such a one-dimensional spatial (cable plus soma) model provides a good approximation for neurons with elongated (apical) dendrites exposed to a uniform extracellular electrical field as long as the dendritic (apical main) cable is not substantially smaller than its electrotonic length [33, chapter 2.5]. Following the somatic doctrine [6], we focused on the effects of the field that are due to the polarization of the membrane voltage at the soma. We analytically calculated the subthreshold voltage response, whose properties are in accordance with electrophysiological observations: the response magnitude scales linearly with the field amplitude [13], as shown by the sensitivity in Fig 5. This sensitivity is of the same order of magnitude as that measured in pyramidal cells [15], i.e., around 0.30 mm for low frequency fields, and decreases with increasing field frequency in a morphology dependent manner [14]. For non-uniform electric fields, e.g., as generated by point source stimulation, however, the sensitivity can be roughly constant for frequencies up to at least 100 Hz [8]. Interestingly, such a behavior can also be observed for a uniform field in case of a rather short dendritic cable (cf. Fig 5B).

While polarization effects due to direct current fields have been extensively studied [34–36], the effects of time-varying fields are less well understood. The response of the subthreshold membrane voltage to time-varying fields has been calculated in [37] for a finite dendritic cable with leaky currents at one end, and in [38] for a spatially non-uniform field. Using a one-dimensional cable model [33] showed that the electrotonic length is a key quantity that determines the neuronal subthreshold response to an electric field. Specifically, elongated neurons are less sensitive to high frequency fields than compact ones. How the voltage response to an input current at a particular location along the cable depends on input frequency is largely determined by the membrane time constant. In case of an electrical field, however, which corresponds to symmetrical stimulation at both ends of the cable, the voltage response is also strongly influenced by currents flowing through the low-resistant intracellular medium. This results in an enhanced high frequency response to an extracellular field when compared to an input current [33, chapter 5].

Nevertheless, a somatic compartment was not considered in these studies. Using the BS model we have shown that the relative size of the soma compared to the dendritic cable substantially affects the neuronal sensitivity to the field.

Further, we found frequency-dependent spike rate modulation (and hence, spike field coherence) caused by the electric field. Unlike neuronal subthreshold sensitivity, spike rate modulation amplitude did not decrease with the field frequency and its precise relationship to field frequency depended on the synaptic input location. Spike rate modulation exhibited a clear resonance in the beta and gamma frequency bands in presence of only somatic inputs (cf. Fig 6 and B in S1 Text), whereas for only distal dendritic inputs, spike rate modulation amplitudes are strongest at much larger frequencies (cf. Fig 7 and C in S1 Text). This can be linked to a theoretical result showing that the response of single-compartment model neurons to high frequency inputs is stronger for larger autocorrelation times of a fluctuating synaptic input current [39]. Since fluctuating synaptic inputs arriving at the distal dendrite are low-pass filtered, the autocorrelation time of the corresponding input current felt by the soma is increased (or rather limited from below). Spike rate resonance frequencies were lower for EIF neurons as compared to LIF neurons, in particular for background inputs only at the distal dendrite. This may be explained by the fact that the presence of the exponential term, describing the spike initiating sodium current, decreases the rate response to high frequency inputs [19] (see also the analytical results in [18]). In all cases, the amplitude of the modulation also depended on the input strength (input mean and noise intensity), but its relationship to field frequency was not strongly affected by the input parameters. Recently it has been shown that Purkinje neurons, due to their large dendritic trees, exhibit spike rate resonance at rather high frequencies in response to somatic input modulations and in the presence of noisy dendritic input [18, Fig 5], which is qualitatively similar to the field-induced resonance effects described here (cf. Figs 7 and 9C). It should be noted, however, that an oscillatory (spatially uniform) external field corresponds to oscillatory input currents with opposite sign at the soma and the distal dendrite, respectively (cf. Eqs 2 and 3). The effects of the field can thus not be easily anticipated from those of an input current modulation at the soma alone. Furthermore, the dendritic membrane surface compared to the somatic one for Purkinje cells [18] is substantially larger than that of pyramidal neurons as considered here, which additionally impedes to directly relate the results. Existing experimental studies on the modulation of neuronal activity by extracellular fields have considered a small number of field frequencies (see [40] for a review). Therefore, our results on spike rate resonance are currently not completely confirmed and may be regarded as predictions. In accordance with our findings weak alternating electric fields (of 30 Hz) have been shown to increase the spiking coherence of pyramidal cells in rat hippocampal slices [41], where this increase was proportional to the subthreshold membrane polarization. Moreover, spatially uniform extracellular fields with high-frequency components entrained spiking activity in ferret primary visual cortex more effectively than fields that only contain low-frequency components [1, Fig. S6]. Our predictions on spike rate modulation by an oscillating electric field are thus in agreement with current knowledge and are informative for future experimental studies. Those results may further be of potential interest for the design of transcranial electrical stimulation protocols.

Regarding the point model extension, we analytically derived an expression for an input current to reproduce the effect of the field as extracted from the biophysically grounded BS model. The amplitude and phase of this input current depend on the parameters of the BS neuron and the electric field. Previously, simple phenomenologically obtained input currents have been used for point neuron network simulations, with either constant amplitudes (across frequencies) [1, 9] or amplitudes fitted to electrophysiological data [7]. Interestingly, the latter study used an input current whose magnitude decreases with increasing frequency, in contrast to the equivalent current we obtained (whose magnitude increases with frequency up to 10 kHz). The neuronal subthreshold sensitivity in that study and the ones shown here, however, are similar. This apparent discrepancy in the currents describing the field effect may be explained by the impedance of the applied model neurons, which naturally influences the equivalent input current. In [7] the model parameters (and thus the impedance) were not fitted to real cells; hence it is unlikely that the model impedance matched with the impedance of the cells from which the current amplitudes were estimated [15]. The successful reproduction of the BS spike rate modulation due to the field by the eP model presented here supports the high-pass properties of the equivalent input current.

In the present study, we derived an extension for point neuron models of the LIF and EIF types. Additional model variables with slow dynamics [42] may also be included in this framework, in order to reflect, for example, effects of slowly deactivating potassium channels that mediate spike rate adaptation and associated characteristic neuronal response properties [43, 44]. In that case, a separation of timescales argument could be used to derive the model extension.

The results we extracted from a canonical spatial neuron model provide insight into the effects of cellular morphology on synaptic input integration and the impact of extracellular electric fields on neuronal activity. In particular, the presented point model extension, which is straightforward to implement and efficient to simulate, shall give rise to comprehensive computational investigations of neuronal population activity entrainment due to transcranial stimulation.

METHODS

The Ball-and-Stick (BS) Neuron Model

Model Derivation

The BS neuron model consists of a finite passive dendritic cable of length L with a lumped somatic compartment at one extremity x = 0 and a sealed-end at the other. We consider this neuron model exposed to synaptic inputs at the soma $I_s(t)$ and the distal dendritic end $I_d(t)$ and to an electric field E(t) (see Fig 1). The electric field is spatially uniform at the scale of the neuron, which is a valid assumption for fields induced by transcranial brain stimulation [6]. Assuming a homogeneous, purely ohmic medium (see [45] for the cable equation in a non-ohmic medium), the subthreshold dynamics of the membrane voltage along the dendritic cable are governed by [20]

$$c_m \frac{\partial V_{BS}}{\partial t} - g_i \frac{\partial^2 V_{BS}}{\partial x^2} + g_m V_{BS} = -g_i \frac{\partial E}{\partial x} = 0 \qquad 0 < x < L, \tag{27}$$

$$E(t) = -\frac{\partial V_{BS,e}}{\partial x}(x, t), \tag{28}$$

where $V_{BS}(x, t) := V_{BS,i}(x, t) - V_{BS,e}(x, t) - V_{rest}$, with intra- and extracellular potentials $V_{BS,i}$ and $V_{BS,e}$, respectively. $c_m = cD_d\pi$ is the membrane capacitance per unit length, $g_i = \varrho_i(D_d/2)^2\pi$ is the internal (axial) conductance per unit length and $g_m = \varrho_m D_d\pi$ is the membrane conductance per unit length. c is the specific membrane capacitance (in F/m^2), ϱ_i is the specific internal conductance (in S/m), ϱ_m is the specific membrane conductance (in S/m^2) and D_d is the cable diameter. Note, that the rightmost equality in Eq 27 is due to our assumption of a spatially uniform electric field E(x, t) ≡ E(t).

At the proximal end of the dendritic cable, x = 0, we consider a lumped soma, assuming that the somatic diameter D_s is small compared to the cable length L. The corresponding boundary condition is given by [46]

$$C_s\frac{\partial V_{BS}}{\partial t} - g_i\frac{\partial V_{BS,i}}{\partial x} + G_s V_{BS} - G_s\Delta_T e^{\frac{V_{BS}-V_T}{\Delta_T}} = I_s(t) \qquad x = 0, \tag{29}$$

and thus

$$C_s\frac{\partial V_{BS}}{\partial t} - g_i\frac{\partial V_{BS}}{\partial x} + G_s V_{BS} - G_s\Delta_T e^{\frac{V_{BS}-V_T}{\Delta_T}} = I_s(t) - g_i E(t) \qquad x = 0, \tag{30}$$

where $C_s = cD_s^2\pi$ and $G_s = \varrho_m D_s^2\pi$ are the somatic membrane capacitance and leak conductance, respectively. At the distal end of the dendritic cable, x = L, we have [46]

$$\frac{\partial V_{BS,i}}{\partial x} = \frac{I_d(t)}{g_i} \qquad x = L, \tag{31}$$

due to the synaptic input $I_d(t)$, and therefore,

$$\frac{\partial V_{BS}}{\partial x} = \frac{I_d(t)}{g_i} + E(t) \qquad x = L. \tag{32}$$

The subthreshold voltage dynamics of the BS model are thus determined by Eqs 27, 30 and 32. The spiking mechanism is implemented by the reset condition 4 with refractory period (see Models in the section Results).

Calculation of the Subthreshold Somatic Response

To analytically calculate the somatic membrane voltage response of the BS model, we consider small variations of the synaptic inputs $I_s(t)$, $I_d(t)$ and a weak oscillatory electric field E(t). This allows us to linearize the exponential term in Eq 30 around a baseline voltage value V_0 to obtain

$$C_s\frac{\partial V_{BS}}{\partial t} - g_i\frac{\partial V_{BS}}{\partial x} + G_s\left(1 - e^{\frac{V_0-V_T}{\Delta_T}}\right)V_{BS} = G_s e^{\frac{V_0-V_T}{\Delta_T}}(\Delta_T - V_0) + I_s(t) - g_i E(t) \tag{33}$$

for x = 0. Note that in case of a purely leaky and capacitive neuronal membrane (i.e., without the exponential term, in the limit $\Delta_T \rightarrow 0$) the linearization above is not required and the response calculated below is also exact for larger (subthreshold) synaptic inputs and electric field magnitudes. The linear partial differential Eq 27 together with the boundary conditions 33and 32 can be solved using separation of variables $V_{BS}(x, t) = W(x)U(t)$ and the temporal Fourier transform

$$\hat{V}_{BS}(x,\omega) = W(x)\hat{U}(\omega) = W(x)\int_{-\infty}^{\infty} U(t)e^{i\omega t}dt, \tag{34}$$

where $\omega = 2\pi f$ denotes angular frequency. We obtain the system of differential equations

$$c_m i\omega\hat{V}_{BS} - g_i\frac{\partial^2\hat{V}_{BS}}{\partial x^2} + g_m\hat{V}_{BS} = 0 \qquad 0 < x < L, \tag{35}$$

$$C_s i\omega\hat{V}_{BS} - g_i\frac{\partial\hat{V}_{BS}}{\partial x} + G_s\left(1 - e^{\frac{V_0 - V_T}{\Delta_T}}\right)\hat{V}_{BS} =$$
$$2\pi\delta(\omega)G_s e^{\frac{V_0 - V_T}{\Delta_T}}(\Delta_T - V_0) + \hat{I}_s(\omega) - g_i\hat{E}(\omega) \qquad x = 0, \tag{36}$$

$$\frac{\partial\hat{V}_{BS}}{\partial x} = \frac{\hat{I}_d(\omega)}{g_i} + \hat{E}(\omega) \qquad x = L, \tag{37}$$

where $\hat{}$ indicates the (temporal) Fourier transform and δ(.) the Dirac delta function. The solution of the second order linear differential Eq 35 is given by

$$\hat{V}_{BS}(x,\omega) = a_1(\omega)e^{z(\omega)x} + a_2(\omega)e^{-z(\omega)x}, \tag{38}$$

where $\pm z(\omega)$ are the roots of the characteristic polynomial $g_i\lambda^2 = g_m + c_m i\omega$ of Eq 35:

$$z(\omega) = \sqrt{\frac{g_m + \sqrt{g_m^2 + \omega^2 c_m^2}}{2g_i}} + \operatorname{sgn}(\omega)i\sqrt{\frac{-g_m + \sqrt{g_m^2 + \omega^2 c_m^2}}{2g_i}}, \tag{39}$$

(same as Eq 8). The coefficients $a_1(\omega)$ and $a_2(\omega)$ are calculated by inserting $\hat{V}_{BS}(x,\omega)$ from Eq 38 in Eqs 36 and 37 to finally obtain

$$\hat{V}_{BS}(0,\omega) = \hat{V}_{BS}^{I_s}(0,\omega) + \hat{V}_{BS}^{I_d}(0,\omega) + \hat{V}_{BS}^{\Delta_T}(0,\omega) + \hat{V}_{BS}^{E}(0,\omega) \tag{40}$$

With

$$\hat{V}_{BS}^{I_s}(0,\omega) = \frac{\hat{I}_s(\omega)}{X(\omega)}; \quad \hat{V}_{BS}^{\Delta_T}(0,\omega) = \frac{2\pi\delta(\omega)G_s e^{\frac{V_0 - V_T}{\Delta_T}}(\Delta_T - V_0)}{X(\omega)}, \tag{41}$$

$$\hat{V}_{\mathrm{BS}}^{I_{\mathrm{d}}}(0,\omega) = \frac{\hat{I}_{\mathrm{d}}(\omega)\mathrm{sech}(z(\omega)L)}{X(\omega)}, \quad \hat{V}_{\mathrm{BS}}^{\mathrm{E}}(0,\omega) = \frac{\hat{E}(\omega)g_{\mathrm{i}}[\mathrm{sech}(z(\omega)L) - 1]}{X(\omega)}, \tag{42}$$

And

$$X(\omega) := C_{\mathrm{s}}\mathrm{i}\omega + G_{\mathrm{s}}\left(1 - e^{\frac{V_0 - V_{\mathrm{T}}}{\Delta_{\mathrm{T}}}}\right) + z(\omega)\, g_{\mathrm{i}}\tanh(z(\omega)L). \tag{43}$$

The function sech(x) = cosh(x)$^{-1}$ refers to the hyperbolic secant. Here, $V_{\mathrm{BS}}^{I_{\mathrm{s}}}$, $V_{\mathrm{BS}}^{I_{\mathrm{d}}}$ and V_{BS}^{E} denote respectively the voltage response components to I_s, I_d and E. $V_{\mathrm{BS}}^{\Delta_{\mathrm{T}}}$ is the voltage "response" caused by the (linearized) exponential term. Since E(t) = E_1 sin(φt), the response to the field can be expressed in the time domain as

$$V_{\mathrm{BS}}^{E}(0,t) = |A(\varphi)|\ \sin(\varphi t + \ \arg(A(\varphi))), \tag{44}$$

$$A(\varphi) = \frac{E_1 g_{\mathrm{i}}\,[\mathrm{sech}(z(\varphi)L) - 1]}{C_{\mathrm{s}}\mathrm{i}\varphi + G_{\mathrm{s}}\left(1 - e^{\frac{V_0 - V_{\mathrm{T}}}{\Delta_{\mathrm{T}}}}\right) + z(\varphi)\, g_{\mathrm{i}} \tanh(z(\varphi)L)}. \tag{45}$$

The Extended Point (eP) Neuron Model

The subthreshold voltage dynamics of the eP model is specified by Eq 5 which is complemented by the reset condition 6 together with a refractory period (see Models in the section Results).

Calculation of the subthreshold response.

We consider again small variations of the synaptic inputs I_s(t), I_d(t) and the current I_E(t) that corresponds to a weak oscillatory electric field E(t). Linearizing the exponential term in Eq 5around the steady-state somatic voltage value, V_0, we obtain

$$C_{\mathrm{eP}}\frac{\partial V_{\mathrm{eP}}}{\partial t} + G_{\mathrm{eP}}\left(1 - \alpha e^{\frac{V_0 - V_{\mathrm{T}}}{\Delta_{\mathrm{T}}}}\right) V_{\mathrm{eP}} = G_{\mathrm{eP}}\alpha e^{\frac{V_0 - V_{\mathrm{T}}}{\Delta_{\mathrm{T}}}}(\Delta_{\mathrm{T}} - V_0) + [L_{\mathrm{s}} * I_{\mathrm{s}}](t) + [L_{\mathrm{d}} * I_{\mathrm{d}}](t) + I_E(t). \tag{46}$$

Note that here again in case of a purely leaky and capacitive membrane ($\Delta_T \rightarrow 0$) the linearization above is not required and the response calculated below is also exact for larger (subthreshold) inputs. Using the Fourier transform on Eq 46 yields

$$C_{eP}i\omega\hat{V}_{eP} + G_{eP}\left(1 - \alpha e^{\frac{V_0 - V_T}{\Delta_T}}\right)\hat{V}_{eP} =$$

$$2\pi\delta(\omega)G_{eP}\alpha e^{\frac{V_0 - V_T}{\Delta_T}}(\Delta_T - V_0) + \hat{L}_s(\omega)\hat{I}_s(\omega) + \hat{L}_d(\omega)\hat{I}_d(\omega) + \hat{I}_E(\omega), \quad (47)$$

which can be easily solved to obtain

$$\hat{V}_{eP}(\omega) = \frac{\hat{L}_s(\omega)\hat{I}_s(\omega) + \hat{L}_d(\omega)\hat{I}_d(\omega) + 2\pi\delta(\omega)G_{eP}\alpha e^{\frac{V_0 - V_T}{\Delta_T}}(\Delta_T - V_0) + \hat{I}_E(\omega)}{C_{eP}i\omega + G_{eP}\left(1 - \alpha e^{\frac{V_0 - V_T}{\Delta_T}}\right)}. \quad (48)$$

Numerical Simulation

Synaptic Input and Electric Field

To generate spike trains we considered in-vivo like noisy synaptic inputs $I_s(t)$, $I_d(t)$. Specifically, $I_x(t)$, $x \in \{s, d\}$ was described by an Ornstein-Uhlenbeck process

$$\frac{dI_x}{dt} = \frac{I_x^0 - I_x}{\tau} + \sigma_x\sqrt{\frac{2}{\tau}}\xi_x(t), \quad (49)$$

with mean I_x^0, correlation time τ and standard deviation σ_x. $\xi_x(t)$ is a Gaussian white noise process, i.e. with zero mean $\langle\xi_x(t)\rangle = 0$ and delta autocorrelation $\langle\xi_x(t)\xi_x(t+t')\rangle = \delta(t')$, where $\langle.\rangle$ denotes the expectation operator. Eq 49 was numerically solved using the method described in [53].

The electric field was described by

$$E(t) = E_1 \sin(\varphi t), \quad (50)$$

(same as Eq 14) with amplitude E_1 and angular frequency $\varphi = 2\pi f$. The values for all parameters are provided in Table 1.

Table 1. Description of parameters and applied values.

Parameter (Unit)	Value (range)	Description
c (F/m^2)	$1 \cdot 10^{-2}$ [47, 48]	Specific membrane capacitance
ϱ_m (S/m^2)	1/2.8 [47]	Specific membrane conductance
ϱ_i (S/m)	1/1.5 [47, 48]	Specific internal (axial) conductance
D_s (m)	$\{5, 10^*, 15\} \cdot 10^{-6}$ [49]	Soma diameter
D_d (m)	$\{0.6, 1.2^*, 1.8\} \cdot 10^{-6}$ [50]	Dendritic cable diameter
L (m)	$\{3.5, 7^*, 10.5\} \cdot 10^{-4}$ [51]	Dendritic cable length
C_s(F)	$cD_s^2\pi$	Somatic membrane capacitance
G_s (S)	$\varrho_m D_s^2 \pi$	Somatic membrane conductance
c_m (F/m)	$cD_d \pi$	Dendritic membrane capacitance per unit length
g_m (S/m)	$\varrho_m D_d \pi$	Dendritic membrane conductance per unit length
g_i (S · m)	$\varrho_i (D_d/2)^2 \pi$	Internal (axial) conductance per unit length
V_s (mV)	$\{10, 20\}$	Spike (or cutoff) voltage
V_r (mV)	0	Reset voltage of BS model
V_T (mV)	10 [52]	Threshold voltage
V_0 (mV)	V_r	Baseline voltage for EIF model extension
Δ_T (mV)	1.5 [52]	Threshold slope factor
T_{Ref} (ms)	1.5	Duration of refractory period
C_{eP} (F)	C_s	Membrane capacitance of eP model
G_{eP} (S)	G_s	Membrane conductance of eP model
V_r' (mV)	5	Reset voltage of eP and P models
I_s^0 (pA)	[4.254, 11.407]	Mean input current at the soma
σ_s (pA)	[8.887, 74.512]	Somatic input noise intensity
I_d^0 (pA)	[6.255, 13.214]	Mean input current at the dendrite
σ_d (pA)	[21.875, 122.363]	Dendritic input noise intensity
τ (ms)	0.5	Synaptic current correlation time
E_1 (V/m)	$\{1, 10\}$	Amplitude of electric field
φ (rad)	$[0, 10^4] \cdot 2\pi$	Angular frequency of electric field
Δ (ms)	3	Spike coincidence precision

* indicates default values.

doi:10.1371/journal.pcbi.1005206.t001

Ball-and-stick Neuron Model

The BS neuron model was numerically solved using the NEURON simulation environment [54]. We applied a finite difference space discretization scheme with 50 segments for the dendritic cable and the implicit (or backward) Euler time discretization scheme. The integration time step was fixed to 0.05 ms when the exponential term was omitted ($\Delta_T \rightarrow 0$) and 0.025 ms otherwise. Decreasing the time step size and increasing the number of segments did not lead to noticeable changes in the membrane voltage time series. The time-varying extracellular potential was included using the built-in "extracellular" mechanism in NEURON.

Point Neuron Model

The point neuron models were numerically solved using the forward Euler time discretization scheme and the same time step as used for the BS model. The solution method was implemented in Python using the library

"Numba" for fast computation. Both point model neurons received the same realization of synaptic input $I_x(t)$, $x \in \{s, d\}$ as the BS model neuron. The linear filters $L_x(t)$, $x \in \{s, d\}$ in the eP model were implemented using the fast Fourier transform (FFT) of $I_x(t)$ and the inverse FFT of $\hat{L}_x(\omega)\hat{I}_x(\omega)$. The membrane capacitance and conductance of the eP model were chosen as equal to the corresponding somatic quantities of the BS model, $C_{eP} = C_s$, $G_{eP} = G_s$. Except stated otherwise, the corresponding parameter values of the P model were determined by fitting the BS model spiking activity in terms of spike coincidences. Specifically, G_P was chosen such that the steady-state (somatic) membrane voltage of the BS model is matched exactly, i.e. $Z^x_P(0) = Z^x_{BS}(0)$, $x \in \{s, d\}$, with impedance $Z^x_m(\omega) := \hat{V}^{I_x}_m(\omega)/\hat{I}_x(\omega)$, $m \in \{BS|_{x=0}, P\}$ using Eqs 7, 12 and 11. The value for C_P was then selected to maximize the coincidence factor Γ (defined below) between 52 s lasting spike trains of the BS and P model neurons for each shown pair (I^0_x, σ_x) of input parameter values. In presence of the exponential term in the models ($\Delta_T > 0$) we used $V_0 = V_r$, which is the steady-state (somatic) voltage in the absence of synaptic input. Several other values, within the range $[V_r, V_s]$, were tested for V_0. This did not lead to a substantial improvement of the reproduction performance.

Analysis methods for the spike Trains

Spike Coincidence Measure

To quantify the similarity between the spike trains of the different model neurons we used the coincidence factor Γ defined by [55]

$$\Gamma_{ref,comp} = \frac{N_{coinc} - \langle N_{coinc} \rangle}{(N_{ref} + N_{comp})/2} \frac{1}{\mathcal{N}}, \tag{51}$$

where N_{coinc} is the number of coincident spikes with precision (i.e., maximal temporal separation) Δ, N_{ref} and N_{comp} are the number of spikes in the reference spike train and in the one being compared to it, respectively. $\langle N_{coinc} \rangle = 2r\Delta N_{ref}$ is the expected number of coincidences generated by a homogeneous Poisson process with the same spike rate $r = N_{comp}/T$ as that shown by the compared spike train, where T is the spike train duration. The factor $N = 1\ 2r\Delta$ normalizes $\Gamma_{ref,comp}$ to a maximum value of 1 which is reached if the spike trains match optimally (with precision Δ). $\Gamma_{ref,comp}$ =

0 on the other hand would result from a homogeneous Poisson process with the same rate as for the reference spike train and thus indicates pure chance.

Spike Rate Resonance and Phase Shift Measure

To examine and compare their suprathreshold responses to an oscillatory electric field E(t) (Eq 50), we simulated the neuron models subject to both a field and a noisy synaptic background current. The latter was located either at the soma or at the distal dendrite $I_x(t)$, $x \in \{s, d\}$ (Eq 49). We considered regimes (in terms of (I_x^0, σ_x)-pairs) where the synaptic drive is sufficiently strong to cause the neuron to spike stochastically with rate r_0. The sinusoidal field then causes a modulation of the spike rate that becomes apparent over many trials (i.e., independent realizations of $I_x(t)$). This quantity can also be thought of as the spike rate averaged over a population of neurons which individually receive a noisy synaptic drive but collectively respond to the same oscillatory field. This population, or trial-averaged, instantaneous spike rate can be expressed as

$$r(t) = r_0 + r_1(\varphi)\sin(\varphi t + \psi(\varphi)), \tag{52}$$

where r_1 and ψ denote respectively the amplitude and phase shift, both depending on the angular frequency φ of the field. Note that in the eP model the field effect is described by the oscillatory current $I_E(t)$. To estimate the spike rate modulation at a given field frequency we first extracted and collected the field phase $\phi_s \in [0, 2\pi)$ for each spike time t_s, such that $E(t_s) = E_0\sin(\phi_s)$. These phases were calculated from 944 trials of at least 26 s duration each, for which the first 2 s were disregarded to avoid transients, and only complete field cycles were considered. We then computed a spike rate histogram from the set $\{\phi_s\}$ using 20 equally sized bins and finally applied a sinusoid of the form $F(\phi) = r_0 + r_1 \sin(\phi + \psi)$ with $\phi \in [0, 2\pi)$ to fit that histogram using the method of least squares, where r_0 was given by the histogram mean value. In this way we obtained r_1 and ψ.

SUPPORTING INFORMATION

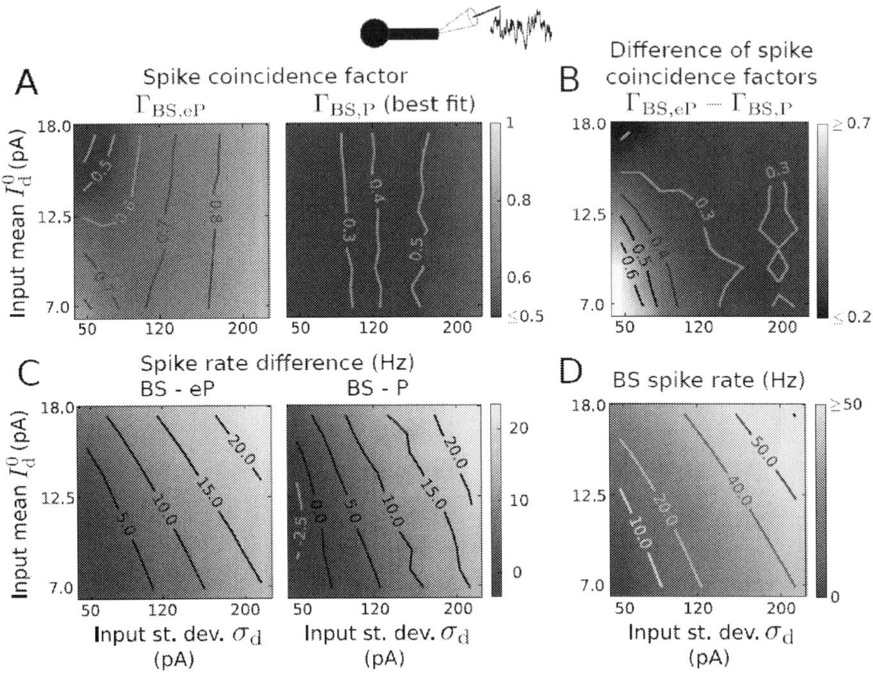

Supplementary Figure A. Reproduction of spiking activity for dendritic inputs using EIF type models A: Coincidence factor for the BS and eP model spike trains, $\Gamma_{BS,eP}$ (left), and for the BS and P model spike trains, $\Gamma_{BS,P}$ (left) as a function of input mean I_d^0 and standard deviation σd. The parameter values of the P model were optimized to maximize ΓBS,P for each input (i.e., (I_d^0, σ_d)-pair) independently. B: Difference $\Gamma_{BS,eP} - \Gamma_{BS,P}$ between the coincidence factors shown in B. C: Spike rate difference of the BS and eP models (left) and of the BS and P models (right) as a function of I_d^0 and σ_d. D: Spike rate of the BS neuron model. Results presented in A-D show averages over 6 noise realizations. The parameter values of the BS model are listed in Table 1.

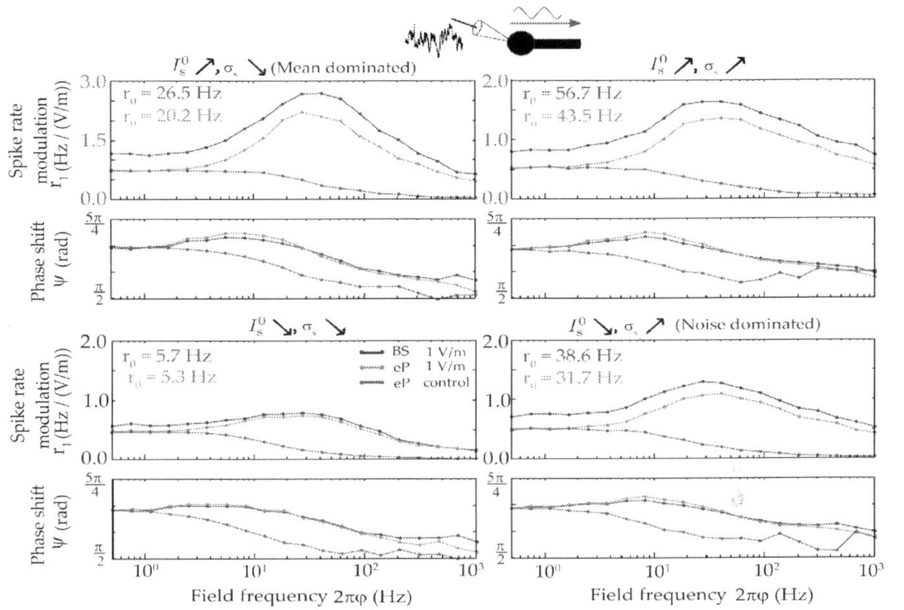

Supplementary Figure B. Spike rate modulation due to an electric field for somatic inputs using neuron models of the EIF type Spike rate modulation of the BS (blue) and the eP (green) models due to an oscillating electric field (E_1 = 1 V/m) as a function of its frequency, for different somatic inputs: I_s^0 = 10.61 pA, σ_s = 24.08 pA (top left), I_s^0 = 10.61 pA, σ_s = 68.21 pA (top right), I_s^0 = 5.05 pA, σ_s = 24.08 pA (bottom left), and I_s^0 = 5.05 pA, σ_s = 68.21 pA (bottom right). Magenta lines show the spike rate modulation of the eP model for which IE was given by $I_E(t) = I_1 \sin(\phi t + \varphi)$ with constant amplitude $I_1 = |B(0.5/(2\pi))|$ and phase shift $\varphi = \arg(B(0.5/(2\pi)))$ with B from Eq. 21 and E_1 = 10 V/m. Note the different amplitude scales in the two top plots.

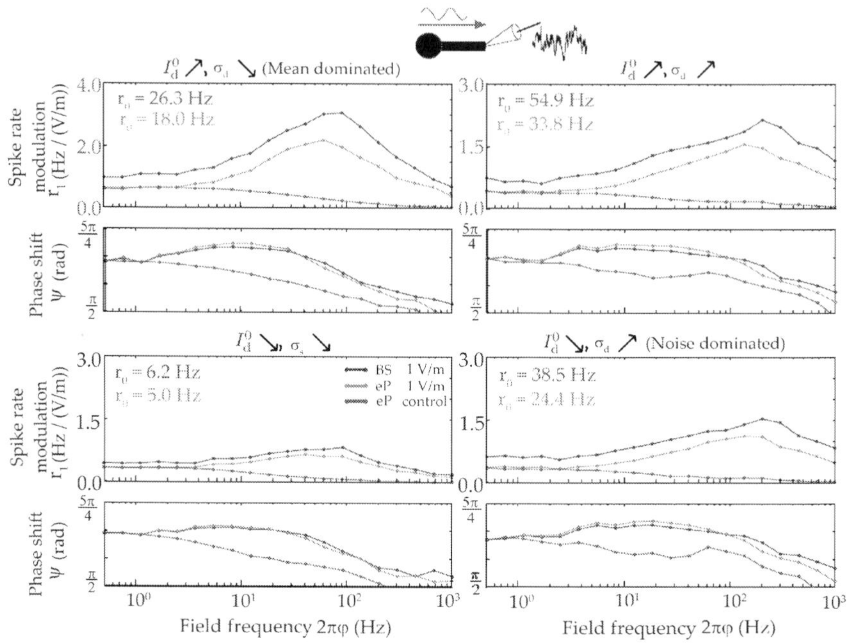

Supplementary Figure C. Spike rate modulation due to an electric field for distal dendritic inputs using neuron models of the EIF type Spike rate modulation of the BS (blue) and the eP (green) models due to an oscillating electric field (E_1 = 1 V/m) as a function of its frequency, for different distal dendritic inputs: I_d^0 = 16.73 pA, σ_d = 57.73 pA (top left), I_d^0 = 16.73 pA, σ_d = 203.41 pA (top right), I_d^0 = 7.56 pA, σ_d = 57.73 pA (bottom left), and I_d^0 = 7.56 pA, σ_d = 203.41 pA (bottom right). Magenta lines show the spike rate modulation of the eP model for which IE was given by $I_E(t) = I_1 \sin(\phi t + \varphi)$ with constant amplitude $I_1 = |B(0.5/(2\pi))|$ and phase shift $\varphi = \arg(B(0.5/(2\pi)))$ with B from Eq. 21 and E_1 = 10 V/m. Note the different amplitude scales in the two top plots.

REFERENCES

1. Fröhlich F, McCormick DA. Endogenous electric fields may guide neocortical network activity. Neuron. 2010;67(1):129–143. pmid:20624597
2. Buzsáki G, Anastassiou CA, Koch C. The origin of extracellular fields and currents—EEG, ECoG, LFP and spikes. Nat Rev Neurosci. 2012;13(6):407–420. pmid:22595786
3. Einevoll GT, Kayser C, Logothetis NK, Panzeri S. Modelling and

analysis of local field potentials for studying the function of cortical circuits. Nat Rev Neurosci. 2013;14(11):770–785. pmid:24135696

4. Neuling T, Wagner S, Wolters CH, Zaehle T, Herrmann CS. Finite-element model predicts current density distribution for clinical applications of tDCS and tACS. Front Psychiatry. 2012;3:83. pmid:23015792
5. Datta A, Bansal V, Diaz J, Patel J, Reato D, Bikson M. Gyri-precise head model of transcranial direct current stimulation: improved spatial focality using a ring electrode versus conventional rectangular pad. Brain Stimul. 2009;2(4):201–207. pmid:20648973
6. Bikson M, Reato D, Rahman A. Cellular and network effects of transcranial direct current stimulation. In: Miniussi C, Paulus W, Rossini PM, editors. Transcranial Brain Stimul. CRC Press; 2012. p. 55–92.
7. Reato D, Rahman A, Bikson M, Parra LC. Low-intensity electrical stimulation affects network dynamics by modulating population rate and spike timing. J Neurosci. 2010;30(45):15067–15079. pmid:21068312
8. Anastassiou CA, Perin R, Markram H, Koch C. Ephaptic coupling of cortical neurons. Nat Neurosci. 2011;14(2):217–223. pmid:21240273
9. Ali MM, Sellers KK, Fröhlich F. Transcranial alternating current stimulation modulates large-scale cortical network activity by network resonance. J Neurosci. 2013;33(27):11262–11275. pmid:23825429
10. Marshall L, Helgadóttir H, Mölle M, Born J. Boosting slow oscillations during sleep potentiates memory. Nature. 2006;444(7119):610–613. pmid:17086200
11. Berenyi A, Belluscio M, Mao D, Buzsaki G. Closed-loop control of epilepsy by transcranial electrical stimulation. Science. 2012;337(6095):735–737. pmid:22879515
12. Herrmann CS, Rach S, Neuling T, Strüber D. Transcranial alternating current stimulation: a review of the underlying mechanisms and modulation of cognitive processes. Front Hum Neurosci. 2013;7:279. pmid:23785325
13. Bikson M, Inoue M, Akiyama H, Deans JK, Fox JE, Miyakawa H, et

al. Effects of uniform extracellular DC electric fields on excitability in rat hippocampal slices in vitro. J Physiol. 2004;557(Pt 1):175–90. pmid:14978199

14. Radman T, Ramos RL, Brumberg JC, Bikson M. Role of cortical cell type and morphology in subthreshold and suprathreshold uniform electric field stimulation in vitro. Brain Stimul. 2009;2(4):215–228. pmid:20161507
15. Deans JK, Powell AD, Jefferys JGR. Sensitivity of coherent oscillations in rat hippocampus to AC electric fields. J Physiol. 2007;583(Pt 2):555–565. pmid:17599962
16. Tiganj Z, Chevallier S, Monacelli E. Influence of extracellular oscillations on neural communication: a computational perspective. Front Comput Neurosci. 2014;8:9. pmid:24570661
17. Brunel N. Dynamics of sparsely connected networks of excitatory and inhibitory spiking neurons. J Comput Neurosci. 2000;8(3):183–208. pmid:10809012
18. Ostojic S, Szapiro G, Schwartz E, Barbour B, Brunel N, Hakim V. Neuronal morphology generates high-frequency firing resonance. J Neurosci. 2015;35(18):7056–7068. pmid:25948257
19. Fourcaud-Trocmé N, Hansel D, van Vreeswijk C, Brunel N. How spike generation mechanisms determine the neuronal response to fluctuating inputs. J Neurosci. 2003;23(37):11628–11640. pmid:14684865
20. Rattay F. Analysis of models for external stimulation of axons. IEEE Trans Biomed Eng. 1986;BME-33(10):974–977. pmid:3770787
21. Richardson MJE. Dynamics of populations and networks of neurons with voltage-activated and calcium-activated currents. Phys Rev E. 2009;80(2):1–16. pmid:19792172
22. Augustin M, Ladenbauer J, Obermayer K. How adaptation shapes spike rate oscillations in recurrent neuronal networks. Front Comput Neurosci. 2013;7:1–11. pmid:23450654
23. Ladenbauer J. The collective dynamics of adaptive neurons: insights from single cell and network models. PhD thesis, Technische Universität Berlin; 2015.
24. Koch C. Biophysics of computation: information processing in

single neurons. Oxford University Press; 2004.

25. Eyal G, Mansvelder HD, de Kock CPJ, Segev I. Dendrites impact the encoding capabilities of the axon. J Neurosci. 2014;34(24):8063–71. pmid:24920612
26. Pinsky PF, Rinzel J. Intrinsic and network rhythmogenesis in a reduced Traub model for CA3 neurons. J Comput Neurosci. 1994;1(1–2):39–60. pmid:8792224
27. Dayan P, Abbott LF. Theoretical neuroscience. vol. 806. Cambridge, MA: MIT Press; 2001.
28. Jolivet R, Lewis TJ, Gerstner W. Generalized integrate-and-fire models of neuronal activity approximate spike trains of a detailed model to a high degree of accuracy. J Neurophysiol. 2004;92(2):959–976. pmid:15277599
29. Wybo WAM, Stiefel KM, Torben-Nielsen B. The Green's function formalism as a bridge between single- and multi-compartmental modeling. Biol Cybern. 2013;107(6):685–694. pmid:24037222
30. Migliore M, Shepherd GM. Emerging rules for the distributions of active dendritic conductances. Nat Rev Neurosci. 2002;3(5):362–370. pmid:11988775
31. Zhou D, Li S, hui Zhang X, Cai D. Phenomenological incorporation of nonlinear dendritic integration using integrate-and-fire neuronal frameworks. PLOS One. 2013;8(1):e53508. pmid:23308241
32. Zhang D, Li Y, Rasch MJ, Wu S. Nonlinear multiplicative dendritic integration in neuron and network models. Front Comput Neurosci. 2013;7:56. pmid:23658543
33. Malik NA. Frequency-dependent response of neurons to oscillating electric fields. PhD thesis, University of Warwick; 2011.
34. Cartee LA, Plonsey R. The transient subthreshold response of spherical and cylindrical cell models to extracellular stimulation. IEEE Trans Biomed Eng. 1992;39(1):76–85. pmid:1572684
35. Plonsey R, Altman KW. Electrical stimulation of excitable cells-a model approach. P IEEE. 1988;76(9):1122–1129.
36. Tranchina D, Nicholson C. A model for the polarization of neurons by extrinsically applied electric fields. Biophys J. 1986;50(6):1139–56. pmid:3801574

37. Monai H, Omori T, Okada M, Inoue M, Miyakawa H, Aonishi T. An analytic solution of the cable equation predicts frequency preference of a passive shunt-end cylindrical cable in response to extracellular oscillating electric fields. Biophys J. 2010;98(4):524–533. pmid:20159148
38. Anastassiou CA, Montgomery SM, Barahona M, Buzsaki G, Koch C. The effect of spatially inhomogeneous extracellular electric fields on neurons. J Neurosci. 2010;30(5):1925–1936. pmid:20130201
39. Brunel N, Chance FS, Fourcaud N, Abbott LF. Effects of synaptic noise and filtering on the frequency response of spiking neurons. Phys Rev Lett. 2001;86(10):2186–2189. pmid:11289886
40. Reato D, Rahman A, Bikson M, Parra LC. Effects of weak transcranial alternating current stimulation on brain activity—a review of known mechanisms from animal studies. Frontiers in human neuroscience. 2013;7:687. pmid:24167483
41. Radman T, Su Y, An JH, Parra LC, Bikson M. Spike timing amplifies the effect of electric fields on neurons: implications for endogenous field effects. J Neurosci. 2007;27(11):3030–3036. pmid:17360926
42. Brette R, Gerstner W. Adaptive exponential integrate-and-fire model as an effective description of neuronal activity. J Neurophysiol. 2005;94(5):3637–3642. pmid:16014787
43. Ladenbauer J, Augustin M, Shiau L, Obermayer K. Impact of adaptation currents on synchronization of coupled exponential integrate-and-fire neurons. PLOS Comput Biol. 2012;8(4):e1002478. pmid:22511861
44. Ladenbauer J, Augustin M, Obermayer K. How adaptation currents change threshold, gain and variability of neuronal spiking. J Neurophysiol. 2014;111(5):939–953. pmid:24174646
45. Bédard C, Destexhe A. Generalized cable theory for neurons in complex and heterogeneous media. Phys Rev E. 2013;88(2):022709. pmid:24032866
46. Tuckwell HC. Introduction to theoretical neurobiology. vol. 1. Cambridge University Press; 1988.
47. Migliore M, Ferrante M, Ascoli GA, Signal GAA. Signal propagation in oblique dendrites of CA1 pyramidal cells. J Neurophysiol. 2005;94(6):4145–4155. pmid:16293591

48. Mainen ZF, Sejnowski TJ. Influence of dendritic structure on firing pattern in model neocortical neurons. Nature. 1996;382(6589):363–366. pmid:8684467
49. Rattay F. The basic mechanism for the electrical stimulation of the nervous system. Neuroscience. 1999;89(2):335–346. pmid:10077317
50. Major G, Larkman AU, Jonas P, Sakmann B, Jack J. Detailed passive cable models of whole-cell recorded CA3 pyramidal neurons in rat hippocampal slices. J Neurosci. 1994;14(8):4613–4638. pmid:8046439
51. Spruston N. Pyramidal neuron. Scholarpedia. 2009;4(5):6130.
52. Badel L, Lefort S, Brette R, Petersen CCH, Gerstner W, Richardson MJE. Dynamic I-V curves are reliable predictors of naturalistic pyramidal-neuron voltage traces. J Neurophysiol. 2008;99(2):656–666. pmid:18057107
53. Gillespie D. Exact numerical simulation of the Ornstein-Uhlenbeck process and its integral. Phys Rev E. 1996;54(2):2084–2091. pmid:9965289
54. Hines ML, Carnevale NT. NEURON: A tool for neuroscientists. Neurosci. 2001;7(2):123–135.
55. Gerstner W, Kistler WM. Spiking neuron models: Single neurons, populations, plasticity. Cambridge University Press; 2002.

4 LEAKY INTEGRATE-AND-FIRE NEURON CIRCUIT BASED ON FLOATING-GATE INTEGRATOR

Vladimir Kornijcuk[1,2], Hyungkwang Lim[1,3], Jun Yeong Seok[1,3], Guhyun Kim[1,3], Seong Keun Kim[1], Inho Kim[1], Byung Joon Choi[2] and Doo Seok Jeong[1]

[1] Center for Electronic Materials, Korea Institute of Science and Technology, Seoul, South Korea
[2] Department of Materials Science and Engineering, Seoul National University of Science and Technology, Seoul, South Korea
[3] Department of Materials Science and Engineering, Seoul National University, Seoul, South Korea

The artificial spiking neural network (SNN) is promising and has been brought to the notice of the theoretical neuroscience and neuromorphic engineering research communities. In this light, we propose a new type of artificial spiking neuron based on leaky integrate-and-fire (LIF) behavior. A distinctive feature of the proposed FG-LIF neuron is the use of a floating-gate (FG) integrator rather than a capacitor-based one. The relaxation time of the charge on the FG relies mainly on the tunnel barrier profile, e.g., barrier height and thickness (rather than the area). This opens up the possibility of large-scale integration of neurons. The circuit simulation results offered biologically plausible spiking activity (<100 Hz) with a capacitor of merely 6 fF, which is hosted in an FG metal-oxide-semiconductor field-effect transistor. The FG-LIF neuron also has the advantage of low operation power (<30 pW/spike). Finally, the proposed circuit was subject to possible types of noise, e.g., thermal noise and burst noise. The simulation results indicated remarkable distributional features of interspike intervals that are fitted to Gamma distribution functions, similar to biological neurons in the neocortex.

KEYWORDS: floating-gate integrator, leaky integrate-and-fire neuron, spiking neural network, synaptic transistor, spatial integration

INTRODUCTION

Ongoing research efforts into spiking neural networks (SNNs) attempt to gain a better understanding of the brain (Gerstner and Kistler, 2002; Markram, 2006) and/or realize its "electronic replicas" that partially imitate brain functionalities such as learning and memory (Mead, 1990; Jeong et al., 2013; Merolla et al., 2014; Qiao et al., 2015). The former generally employs computational SNNs; a vast number of spiking neurons are simulated on computers in search of their behaviors relating to neuronal representation at both low and high levels (Markram, 2006). By contrast, the latter relates physically working hardware SNNs and their components, e.g., spiking neurons and synapses, in favor of real-time interaction with environments, which is referred to as neuromorphic engineering (Mead, 1990). When emulating an SNN with a vast number of neurons, the hardware SNN largely outperforms the computational SNN in terms of runtime, given the latter's need for substantial computational resources. The larger the SNN, the greater the severity of its need for computational resources. The hardware SNN is thus perhaps a good solution to this practical problem, if the artificial neurons and synapses capture their biological counterparts with high precision (Indiveri et al., 2011; Azghadi et al., 2014). The components with limited precision—only capturing the essence of their biological counterparts—can be engaged in neuromorphic systems that are endowed with several brain functionalities such as spatiotemporal recognition despite the component-wise disparity in detailed behavior (Eliasmith and Anderson, 2004; Eliasmith et al., 2012).

Essentially, neurons in a biological network communicate by spikes. The membrane potential of each neuron rises amid incident presynaptic spikes that cause excitatory postsynaptic currents (EPSCs) through the dendrites. That is, the membrane integrates the EPSCs until the membrane potential reaches a threshold for spiking. This procedure is referred to as integrate-and-fire (IF; Burkitt, 2006). This IF procedure lays the foundations of computational neuron models, e.g., the leaky IF neuron (LIF; Burkitt, 2006), Hodgkin-Huxley neuron (Hodgkin and Huxley, 1952), and Izhikevich neuron models (Izhikevich, 2003), and the corresponding hardware models (Mead, 1989; Indiveri et al., 2011; Lim et al., 2015). Among the models, the LIF neuron is one of the most widely used models in light of its simplicity.

In view of real-time interaction with physical environments, it is desirable that the hardware neuron spikes at a rate similar to that of the biological neuron (ca. <100 Hz) given that each spike consumes a certain amount of power. The higher the activity in a given period of time, the more power the neuron consumes. Toward this end, the interspike interval (ISI) between neighboring spikes in time reaches a few tens of milliseconds, which requires a comparable R-C time constant within the framework of the LIF neuron. To put it precisely, a linear low-pass filter, i.e., integrator, in the LIF neuron needs to be endowed with a cutoff frequency below the minimum activity of the biological neuron. Signal integration can be realized in different integrators, e.g., Tau-cell (Edwards and Cauwenberghs, 2000), the subthreshold log-domain integrator by Arthur and Boahen (2004), and a differential pair integrator (Bartolozzi and Indiveri, 2007), and they are nicely reviewed in a paper written by Indiveri et al. (2011). For these integrators, a capacitor causes a delay in the response to an input signal so that the capacitance significantly alters the time delay, partly akin to an R-C delay in a simple R-C circuit.

The FG-based metal-oxide-semiconductor field-effect transistor (MOSFET), FG-MOSFET for short, is one of the most successfully commercialized nonvolatile memory bits in flash memory (Jeong et al., 2012). Remarkable progress in flash memory technology has been made, ranging from the charge trap flash as a variation of the FG-MOSFET to vertical NAND memory. The high maturity level of flash memory technology offers great opportunities for neuromorphic engineering; in particular, FG-MOSFETs are promisingly utilized as programmable synapses that work as local memories within a neuromorphic circuit (Hasler et al., 1994; Gordon et al., 2004; Tenore et al., 2006; Brink et al., 2013; Ramakrishnan et al., 2013). To date, diverse FG-MOSFET-based synapse circuits have been proposed with different precision; the simplest case is the single-transistor synapse device that can maintain the programmed synaptic weight for sufficiently long time periods and implement the spike-timing-dependent plasticity protocol (Hasler et al., 1994; Ramakrishnan et al., 2013). In addition, FG-MOSFETs are also employed as the core part of a synapse circuit (Tenore et al., 2006). Although synapse circuits containing FG-MOSFETs are diverse, it is common that the FG-MOSFETs are responsible for the memory of a programmed synaptic weight.

In this study, we propose an LIF neuron circuit based on a floating-gate (FG) integrator as a replacement for a capacitor integrator. Compared with FGs in synapse circuits, the role of an FG in this type of integrator is counterintuitive given that the FG is deliberately designed to retain the charge on the FG for a few seconds, at most. This poor charge retention is not acceptable in the FGs in synaptic circuits. Circuit simulations were conducted using LTspice IV in support of the proposed circuit. The kinetics of filling the floating gate with charge (charging) and emptying it (discharging) resembles the charging and discharging of a capacitor. However, a significant difference lies in the mechanism for charging and discharging. Charge transfer into and out of the FG is mainly determined by area-independent properties of the tunnel barrier, e.g., barrier height and thickness. Thus, the characteristic time constant—corresponding to that in a capacitor-based integrator—can be tweaked irrespective of the area of the FG, unlike the capacitor-based integrator. As a result, the circuit has excellent potential for scalability and very low power consumption.

MATERIALS AND METHODS

Circuit SIMULATIONS

The circuit simulations were performed using LTspice IV. The LIF neuron circuit was designed by adopting 65-nm complementary metal-oxide-semiconductor (CMOS) technology that was implemented by using the BSIM 4.6.0 model (a built-in model in LTspice IV; Dunga et al., 2006). The parameters for all devices in this work can be found in Table 1. Quantum mechanical elastic tunneling through the tunnel barrier in a tunnel junction is a key phenomenon in the FG integrator; we utilized the tunneling equation included in the BSIM 4.6.0 model (Cao et al., 2000; Lee and Hu, 2001; Dunga et al., 2006). The tunneling equation is based on the Fowler-Nordheim tunneling within the framework of the Wentzel-Kramers-Brillouin approximation. The tunneling equation in the BSIM 4.6.0 model is semi-empirical with regard to the use of an auxiliary function that improves the accuracy of the original Fowler-Nordheim tunneling equation (Ranuárez et al., 2006).

Table 1. Size of MOSFETs in use.

	Transistor number	Channel length (nm)	Channel width (nm)	Gate oxide thickness (nm)
FG integrator	MT1–MT2	60	120	1.3
	M1	60	120	2.5
Amp1	M2–M5	60	240	2.5
Amp2	M6, M8	120	120	2.5
	M7, M9	60	120	2.5
Polarity inverter	M10	600	120	2.5
	M11	60	120	2.5
	M12, M13	120	120	2.5

Noise Implementation

Consecutive random number generation is required for simulating time-varying noise to be applied to the FGLIF neuron circuit. White voltage noise, e.g., thermal voltage noise, was simulated by generating an identical and independently distributed (i.i.d.) random number whose probability follows a normal distribution. The probability distribution function (PDF) is centered at zero with a standard deviation corresponding to the root-mean-square (RMS) amplitude of voltage noise (ΔV_{RMS}). A new random number was repeatedly generated at each time bin (Δt) from this PDF. The gate terminals of all MOSFETs in the neuron circuit were subject to such white noise.

Burst noise and flicker noise as a group of individual burst noises are nonwhite noise. Burst noise in an *n*-channel MOSFET, *n*MOS for short, is estimated to originate from repeated localization and delocalization of electrons by traps at the gate oxide/semiconductor interface (Hung et al., 1990). Both localization (trapping) and delocalization (detrapping) are stochastic and renewal processes with regard to the exponentially decaying PDF with the duration of an empty trap (τ_t) and a filled trap (τ_{det}) (Yonezawa et al., 2013). That is, the interaction between an electron and a trap is a Poisson process. First we defined the electron-trapping (detrapping) rate that evaluates the number of trapping (detrapping) events per unit time as $r_t = 1/\tau_t$ ($r_{det} = 1/\tau_{det}$). A uniform random number in the range between 0 and 1 was generated at each time step and compared with $r_t \cdot \Delta t$ to determine the occurrence of an electron-trapping event; the electron is trapped at the time step if the random number is smaller than $r_t \cdot \Delta t$. The same holds for an electron-detrapping event except that the random number is compared with $r_{det} \cdot \Delta t$. The time bin Δt was sufficiently small (100 μs). For simplicity, it was assumed that $r_t = r_{det}$.

The traps were assumed to be neutral if they were empty and located at the gate oxide/Si interface. Additionally, each trap was assumed to interact with only one electron. Each filled trap induces a change in the flat band voltage on average $\langle \Delta V_{fb} \rangle$ by $\langle \Delta V_{fb} \rangle = \alpha \cdot tox \cdot N_a^{0.6} / \sqrt{L_{eff} \cdot W_{eff}}$, where α, t_{ox}, N_a, L_{eff}, and W_{eff} denote an empirical constant, gate oxide thickness, acceptor density in the channel, and effective channel length and width, respectively (Fukuda et al., 2007). For all MOSFETs, α was set to 1.5×10^{-12}(Fukuda et al., 2007) and N_a to 1.7×10^{17} cm^{-3} (default value in the BSIM 4.6.0 model). ΔV_{fb} is generally distributed following an exponential PDF; the abovementioned $\langle \Delta V_{fb} \rangle$ denotes the expected, i.e., mean, value given the exponential PDF (Fukuda et al., 2007). In this regard, each trap was endowed with a particular random ΔV_{fb} value that was obtained from the exponential distribution. Note that ΔV_{fb} upon an electron-trapping event is positive given the appearance of a negative point charge at the gate oxide/Si interface.

Given the stochastic electron-trapping and detrapping processes, V_{fb} fluctuates with time following a Poisson process. The fluctuation in V_{fb} (ΔV_{fb}) consequently alters the drain current (I_d) at a given gate voltage (V_g) with regard to the consequent change in the threshold voltage V_{th} (ΔV_{th}), and $\Delta V_{th} = \Delta V_{fb}$. The transconductance g_m is a function of the difference between V_g and $V_{th0} + \Delta V_{th}$, i.e., $g_m = f\{V_g - (V_{th0} + \Delta V_{th})\} = f\{(V_g - \Delta V_{fb}) - V_{th0}\}$, where V_{th0} denotes V_{th} in the absence of a trap. Imposing ΔV_{fb} on V_{fb} is thus equivalent to V_g subject to a fluctuation by $-\Delta V_{fb}$ with a noise-free V_{th}, i.e., V_{th0}. In our circuit simulations, the stochastic change in V_{fb} and the resulting change in g_m were simulated by changing V_g by $-\Delta V_{fb}$ at each time step. The aforementioned thermal noise was applied to V_g on top of this burst noise.

To simulate flicker noise, a group of *n* traps was assumed to simultaneously interact with electrons, and the interaction of each trap with an electron was independent of the others. The noise generation algorithm was analogous to the abovementioned burst noise generation except that the independent interactions of all *n* traps with electrons were simultaneously considered. The occurrence of trapping or detrapping at each trap at each time step was determined by comparing $r_t \cdot \Delta t$ or $r_{det} \cdot \Delta t$ with an i.i.d. random number. r_t for each trap was randomly chosen as for the abovementioned single-trap case. This comparison was repeated *n* times with *n* i.i.d. random numbers.

The same method was applied to *p*-channel MOSFETs, *p*MOS for short, regarding the interaction between a trap and a hole (rather than an electron). The donor density in the channel (N_d) was 1.7×10^{17} cm^{-3}, which is the default value in the BSIM 4.6.0 model. The main difference from the *n*MOS case lies in the charge of a filled trap, which is positive and negative for a *p*MOS and *n*MOS, respectively. Consequently, ΔV_{fb} for a *p*MOS upon a hole-trapping event is negative because of the appearance of a positive point charge at the gate oxide/Si interface, which is the opposite of the *n*MOS case. Thus, the equivalent change in V_g is also the opposite of the *n*MOS case.

RESULTS

Circuit Configuration

The proposed floating-gate-based leaky integrate-and-fire (FGLIF) neuron circuit is depicted in Figure 1. The circuit consists of 12 MOSFETs (M2–M13), a single FG transistor (M1 + C_{FG}), and a capacitor C1. The FG transistor has separate terminals MT1 and MT2 (tunnel junctions) for programming charge in the FG through quantum mechanical tunneling. Such an FG transistor is often referred to as a synapse transistor (Diorio et al., 1998; Rahimi et al., 2002). As shown in Figure 1, the circuit is divided into four subcircuits on functional grounds: (a) charge integrator, (b) non-inverting common source amplifier (Amp1), (c) non-inverting common-source amplifier (Amp2) with positive feedback, and (d) polarity inverter.

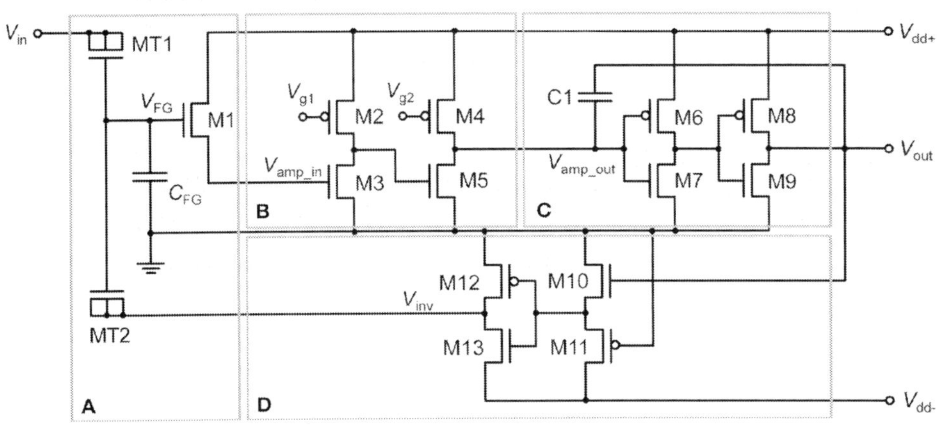

Figure 1. FGLIF neuron circuit. The circuit consists of four subcircuits: **(A)** FG integrator, **(B)**voltage amplifier, **(C)** voltage amplifier with positive feedback, and **(D)** polarity inverter.

To begin, it is worth noting the synapse transistor (Figure 1A) that plays the key role in the proposed LIF neuron circuit. The additional terminals on MT1 and MT2 exclusively control the charge in the FG by means of quantum mechanical tunneling through the tunnel barriers in MT1 and MT2 given the use of a relatively thick gate oxide layer (2.5 nm) in FG transistor M1, which hinders the charge transfer through it. To put it precisely, the FG charge is controlled in a vertical tunnel junction of electrode-barrier-FG-barrier-electrode, following the prototypical synapse transistor proposed by Hasler et al. (1994). In the circuit simulator, the tunnel junctions are implemented by shorting the source, drain, and body terminals of a MOSFET, as shown in Figure 1A. Presynaptic currents are incident on the shorted terminal. All dielectrics in use in the circuit are SiO_2 unless otherwise stated.

Integrator Circuit

The integration in the circuit is realized by the FG integrator in Figure 1A. A charge is injected into the FG upon an incident presynaptic current on the tunnel junction and temporarily remains trapped. FG potential V_{FG} evolves in due course, which consequently changes the channel conductance of M_1. V_{FG} thus parameterizes the integration. The injected charge simultaneously decays with time amid the presynaptic current injection, implying leaky integration that lays the foundation of LIF behavior in the FGLIF neuron. The retention time of an FG charge is mainly determined by the tunnel barrier thickness d_{tun_ox} and area, and the capacitance of C_{FG}. The thick gate oxide of M_1 barely intervenes in the charge ejection on the timescale of our interest. In contrast to an FG transistor as a nonvolatile memory element, our approach aims at the active use of charge-ejection dynamics; therefore, a charge retention time of approximately several seconds is desired. Accordingly, the circuit design and programming voltage are in need of tweaks. Several device parameters of the FG integrator in this work are listed in Table 1. The parameters in Table 2were used for the circuit simulations unless otherwise stated.

Table 2. Circuit parameters used for the circuit simulations.

V_{dd+} (V)	V_{dd-} (V)	V_{g1} (V)	V_{g2}(V)	C_{FG} (fF)	C1 (fF)
0.5	−0.5	0.7	0.65	6	0.15

The synapse transistor has an advantage of employing multiple input tunnel junctions, enabling spatial integration (Polsky et al., 2004) that denotes the simultaneous integration of synaptic currents through different synapses. Figure 2A displays an FG integrator with n identical input terminals ($MT1_1$–$MT1_n$). For $n = 10$, a time-varying V_{FG} in response to an incident spike on a single terminal at 0 s was simulated with the other nine terminals being grounded. For comparison, the same simulation was conducted for $n = 1$. The input spike amplitude (V_{in}) and width (t_{sp}) were 0.5 V and 10 μs, respectively. Figure 2B relates the simulation results that uncover a decaying V_{FG} with different time constants, i.e., relaxation times, which depend on n and d_{tun_ox}. The relaxation time is defined as the requisite time for V_{FG} to reach $1/e$ of V_{FG} at 0 s. The higher the number of input terminals, the shorter the relaxation time since each terminal works as a charge leakage path. Notably, a time constant of ca. 2.7 s for $n = 1$ is significantly reduced to 0.3 s for $n = 10$, as seen in Figure 2B. A workaround solution to such a reduction is to make use of thicker tunnel barriers, which offers a larger relaxation time (Figure 2B). The use of a thicker tunnel barrier trades off V_{FG} at 0 s for a larger relaxation time in light of the difficulty in charge injection upon spike arrival. Thus, one should carefully choose the tunnel barrier thickness that reconciles the charge relaxation (ejection) kinetics with the charge injection kinetics.

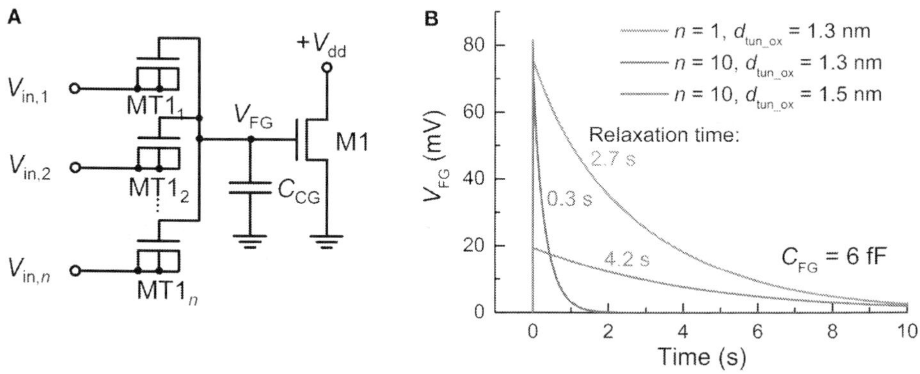

Figure 2. FG integrator realizing spatial integration. (A) FG integrator circuit. **(B)** Different relaxation behaviors of V_{FG}—parameterizing integration—of the FG integrator for different n (1 and 10) and d_{tun_ox} values (1.3 and 1.5 nm). For all cases, the FG integrator was subject to an incident spike ($V_{in} = 0.5$ V, $t_{sp} = 10$ μs) on a single tunnel junction at 0 s while the others were grounded.

Amplifiers

MOSFETs M2–M5 in Figure 1B form a signal amplifier (Amp1) of two inverting common-source stages with *p*MOS loads. The input into Amp1 (V_{amp_in}) is controlled by a voltage drop across the channel of M1, which is determined by V_{FG}. That is, V_{FG} determines the output of Amp1. The output is subsequently relayed to the next amplifier (Amp2) in Figure 1C. The voltage-transfer characteristic (VTC) of each stage in Amp1 is controlled by constant gate voltages (V_{g1} and V_{g2}); they are important in designing the output spike width and spiking threshold. We will set aside this issue until Section Adjustment of Circuit Parameters.

The signal exiting from Amp1 (V_{amp_out}) enters Amp2, which directly elicits a spike at output terminal V_{out} when V_{amp_out} exceeds the transition region in the VTC of Amp2. The transition region is determined by two identical inverters (M6–M7 and M8–M9). The midpoint voltage, where $V_{out} = V_{amp\_out}$, is taken as a threshold voltage of Amp2 for spiking. Capacitor C1 realizes a positive capacitive feedback to the input to Amp2, allowing the output to remain high until the reset of the FG integrator. The reset occurs in rapid succession following the onset of spiking at V_{out} by negative feedback that is achieved by the polarity inverter in Figure 1D.

Polarity Inverter

The FG integrator is in need of a reset in order to complete a single spike. The reset is equivalent to emptying a charge in the FG by means of an electric field. The application of a negative voltage to terminal MT2 lets the previously injected charge vanish. Thus, a subcircuit that inverts V_{out} and relays it to the FG integrator is necessary. The subcircuit in Figure 1D, in conjunction with negative V_{dd} (V_{dd-}), flips the polarity of V_{out}, resulting in a negative output at V_{inv} in Figure 1D. To highlight this polarity inverter, the subcircuit is separately illustrated in Figure 3A, and the simulated VTC at a V_{dd-} of −0.5 V is plotted in Figure 3B. Note that V_{in} in Figure 3A corresponds to V_{out} in Figure 1. The VTC evidences a polarity reversal for V_{in} larger than the midpoint voltage. A negative V_{inv} pulse is accordingly elicited from the polarity inverter in response to a positive V_{in} pulse (0.5 V in amplitude and 25 μs in width), as seen in Figure 3B.

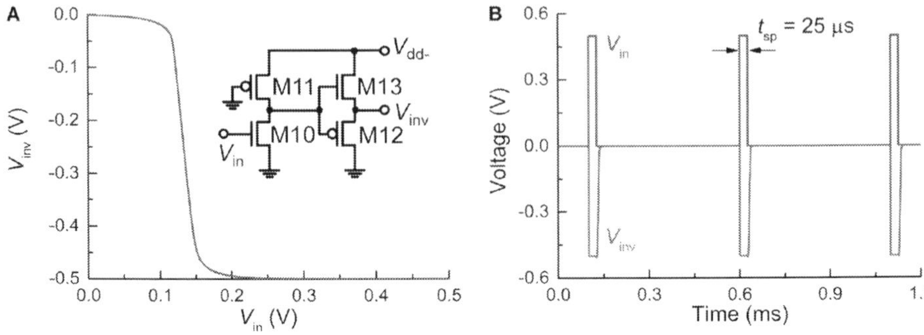

Figure 3. Polarity inverter. (A) VTC of the polarity inverter whose circuit is redrawn in the inset. **(B)** Output V_{inv} (blue solid line) in response to three input spikes (red solid line) in close succession (V_{in} = 0.5 V, t_{sp} = 25 μs).

Provided that such a negative voltage pulse resets the FG integrator, V_{FG} consequently falls below zero. This reset process continues until V_{FG} becomes sufficiently low to let an input to either amplifier fall below the threshold for amplification.

CIRCUIT OPERATION

Dc Input Mode

First, we verify the spiking dynamics of the FGLIF neuron circuit (n = 1) under a constant voltage, which is equivalent to controlled neurophysiology experiments. Applying a constant voltage to MT1 continuously elevates V_{FG} (integration) in light of the positive charge injection into the FG. With that said, the detailed balance (Riggert et al., 2014) eventually reconciles the charge injection with the ejection at a particular V_{FG} level; therefore, the rate of a V_{FG} increase largely declines when it is close to this level. By contrast, an R-C integrator maintains such a balance through charging on C and simultaneous discharging through R. Figure 4B shows the V_{FG} variation in time at 0.26 V amid output spiking. V_{FG} rises in the first place, and thus so does the channel conductance of M1. V_{amp_in} and the resulting V_{amp_out} consequently increase until V_{amp_out} reaches the threshold (ca. 0.29 V) for high V_{out} through Amp2, as plotted in Figure 4C. The high V_{out} is temporally maintained in view of the positive feedback through C1. The high V_{out} simultaneously triggers the polarity inverter that activates negative feedback to the FG integrator, resetting the FG integrator. V_{amp_out} therefore falls below the threshold, leading to the termination of the high V_{out} and negative feedback. This procedure produces a single spike and is repeated for the next spike within the ISI. The reset rate determines the spike width: the faster the reset, the narrower the spike width. This relationship will be addressed in detail in Section Adjustment of Circuit Parameters. The simulation with the parameters in Table 1 uncovered an output spike width of 25 μs and an output activity a_{out} (spiking frequency) of ~23 Hz (Figure 4D).

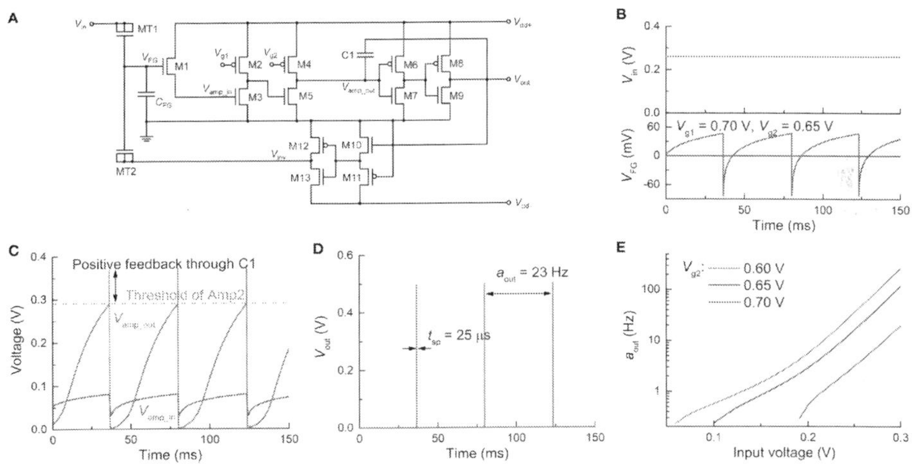

Figure 4. Spiking characteristics in dc input mode. (A) FGLIF neuron circuit (identical to Figure 1). **(B)** Input dc voltage V_{in} (upper panel) and resulting V_{FG} evolution (bottom panel). **(C)**Responses of V_{amp_in} and V_{amp_out} to V_{in}, which correspond to the input and output of Amp1, respectively. The gray dashed line denotes the threshold of Amp2. The increase in V_{amp_out} owing to positive feedback through C1 is indicated. **(D)** Output spikes (t_{sp} = 25 μs, a_{out} = 23 Hz). **(E)** Neuronal gain function for three different V_{g2} values (0.6, 0.65, and 0.7 V).

The neuronal gain function is the substrate of encoding neuronal information (Gerstner and Kistler, 2002; Eliasmith and Anderson, 2004); different inputs are encoded to represent distinguishable outputs. In an attempt to verify a gain function of the FGLIF neuron, the neuronal activity was evaluated at different constant voltages (Figure 4E). A higher input voltage significantly speeds up FG charging within the ISI, reducing the ISI to a large extent. Thus, the activity largely increases with the input voltage. Figure 4E also uncovers a threshold input for spiking

(60, 100, and 190 mV for V_{g2} = 0.60, 0.65, and 0.70 V, respectively), which is determined by the threshold of Amp2. Furthermore, there exists a minimum activity of ~0.2 Hz, and it features the type-II excitability of the Hodgkin-Huxley model (Dayan and Abbott, 2001). The minimum activity is fairly negligible. Note that the exponential change in activity upon voltage (Figure 4E) arises from the tunnel current that exponentially varies upon voltage (Jeong and Hwang, 2005; Soni et al., 2014).

Spiking Input Mode

A spiking input mode realizes practical circumstances for the operation of an FGLIF neuron in an SNN. As a whole, the response of the FGLIF neuron is comparable to the dc input mode, although there are differences to some extent. The subcircuit-wise responses to an input spike train (activity: 100 Hz; spike amplitude: 0.42 V; spike width: 25 μs) are plotted in Figure 5 in the same order as Figure 4. The main difference between the two modes lies in the integration: the input spikes cause stepwise evolution of V_{FG}, unlike the former case (Figure 5A). However, irrespective of an input type, the FG integrator successfully features leaky integration. Analogous to the dc input mode, V_{amp_in} is amplified through Amp1 (Figure 5B), and a spike is elicited from V_{out} when V_{amp_out} crosses the threshold of Amp2 (Figure 5C). Likewise, the output activity differs for different spike amplitudes and activities, as shown in Figure 5D; the larger the amplitude or/and activity, the more frequently the neuron spikes. That is, when information transmission between pre- and postsynaptic neurons is invoked in an SNN, the postsynaptic FGLIF neuron is able to represent the presynaptic neuron›s activity, i.e., input activity for the postsynaptic neuron, by outputting a distinguishable activity.

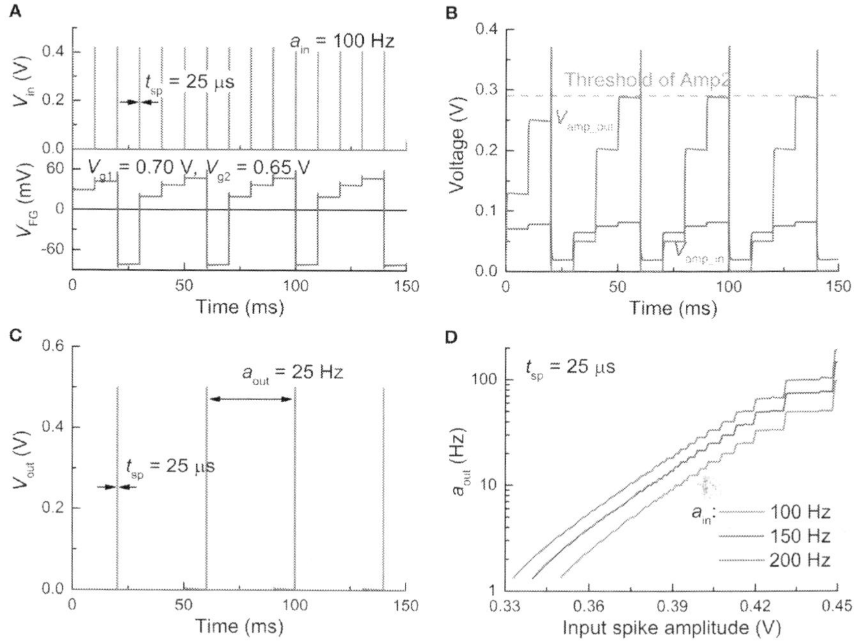

Figure 5. Spiking characteristics in spiking input mode. (A) Input signal profile (V_{in} = 0.42 V, t_{sp} = 25 μs) (upper panel) and resulting V_{FG} evolution (bottom panel). The input activity (a_{in}) was set to 100 Hz. **(B)** Responses of V_{amp_in} and V_{amp_out} to the input spikes. **(C)** Output spikes (t_{sp} = 25 μs, a_{out} = 25 Hz). **(D)** Neuronal gain function for three a_{in} values (100, 150, and 200 Hz).

Adjustment of Circuit Parameters

The proposed circuit provides a means of tweaking neuronal behavior such as relaxation time, spiking threshold, and spike width and amplitude. Recalling the change of the relaxation time upon the tunnel barrier thickness (Section Circuit Simulations), a thicker tunnel barrier, e.g., 1.5 nm for n = 10, is *a priori*preferred in favor of a relaxation timescale that is biologically plausible. The consequent decrease in maximum V_{FG} can be compensated for by increasing the spike width. For instance, for d_{tun_ox} = 1.5 nm, the use of a wider spike (100 μs) raises the maximum V_{FG} by approximately one order of magnitude, per our simulation (not shown).

In addition, gate voltages V_{g1} and V_{g2} of the loads in Amp1 alter the VTC. For instance, Figure 6A shows the VTC of Amp1 for three different V_{g2} values at the same V_{g1} (0.7 V) where significant changes in the VTC are seen. The same holds for V_{g1} as shown in Figure 6B; however, the VTC merely shifts relying on V_{g1}. Provided that high V_{out} (spiking) is triggered only if V_{amp_out} reaches the threshold of Amp2 (denoted by a dashed line in Figure 6A), V_{amp_in} for spiking substantially varies upon V_{g2}. In Figure 6A, a higher V_{g2} leads to a higher V_{amp_in} for spiking; therefore, spiking requires a higher V_{FG}. A higher V_{FG} is in need of a longer integration time at a given input voltage in the dc input mode, or equivalently at a given input activity (if sufficiently high to evoke a spike) in the spiking input mode. Thus, output activity a_{out} declines with V_{g2} (Figure 6C). On the same grounds, a higher V_{FG} requires a higher V_{in} to output the same activity. As a consequence, the threshold for spiking in the dc input mode increases with V_{g2}, as shown in Figure 6D. Therefore, the neuronal gain function can be easily modified by tweaking V_{g2}.

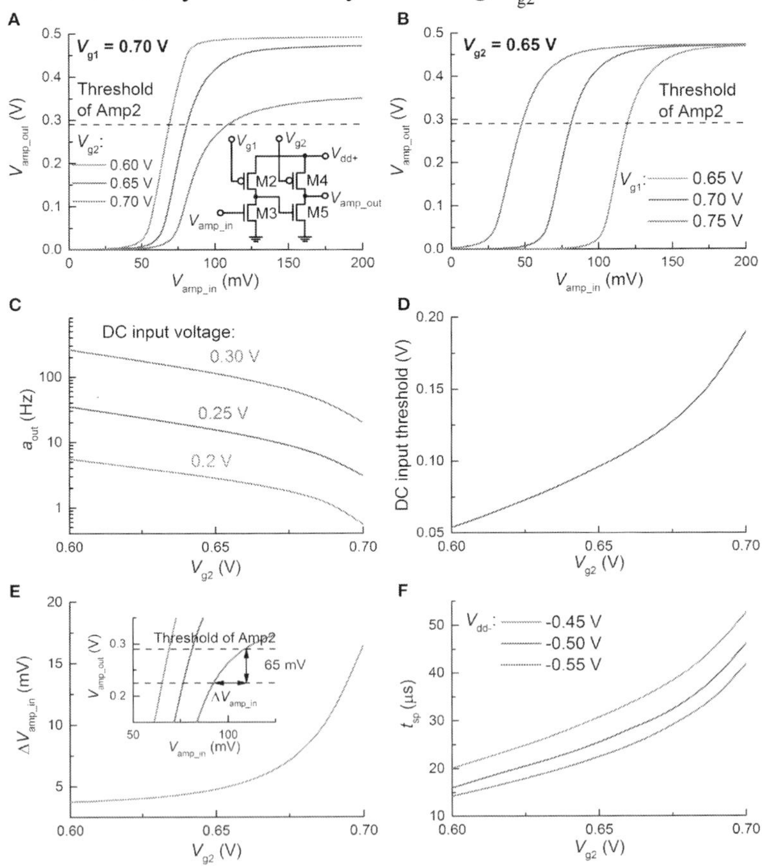

Figure 6. Change in spiking dynamics upon circuit parameters. Change in VTC of Amp1 **(A)** for three V_{g2} values and the same V_{g1} (0.7 V), and **(B)** for

three V_{g1} and the same V_{g2} (0.65 V). **(C)** Change in a_{out} upon V_{g2} at 0.7 V V_{g1} for different input dc voltages (0.2, 0.25, and 0.3 V). **(D)**Threshold of V_{in} for spiking with V_{g2}. **(E)** Requisite change in V_{amp_in} for ceasing spiking (ΔV_{amp_in}) with V_{g2}. The inset provides a graphical view of ΔV_{amp_in} for the VTCs **(A)** in case of an overshoot of 65 mV by the positive feedback. **(F)** Output spike width with V_{g2} for different V_{dd-} values (−0.45, −0.5, and −0.55 V).

Triggering the positive feedback through C1 elevates V_{amp_out} over the threshold of Amp2, as seen in Figures 4C, 5B; the overshoot (ca. 65 mV in this work) is mainly determined by the capacitance of C1 and V_{dd+}. The overshoot is independent of V_{g2}. Then V_{amp_in} immediately declines upon the onset of the negative feedback to the FG integrator through the polarity inverter. V_{amp_in} eventually falls below the threshold of Amp2. The requisite time for this process is equivalent to the spike width. The V_{amp_out}-V_{amp_in} relations in Figure 6A indicate that the higher V_{g2} is given, the larger decrease in V_{amp_in}(ΔV_{amp_in}) needs to be made by the negative feedback to drag V_{amp_out} below the threshold. ΔV_{amp_in} is elucidated in the inset of Figure 6E, where the VTCs in Figure 6A are zoomed in. Notably, the requisite ΔV_{amp_in} increases with V_{g2}, and thus it takes longer for a higher V_{g2} to decrease V_{amp_out} by 65 mV, i.e., to reset the FG integrator with the same negative V_{inv}. As a consequence, a higher V_{g2} offers a wider spike, as shown in Figure 6F. In addition, given that a higher $|V_{dd-}|$evokes a higher $|V_{inv}|$, resetting the FG integrator takes less time with a higher $|V_{dd-}|$, rendering the output spike width narrower (Figure 6F).

Power Consumption

Regarding the principles of neuromorphic engineering, low power consumption is strongly desired. Toward this end, we evaluated the power consumption of the FGLIF neuron that elicits different output activities amid the application of a constant input voltage. The average power consumption was acquired for various output activities by evaluating the consumed energy during the period of a single spike and dividing it by the period. The results are plotted in Figure 7, in which the average power consumption is gently proportional to the output activity. Notably, the proposed FGLIF neuron circuit consumes power less than 30 pW in the entire activity range. This is mainly ascribed to the subthreshold operation of Amp1 and Amp2, allowing a low current flow through the channels in series.

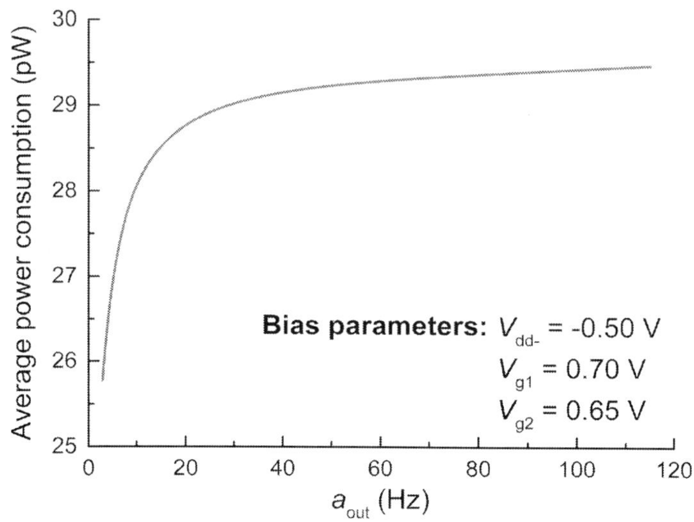

Figure 7. Average power consumption of FGLIF neuron with respect to a_{out}.

Power consumption of the proposed circuit was compared to the values achievable in the alternative VLSI neuron designs, which are listed in Table 3. The FGLIF neuron circuit provides power consumption that is several orders of magnitude lower than those of other models, while the number of transistors in use is comparable. Of course, a precise comparison between the circuits is still difficult since power consumption also strongly depends on several other factors such as CMOS technology in use, neuron spike width, and firing rate. Therefore, the results in Table 3 provide only an approximate overview and can be expected to change.

Table 3. Power consumption comparison between the proposed neuron circuit and other VLSI neuron models.

Neuron model	Number of transistors used	Power consumption	References
Conductance-based	27–30+	60 µW	Mahowald and Douglas, 1991
Integrate-and-Fire	18–20	~10 µW for output firing rate of 100 Hz	Indiveri et al., 2006
Hindmarsh-Rose	90	153.4 µW	Lee et al., 2004
Quadratic Integrate-and-Fire	14	8–40 µW	Wijekoon and Dudek, 2008
Log-domain low pass filter neuron	16	50–1000 nW	Arthur and Boahen, 2007
Log-domain Izhikevich neuron	17+	2.6 µW at rest state	Van Schaik et al., 2010
FGLIF	13	<30 pW	

Spiking in the Presence of Noise

The operation of the FGLIF neuron circuit appears to be markedly susceptible to noise because of the subthreshold operation of the MOSFETs in the circuit. Thus, it is important to identify the effect of noise on the operation of the neuron circuit and analyze it in comparison with its biological counterpart. Types of noise in a MOSFET are (i) thermal noise, (ii) shot noise, (iii) burst noise, and (iv) flicker noise. The first two are white noise, whereas the last two are nonwhite noise whose power spectral density (PSD) relies on frequency (Chong and Sansen, 2013; Hamilton et al., 2014). The thermal voltage noise across a MOSFET channel is analogous to that across a resistor, and its PSD is $S_v = 4kTR_d$ (Nyquist, 1928). Here R_d denotes the resistance of the channel. In the FGLIF circuit, the thermal noise in each CMOS stage (a pair of *p*MOS and *n*MOS channels) is filtered by the capacitance of the following stage. Therefore, the thermal noise endows each voltage input node with ΔV_{RMS}:

$$\triangle V_{RMS} = \sqrt{k_b T/C}, \quad (1)$$

where k_b, T, and C denote the Boltzmann constant, temperature, and equivalent MOS capacitance of the following stage, respectively. Note that the MOS capacitance relies on the gate oxide capacitance and gate voltage. Given the subthreshold operation of the MOSFETs in the neuron circuit, C is smaller than that of above-threshold-working MOSFETs, and thus the thermal noise effect is perhaps prominent with regard to Equation (1). For simplicity, C was evaluated at the average V_g values that are applied to the input of each CMOS stage during the circuit's operation. ΔV_{RMS} was evaluated at each node that precedes a given capacitance value, and it varied from 0.8 to 5.4 mV for different nodes. This thermal noise was taken into account in the following circuit simulations, as detailed in Section Materials and Methods. Additionally, it should be mentioned that Equation (1) is in fact not limited to thermal noise but is rather universal for the entire white noise of the system in the presence of a filtering capacitor (Sarpeshkar et al., 1993). As a result, no extra terms are needed to evaluate the shot noise in the circuit.

Burst noise (also known as random telegraph noise) appears to markedly affect the operation of the neuron circuit, given the fluctuation of V_{fb}, i.e., ΔV_{fb}, upon the interaction of a charge trap with an electron. The contribution of each trap interacting with an electron to $\langle \Delta V_{fb} \rangle$ differs for the MOSFETs in the circuit owing to the different channel areas and gate oxide thicknesses (see Section Materials and Methods). In the following simulations, $\langle \Delta V_{fb} \rangle$ for each MOSFET was evaluated, and the fluctuation of V_g(equivalent to ΔV_{fb} that was randomly generated from an exponential distribution function with $\langle \Delta V_{fb} \rangle$) was applied to each MOSFET following the Poisson process that is elucidated in Section Materials and Methods. The same held for M1 (FG-MOSFET) except that ΔV_{fb} was imposed on V_{FG} rather than V_g. The electron-trapping rate r_t was also an i.i.d. random variable (see Section Materials and Methods).

The topmost panel in Figure 8A shows a typical fluctuation in V_g for *n*MOS (channel area: 120 × 60 nm^2; gate oxide thickness: 2.5 nm) attributed to a single trap at the gate oxide/Si interface. ΔV_{fb} was 1.0 mV. ΔV_g is thus toggled between 0 and −1.0 mV upon the interaction between the trap and an electron; V_g is −1.0 mV when the trap is filled with an electron, and V_g is 0 otherwise. The corresponding PSD is plotted in Figure 8B, in which the PSD dispersion in the double logarithmic plot follows a power low with an exponent of −2 (Lorentzian PSD). The same holds for *p*MOSs with the same parameters, except that ΔV_g is toggled between 0 and 1.0 mV, and ΔV_g is 1.0 mV when the trap is filled with a hole.

The number of traps underneath the gate oxide is not necessarily one. Interfacial trap density D_{it} evaluates the average number of traps per unit area in the *n*-channel area; the larger D_{it}, the more traps are likely present at the interface. The number of traps (7, 14, and 72 traps) markedly alters the fluctuation in V_g, as shown in Figure 8A. An i.i.d. random ΔV_{fb} and r_t were assigned to each trap; both were drawn from exponential distribution functions with $\langle \Delta V_{fb} \rangle$ differing for different MOSFETs and a $\langle \tau_t \rangle$ of 10 ms, respectively. Additionally, each trap was assumed to interact with a single carrier. The method of noise generation is detailed in Materials and Methods.

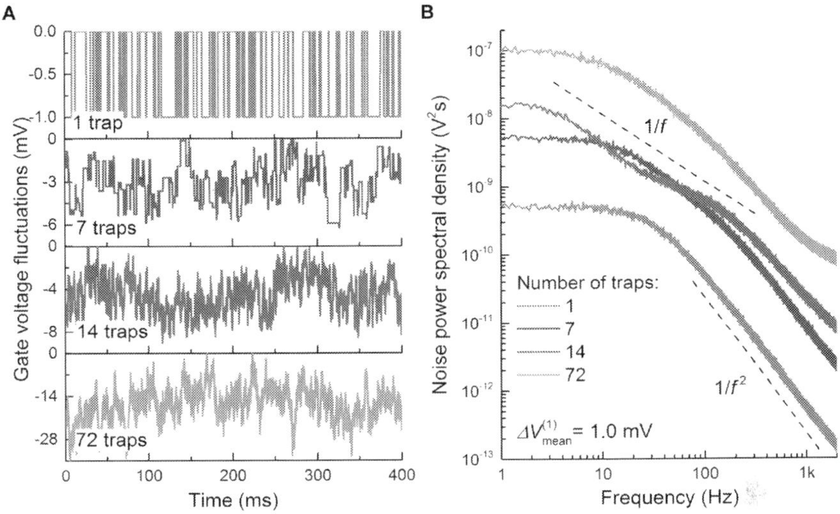

Figure 8. Burst noise in MOSFET. (A) Gate voltage fluctuations in an *n*MOS caused by electron trapping and detrapping events through interfacial charge traps. Average trapping and detrapping times for single-trap case (upper graph) were chosen to be 10 ms each, while in other cases they were taken from exponential distribution with a mean value of 10 ms. ΔV_{fb} owing to a single trap was also taken from exponential distribution with a mean value of 1.0 mV. **(B)** Noise PSD, evaluated for the temporal patterns in **(A)**.

As shown, the V_g deviation from 0 increases with the number of traps because ΔV_{fb} proportionally increases with the number of filled traps. Notably, the PSD becomes close to flicker noise with the number of traps; for instance, 72 traps (orange PSD in Figure 8B) lead to a PSD approximately following an exponent of -1 in the double logarithmic plot. This is a consequence of the presence of multiple Lorentzian PSD functions (72 in total), i.e., superposition of these many PSD functions results in a $1/f$ PSD function (Campbell et al., 2009). Burst noise is thus a subset of flicker noise with regard to their origins. Long-channel MOSFETs often exhibit flicker noise attributed to the probable large number of traps at a given D_{it} (Uren et al., 1985). By contrast, deep-submicron MOSFETs likely include a few traps (or even a single trap); therefore, burst noise is predominantly observed (Hung et al., 1990).

Eventually, the FGLIF neuron circuit was simulated taking into account the aforementioned types of noise. The circuit simulation was done for both dc input mode (V_{in}: 0.26 V) and spike input mode (a_{in}: 100

Hz, V_{in}: 0.42 V) for direct comparison with the behaviors without noise shown in Figures 4, 5. Figure 9displays the simulation results, including output spikes in time and ISI distribution. Note that D_{it} for all MOSFETs was set to 10^{11} cm^{-2} in the simulations. The ISI (τ) distribution apparently arises from the present noise as compared with the perfect periodicity of spikes (single ISI) for the noise-free neuron circuit. The ISI (τ) histogram for both input modes can be fitted well to a Gamma distribution function given by

$$p(\tau) = \tau^{k-1} e^{-\tau/\theta} \cdot \left(\theta^k \Gamma(k)\right)^{-1}, \tag{2}$$

where k and θ are fitting parameters that determine the shape and scale of the distribution, respectively (Maimon and Assad, 2009). As shown in Figure 8, the noise characteristic markedly varies upon the number of traps; therefore, the ISI distribution may change accordingly. To identify this effect, additional ISI histograms were obtained with different D_{it} values (10^{11}, 2×10^{11}, and 4×10^{11} cm^{-2}), and the results are shown in Figure 10. The higher number of traps at the interface, the more widely the ISI is distributed. The decrease in the shape parameter k denotes an increase in spiking irregularity (Maimon and Assad, 2009), which is captured by the ISI widening. Note that in these simulations the contribution of each noise source to the overall noise effect cannot be distinguished mostly due to the physical origin of flicker and burst noise as mentioned earlier. Nevertheless, the power spectral density arising from interfacial charge traps becomes close to flicker noise with the number of charge traps (see Figure 8B), and thus the change in the ISI distribution with trap density in Figure 10 most likely reflects the distributional change upon the transition of a noise mechanism from burst-like to flicker-like noise. This transition widens a range of trapping and detrapping probabilities that are proportional to trapping and deptrapping rates as detailed in Section Materials and Methods. Consequently, a wide range of such rates is intertwined in the overall noise dynamics. Therefore, spiking regularity—parameterized by k—increases with trap density.

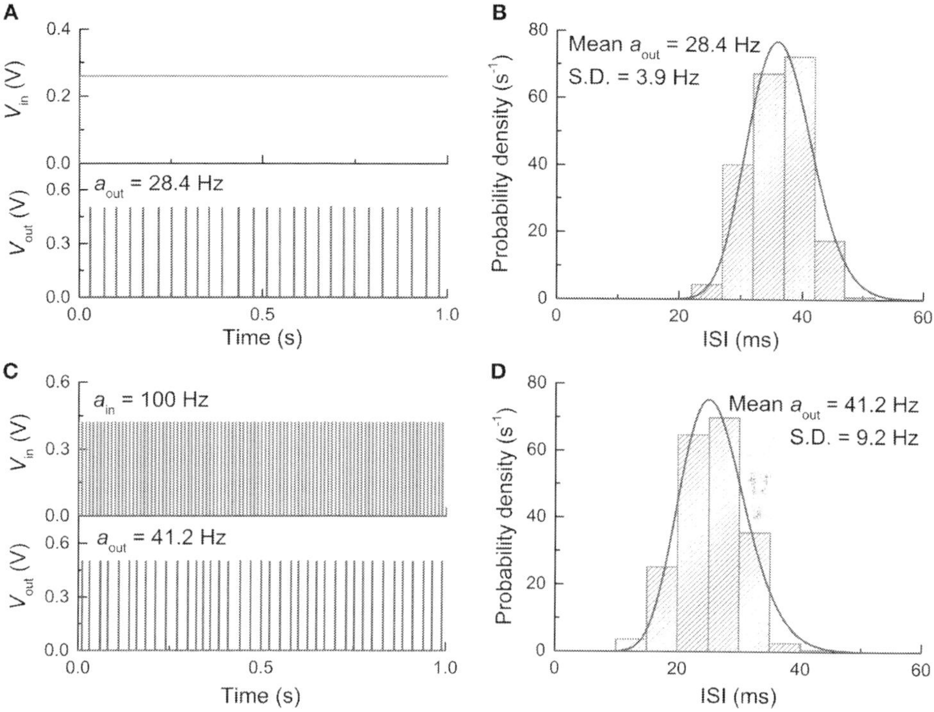

Figure 9. Neuron firing in the presence of noise. (A) Neuron firing when stimulated by a dc input voltage of 0.26 V. **(B)** ISI histogram for dc input and Gamma distribution function fitted to the ISI histogram. **(C)** Neuron firing when stimulated by a periodic spike train with a firing rate of 100 Hz and amplitude of 0.42 V. **(D)** ISI histogram for spiking input condition and Gamma distribution function fitted to the histogram.

To identify the relative contributions of thermal and burst noise to the overall ISI distribution, further simulations were conducted by controlling the noise sources. Three cases were considered: thermal noise only, burst noise (subset of flicker noise) only, and simultaneous thermal and burst noise. Each case was examined from 10 independent simulations.

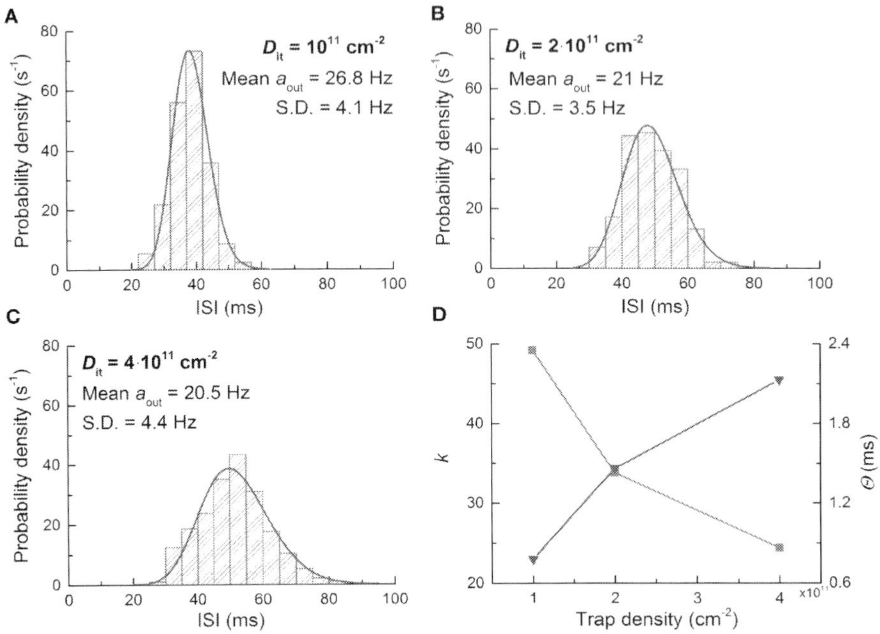

Figure 10. ISI distribution with interfacial trap density. (A–C) Neuron output ISI distributions for three different D_{it} values when it is subjected to a dc input of 0.26 V. **(D)** Gamma distribution parameters used for fitting histograms in **(A–C)**. The gray and red symbols denote *k* and Θ, respectively.

The resulting ISI distributions (fitted to Gamma distribution functions) are shown in Figures 11A–C. As such, the thermal noise is determined by capacitance (see Equation 1) so that it is almost invariant through the trials. This is featured by the negligible trial-to-trial variation in the ISI distribution (Figure 11A). Notably, the average ISI shifts toward a lower value due to the thermal noise as seen in comparison with the noise-free case (ca. 43.5 ms) indicated by a dashed vertical line. By contrast, the burst noise markedly alters the ISI distribution upon trial (Figure 11B) because ΔV_{fb} and r_t values were sampled at random for each trial from exponential distributions (Yonezawa et al., 2013). The center of the distribution for each trial is scattered around the ISI of the noise-free case. In the present of simultaneous thermal and burst noise, the effects are superimposed (Figure 11C); the scattered positions of distributional centers relate to the burst noise, and their shift below 43.5 ms relates to the thermal noise. The noise-sensitive ISI distribution is mostly dictated by the noise in Amp1 (Figure 11B) whose input (V_{amp_in}) directly evokes a spike. The thermal noise endows V_{amp_in} with a fluctuation around the noise-free V_{amp_in} so that the output from the input stage (M2–M3), i.e., V_{M3} in the inset of Figure 11D, fluctuates around the noise-free VTC (gray zone in Figure 11D). This fluctuation, in turn, affects the input into the following stage (M4–M5) in conjunction with its own noise. A typical reduction in ISI due to thermal noise is shown in Figure 11G. The output overshoot accelerates spiking (orange line) that precedes the noise-free spiking (gray line) by ~10 ms. Therefore, the thermal noise generally shorten the ISI.

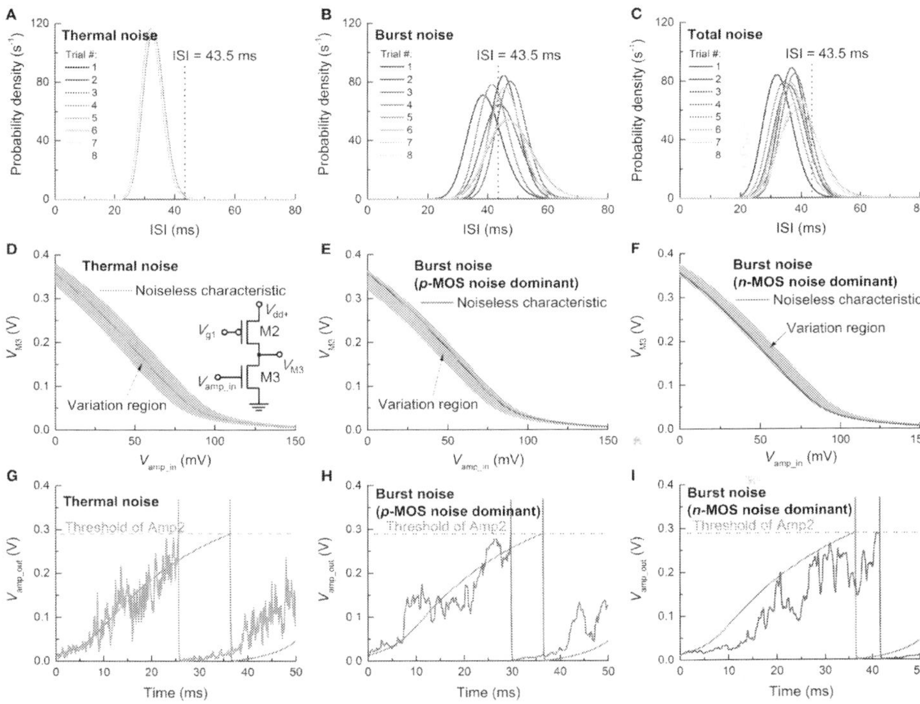

Figure 11. ISI variation upon noise-type. Output ISI distributions in the presence of **(A)** only thermal noise, **(B)** only burst noise, and **(C)** simultaneous thermal and burst noise at a constant dc input of 0.26 V. The dashed vertical lines at 43.5 ms indicate the ISI of noise-free spiking. The VTC of the input stage (M2–M3) of Amp1—shown in the inset—in the presence of **(D)** only thermal noise and only burst noise in **(E)** *p*MOS (M2) and **(F)** *n*MOS (M3). The shaded regions denote VTC variation ranges. The evolution of output of Amp1 (V_{amp_out}) for cases of **(D–F)** is exemplified in **(G–I)**, respectively.

The effect of burst noise on Amp1 differs for *p*MOS and *n*MOS. For *p*MOS, the trapped holes at the gate oxide/Si interface leads to a negative shift in V_{fb}, i.e., $\Delta V_{fb} < 0$; therefore, when the burst noise in the *p*MOS is solely present in the input stage, the VTC shifts as a whole as shown in Figure 11E. Such a shift causes a reduction in Vamp_in to output the same V_{M3} as for the noise-free case so that the ISI is reduced—similar to the thermal noise effect. An example is shown in Figure 11H. By contrast, for *n*MOS, the electrons trapped at the channel interface elevate V_{fb}, i.e., $\Delta V_{fb} > 0$, indicating a right shift in the VTC as a whole (Figure 11F). Consequently, Amp1 needs higher V_{amp_in} than the noise-free case to output the same V_{M3} as for the noise-free case, implying larger ISI. An example of this case is plotted in Figure 11I.

DISCUSSION

Recall that unlike capacitor-based integrators, the FG integrator in the proposed FGLIF neuron has the characteristic relaxation time defined by quantum mechanical tunneling dynamics through the tunnel barrier. The different basis of charge integration largely mitigates a severe need for high capacitance in favor of a biologically plausible timescale. In order to highlight the scalability of the proposed FGLIF neuron circuit (particularly the FG integrator), the FG integrator was compared with a switched-capacitor integrator comprising *n* MOSFET switches (MR1–MR*n*) and one capacitor C_{mem}. The switched-capacitor integrator is illustrated in Figure 12A. Note that the MOSFETs have the gate and drain shorted in order to realize "fast charging and slow discharging." They are switched on upon the application of a voltage pulse to the shorted terminal.

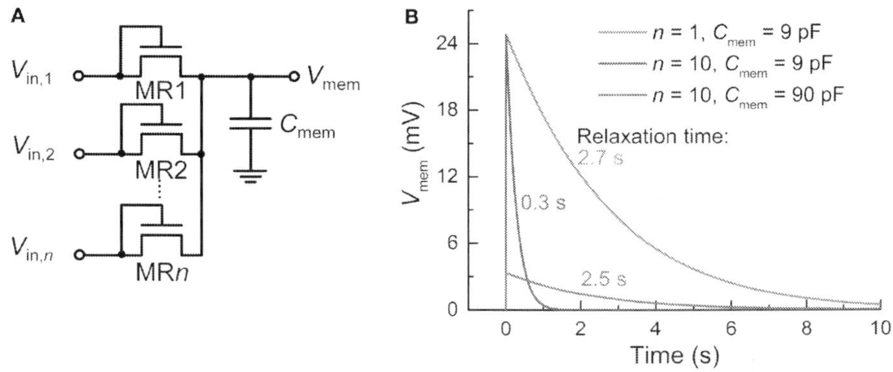

Figure 12. Comparison with switched-capacitor integrator. (A) Switched-capacitor integrator circuit with n input MOSFET terminals (MR1–MRn). **(B)** Relaxation of V_{mem}—parameterizing integration—with time for different n's (1 and 10) and capacitance values of C_{mem} (9 and 90 pF). An input spike (V_{in} = 0.5 V, t_{isp} = 10 μs) was applied to a single tunnel junction at 0 s and the others were grounded.

All MOSFETs in this integrator were of the same size (channel length/width: 60/120 nm) as the tunnel junctions and FG transistor M1 in the FG integrator. The circuit simulation results indicate a need for remarkably high capacitance—more than three orders of magnitude higher than that of C_{FG} (6 fF)—to achieve a relaxation time of a few seconds for the FG integrator (compare Figure 12B with Figure 2B). For a fair comparison, the input voltage pulse was identical to that of the FG integrator shown in Figure 2(amplitude: 0.5 V, width: 10 μs). Furthermore, the use of multiple terminals (n = 10) significantly reduces the relaxation time by approximately one order of magnitude in light of a decrease in the equivalent resistance owing to the MOSFETs being in parallel (Figure 12B). The same holds for the FG integrator as addressed in Section Circuit Simulations, and a relaxation time of a few seconds was recovered by introducing a slightly thicker tunnel barrier. By contrast, the switched-capacitor integrator requires a higher capacitance to avoid such a large decrease in the relaxation time, e.g., 90 pF to endow the integrator with a relaxation time of ~2.5 s, as shown in Figure 12B. However, the high capacitance value retards not only the discharging (relaxation) but also the charging; therefore, the V_{mem} maximum (ca. 24.8 mV) was not reached during the period of a single spike.

Such high capacitance in this simple integrator is hardly affordable in integrated circuits, delimiting the scalability. It is revealed that realizing a capacitance of a few pF requires a few hundred μm^2, which cannot fit in with the framework for scaling down. As addressed in Section Circuit Simulations, C_{FG} of 6 fF in capacitance was placed on M1 (channel length/width: 60/120 nm), and thus the FG integrator scheme offers great scalability, which is a definite advantage over switched-capacitor integrators. Of course, several strategies for reducing the capacitor area are likely available, e.g., introducing high-*k* dielectrics, or HfO_2 and/or three-dimensional capacitors (Kim et al., 2010). However, such attempts may cause additional complexity in chip fabrication and a consequent fabrication cost that may outweigh the benefits. Alternatively, maximizing the resistance of the switch in a switched-capacitor integrator equally enables a long relaxation time, as demonstrated by Noack et al. (2015). Another strategy based on capacitor-based integration by Qiao et al. (2015) accomplished a time constant of a few tens of milliseconds using a 1 pF capacitor. Nevertheless, our FG-based strategy may be a potential alternative to the capacitor-based integration for further scaling down neurons.

One of the main concerns for flash memories is reliability, particularly in the endurance of the FG transistor. Poor endurance—although sufficient for flash memory applications with a write endurance of 10^4–10^5 times—is mainly ascribed to a high write voltage in exchange for better data retention (Cappelletti and Modelli, 1999). Given the circumstances for the operation of the FG integrator (receiving a train and/or burst of a number of spikes), the poor endurance is perhaps a significant obstacle. Low neuronal activity (as low as biological neurons) is preferred partly on these grounds. Fortunately, unlike with flash memory, we deliberately allow data (charge on the FG) loss for a relaxation time of a few seconds, so that the requisite write and erase voltages are much lower, i.e., ±0.5 V in the proposed circuit. This low voltage is mostly applied to the tunnel junction; the voltage across C_{FG} is merely 47 mV (maximum V_{FG} shown in Figure 4B) at most, and the gate oxide is subject to even less voltage with regard to the additional voltage drop in the Si channel underneath. The simulation results show ~90 electrons that tunnel through the gate oxide in a single spiking period, i.e., one output spike (amplitude: 0.5 V, width: 25 μs) and one ISI, which is equivalent to a charge density of ~2

$\times 10^{-7}$ C/cm^2. Thus, such low voltage likely alleviates the burden on the FG, rendering the FG fairly endurable (Wann and Hu, 1995).

Significant improvement in the reliability of flash memory occurred by replacing the standard silicon FG with an insulating SiN_x one, forming so-called silicon-oxide-nitride-oxide-silicon (SONOS; Chan et al., 1987). The insulating SiN_x tolerates shorting paths embedded in the tunnel barriers to a greater degree than the standard SOSOS (silicon instead of SiN_x in SONOS; Chan et al., 1987). However, provided that the synapse transistor stretches the FG to the separate tunnel junctions, the conductance of the FG is of significant concern. To put it precisely, the highly resistive SiN_x FG barely alters V_{FG} in M1 upon a charge transfer through the separate tunnel junctions. The standard SOSOS is therefore suitable for the FG integrator. The aforementioned low write and erase voltages that suffice for the FG integrator likely support the reliability of SOSOS.

As such, the number of the interfacial charge traps is most likely proportional to the channel size, and the effect of a single trap on ΔV_{fb} is predicted to become more significant as the channel size shrinks (Fukuda et al., 2007; Miki et al., 2012). The continuum-based model in this study may not precisely elucidate the effect of interfacial traps on ΔV_{fb} in a state-of-the-art nanoscale MOSFET. However, discrete models that are more suitable for dealing with a few traps also indicate the same tendency as for the continuum model (Fukuda et al., 2007; Miki et al., 2012). Experimental studies have pointed to burst noise as a dominant type of noise in nanoscale MOSFETs (Tega et al., 2009; Miki et al., 2012).

Noise in integrated circuits is inevitable as such, and the noise induces irregular spiking patterns (Figures 9, 10). The ISI histograms are nicely fitted to Gamma distribution functions with different fitting parameters akin to biological neurons, e.g., parietal neurons (Bair et al., 1994; Maimon and Assad, 2009). The integrated FGLIF neuron circuit in practice may differ in noise characteristics from our theoretical one. Even among integrated FGLIF neuron circuits, the difference in noise is most likely evident because the variables that relate to noise, e.g., D_{it}, ΔV_{fb}, and r_t, are markedly dependent on the details of fabrication methods. It is obvious that traps are incorporated in a MOSFET to some extent, and they endow the MOSFET with noise. A fundamental question arises as to whether such irregular spiking helps neural processing or is an obstacle to neural processing. Within the framework of stochastic electronics (Chen et al., 2010; Hamilton et al., 2014), such irregular spiking is necessary, and imperfections such as charge traps are required for the disorder. For instance, it has turned out that the noise given to a VLSI Hodgkin-Huxley neuron enhances signal modulation (Chen et al., 2010). Nevertheless, answering this fundamental question is beyond the scope of this paper, so we leave the question open.

CONCLUSION

We proposed an FGLIF neuron circuit, which likely achieves input-dependent output spiking activity within a biologically plausible range with a capacitor of merely 6 fF. Such low capacitance offers a great opportunity for scaling down the FGLIF neuron circuit while maintaining the activity scale. In addition, given the subthreshold operation of most MOSFETs in the circuit, spiking consumes less than 30 pW of power irrespective of spiking activity, rendering the FGLIF neuron very suitable for large-scale SNNs. In addition, the FGLIF neuron circuit is fully compatible with standard CMOS technology, which is of great benefit. Unavoidable noise in the circuit leads to distributional features of ISIs. The ISI distribution was fitted to a Gamma distribution function that has often described the ISI distribution of neocortical neurons.

Author Contributions

VK conceived the idea and performed the circuit calculations together with HL, JS, and BC, GK, SK, and IK also conceived the idea and support the calculations. DJ initiated and supervised this study and wrote the manuscript. All authors discussed the results and contributed to the refinement of the manuscript.

Conflict of Interest Statement

The authors declare that the research was conducted in the absence of any commercial or financial relationships that could be construed as a potential conflict of interest.

Acknowledgments

The authors acknowledge a Korea Institute of Science and Technology grant (grant number 2E26691).

REFERENCES

1. Arthur, J. V., and Boahen, K. (2004). Recurrently connected silicon neurons with active dendrites for one-shot learning, in *IEEE International Joint Conference on Neural Networks*, Vol. 3 (Budapest), 1699–1704.
2. Arthur, J. V., and Boahen, K. (2007). Synchrony in silicon: the gamma rhythm. *IEEE Trans. Neural Netw.* 18, 1815–1825. doi: 10.1109/TNN.2007.900238
3. Azghadi, M. R., Iannella, N., Al-Sarawi, S. F., Indiveri, G., and Abbott, D. (2014). Spike-based synaptic plasticity in silicon: design, implementation, application, and challenges. *Proc. IEEE* 102, 717–737. doi: 10.1109/JPROC.2014.2314454
4. Bair, W., Koch, C., Newsome, W., and Britten, K. (1994). Power spectrum analysis of bursting cells in area MT in the behaving monkey. *J. Neurosci.* 14, 2870–2892.
5. Bartolozzi, C., and Indiveri, G. (2007). Synaptic dynamics in analog VLSI. *Neural Comput.* 19, 2581–2603. doi: 10.1162/neco.2007.19.10.2581
6. Brink, S., Nease, S., and Hasler, P. (2013). Computing with networks of spiking neurons on a biophysically motivated floating-gate based neuromorphic integrated circuit. *Neural Netw.* 45, 39–49. doi: 10.1016/j.neunet.2013.02.011
7. Burkitt, A. N. (2006). A review of the integrate-and-fire neuron model: I. Homogeneous synaptic input. *Biol. Cybern.* 95, 1–19. doi: 10.1007/s00422-006-0068-6
8. Campbell, J. P., Qin, J., Cheung, K. P., Yu, L. C., Suehle, J. S., Oates, A., et al. (2009). Random telegraph noise in highly scaled nMOSFETs, in *Reliability Physics Symposium, 2009 IEEE International* (Montreal, QC), 382–388.
9. Cao, K. M., Lee, W. C., Liu, W., Jin, X., Su, P., and Fung, S. K. H. (2000). BSIM4 gate leakage model including source-drain partition,

in *Electron Devices Meeting, 2000. IEDM 00. Technical Digest. International* (San Francisco, CA), 815–818.

10. Cappelletti, P., and Modelli, A. (1999). *Flash Memory Reliability*. New York, NY: Springer.
11. Chan, T. Y., Young, K. K., and Hu, C. (1987). A true single-transistor oxide-nitride-oxide EEPROM device. *Electron Device Lett.* 8, 93–95.
12. Chen, H., Saïghi, S., Buhry, L., and Renaud, S. (2010). Real-time simulation of biologically realistic stochastic neurons in VLSI. *IEEE Trans. Neural Netw.* 21, 1511–1517. doi: 10.1109/TNN.2010.2049028
13. Chong, Z. Y., and Sansen, W. (2013). *Low-Noise Wide-Band Amplifiers in Bipolar and CMOS Technologies*, Vol. 117. New York, NY: Springer Science & Business Media.
14. Dayan, P., and Abbott, L. F. (2001). *Theoretical Neuroscience*, Vol. 806. Cambridge, MA: MIT Press.
15. Diorio, C., Hasler, P., Minch, B. A., and Mead, C. (1998). Floating-gate MOS synapse transistors, in *Neuromorphic Systems Engineering*, ed T. S. Lande (Norwell, MA: Kluwer Academic Publishers), 315–337.
16. Dunga, M. V., Xi, X., He, J., Liu, W., Cao, K. M., Jin, X., et al. (2006). *BSIM 4.6.0 MOSFET Model.* Berkeley, CA: University of California.
17. Edwards, R. T., and Cauwenberghs, G. (2000). Synthesis of log-domain filters from first-order building blocks, in *Research Perspectives on Dynamic Translinear and Log-Domain Circuits*, eds W. A. Serdijn and J. Mulder (New York, NY: Springer), 71–80.
18. Eliasmith, C., and Anderson, C. H. (2004). *Neural Engineering: Computation, Representation, and Dynamics in Neurobiological Systems*. Cambridge, MA: MIT press.
19. Eliasmith, C., Stewart, T. C., Choo, X., Bekolay, T., DeWolf, T., Tang, Y., et al. (2012). A large-scale model of the functioning brain. *Science* 338, 1202–1205. doi: 10.1126/science.1225266
20. Fukuda, K., Shimizu, Y., Amemiya, K., Kamoshida, M., and Hu, C. (2007). Random telegraph noise in flash memories-model and

technology scaling, in *Electron Devices Meeting, 2007. IEDM 2007. IEEE International* (Washington, DC), 169–172.

21. Gerstner, W., and Kistler, W. M. (2002). *Spiking Neuron Models: Single Neurons, Populations, Plasticity*. Cambridge: Cambridge University Press.
22.
23. Gordon, C., Farquhar, E., and Hasler, P. (2004). A family of floating-gate adapting synapses based upon transistor channel models, in *Proceedings of the 2004 International Symposium on Circuits and Systems, 2004. ISCAS 04*, Vol. 1 (Vancouver, BC), I-317–I-320.
24. Hamilton, T. J., Afshar, S., van Schaik, A., and Tapson, J. (2014). Stochastic electronics: a neuro-inspired design paradigm for integrated circuits. *Proc. IEEE* 102, 843–859. doi: 10.1109/JPROC.2014.2310713
25. Hasler, P., Diorio, C., Minch, B. A., and Mead, C. (1994). Single transistor learning synapses, in *Advances in Neural Information Processing Systems 7*, eds G. Tesauro and D. S. Touretzky (Cambridge, MA: MIT Press), 817–824.
26. Hodgkin, A. L., and Huxley, A. F. (1952). A quantitative description of membrane current and its application to conduction and excitation in nerve. *J. Physiol.* 117, 500–544.
27. Hung, K. K., Chenming, H., and Cheng, Y. C. (1990). Random telegraph noise of deep-submicrometer MOSFET›s. *IEEE Electron Device Lett.* 11, 90–92.
28. Indiveri, G., Chicca, E., and Douglas, R. (2006). A VLSI array of low-power spiking neurons and bistable synapses with spike-timing dependent plasticity. *IEEE Trans. Neural Netw.* 17, 211–221. doi: 10.1109/TNN.2005.860850
29. Indiveri, G., Linares-Barranco, B., Hamilton, T. J., Van Schaik, A., Etienne-Cummings, R., Delbruck, T., et al. (2011). Neuromorphic silicon neuron circuits. *Front. Neurosci.* 5:73. doi: 10.3389/fnins.2011.00073
30. Izhikevich, E. M. (2003). Simple model of spiking neurons. *IEEE Trans. Neural Netw.* 14, 1569–1572. doi: 10.1109/TNN.2003.820440
31. Jeong, D. S., and Hwang, C. S. (2005). Tunneling current from a

metal electrode to many traps in an insulator. *Phys. Rev.* B71:165327. doi: 10.1103/PhysRevB.71.165327

32. Jeong, D. S., Kim, I., Ziegler, M., and Kohlstedt, H. (2013). Towards artificial neurons and synapses: a materials point of view. *RSC Adv.* 3, 3169–3183. doi: 10.1039/C2RA22507G
33. Jeong, D. S., Thomas, R., Katiyar, R. S., Scott, J. F., Kohlstedt, H., Petraru, A., et al. (2012). Emerging memories: resistive switching mechanisms and current status. *Rep. Prog. Phys.* 75:076502. doi: 10.1088/0034-4885/75/7/076502
34. Kim, S. K., Lee, S. W., Han, J. H., Lee, B., Han, S., and Hwang, C. S. (2010). Capacitors with an equivalent oxide thickness of <0.5 nm for nanoscale electronic semiconductor memory. *Adv. Funct. Mater.* 20, 2989–3003. doi: 10.1002/adfm.201000599
35. Lee, W. C., and Hu, C. (2001). Modeling CMOS tunneling currents through ultrathin gate oxide due to conduction-and valence-band electron and hole tunneling. *Electron Devices IEEE Trans.* 48, 1366–1373. doi: 10.1109/16.930653
36. Lee, Y. J., Lee, J., Kim, Y. B., Ayers, J., Volkovskii, A., Selverston, A., et al. (2004). Low power real time electronic neuron VLSI design using subthreshold technique, in *Proceedings of the 2004 International Symposium on Circuits and Systems, 2004. ISCAS 04*, Vol. 4 (Vancouver, BC), IV-744–IV-747.
37. Lim, H., Kornijcuk, V., Seok, J. Y., Kim, S. K., Kim, I., Hwang, C. S., et al. (2015). Reliability of neuronal information conveyed by unreliable neuristor-based leaky integrate-and-fire neurons: a model study. *Sci. Rep.* 5, 9776. doi: 10.1038/srep09776
38. Mahowald, M., and Douglas, R. (1991). A silicon neuron. *Nature* 354, 515–518.
39. Maimon, G., and Assad, J. A. (2009). Beyond poisson: increased

spike-time regularity across primate parietal cortex. *Neuron* 62, 426–440. doi: 10.1016/j.neuron.2009.03.021

40. Markram, H. (2006). The blue brain project. *Nat. Rev. Neurosci.* 7, 153–160. doi: 10.1038/nrn1848
41. Mead, C. (1989). *Analog VLSI and Neural Systems*. Reading, MA: Addison-Wesley.
42. Mead, C. (1990). Neuromorphic electronic systems. *Proc. IEEE* 78, 1629–1636.
43. Merolla, P. A., Arthur, J. V., Alvarez-Icaza, R., Cassidy, A. S., Sawada, J., Akopyan, F., et al. (2014). A million spiking-neuron integrated circuit with a scalable communication network and interface. *Science* 345, 668–673. doi: 10.1126/science.1254642
44. Miki, H., Tega, N., Frank, D. J., Bansal, A., Kobayashi, M., Cheng, K., et al. (2012). Statistical measurement of random telegraph noise and its impact in scaled-down high-k/metal-gate MOSFETs, in *Electron Devices Meeting, 2012. IEDM 2012. IEEE International* (San Francisco, CA), 450–453.
45. Noack, M., Partzsch, J., Mayr, C. G., Hänzsche, S., Scholze, S., Höppner, S., et al. (2015). Switched-capacitor realization of presynaptic short-term-plasticity and stop-learning synapses in 28 nm CMOS. *Front. Neurosci.* 9:10. doi: 10.3389/fnins.2015.00010
46. Nyquist, H. (1928). Thermal agitation of electric charge in conductors. *Phys. Rev.* 32, 110–113.
47. Polsky, A., Mel, B. W., and Schiller, J. (2004). Computational subunits in thin dendrites of pyramidal cells. *Nat. Neurosci.* 7, 621–627. doi: 10.1038/nn1253
48. Qiao, N., Mostafa, H., Corradi, F., Osswald, M., Stefanini, F., Sumislawska, D., et al. (2015). A reconfigurable on-line learning spiking neuromorphic processor comprising 256 neurons and 128K synapses. *Front. Neurosci.* 9:141. doi: 10.3389/fnins.2015.00141
49. Rahimi, K., Diorio, C., Hernandez, C., and Brockhausen, M. D. (2002). A simulation model for floating-gate MOS synapse transistors, in *IEEE International Symposium on Circuits and Systems, 2002. ISCAS 2002*, Vol. 2 (Phoenix-Scottsdale, AZ), II-532–II-535.

50. Ramakrishnan, S., Wunderlich, R., Hasler, J., and George, S. (2013). Neuron array with plastic synapses and programmable dendrites. *IEEE Trans. Biomed. Circuits Syst.* 7, 631–642. doi: 10.1109/BioCAS.2012.6418412
51. Ranuárez, J. C., Deen, M. J., and Chen, C. H. (2006). A review of gate tunneling current in MOS devices. *Microelectronics Reliab.* 46, 1939–1956. doi: 10.1016/j.microrel.2005.12.006
52. Riggert, C., Ziegler, M., Schroeder, D., Krautschneider, W. H., and Kohlstedt, H. (2014). MemFlash device: floating gate transistors as memristive devices for neuromorphic computing. *Semiconductor Sci. Tech.* 29, 104011–104019. doi: 10.1088/0268-1242/29/10/104011
53. Sarpeshkar, R., Delbruck, T., and Mead, C. (1993). White noise in MOS transistors and resistors. *Circuits Devices Mag. IEEE* 9, 23–29. doi: 10.1109/101.261888
54. Soni, R., Petraru, A., Meuffels, P., Vavra, O., Ziegler, M., Kim, S. K., et al. (2014). Giant electrode effect on tunnelling electroresistance in ferroelectric tunnel junctions. *Nat. Commun.* 5, 5414. doi: 10.1038/ncomms6414
55. Tega, N., Miki, H., Ren, Z., D›Emic, C. P., Zhu, Y., Frank, D. J., et al. (2009). Reduction of random telegraph noise in high-κ/metal-gate stacks for 22 nm generation FETs, in *Electron Devices Meeting, 2009. IEDM 2009. IEEE International* (Baltimore, MD), 771–774.
56. Tenore, F., Vogelstein, R. J., Etienne-Cummings, R., Cauwenberghs, G., and Hasler, P. (2006). A floating-gate programmable array of silicon neurons for central pattern generating networks, in *2006 IEEE International Symposium on Circuits and Systems, 2006. ISCAS 2006. Proceedings* (Island of Kos), 3157–3160.
57. Uren, M. J., Day, D. J., and Kirton, M. (1985). 1/*f* and random telegraph noise in silicon metal–oxide–semiconductor field–effect transistors. *Appl. Phys. Lett.* 47, 1195–1197.
58. Van Schaik, A., Jin, C. T., McEwan, A. L., and Hamilton, T. J. (2010). A log-domain implementation of the Izhikevich neuron model, in *Proceedings of 2010 IEEE International Symposium on Circuits and Systems (ISCAS)* (Paris), 4253–4256.

59. Wann, C. H., and Hu, C. (1995). High endurance ultra-thin tunnel oxide for dynamic memory application, in *Electron Devices Meeting, 1995. IEDM'95., International (IEEE)* (Washington, DC), 867–870
60. Wijekoon, J. H., and Dudek, P. (2008). Compact silicon neuron circuit with spiking and bursting behaviour. *Neural Netw.* 21, 524–534. doi: 10.1016/j.neunet.2007.12.037
61. Yonezawa, A., Teramoto, A., Obara, T., Kuroda, R., Sugawa, S., and Ohmi, T. (2013). The study of time constant analysis in random telegraph noise at the subthreshold voltage region, in *2013 IEEE International Reliability Physics Symposium* (Anaheim, CA), XT.11.1–XT.11.6. doi: 10.1109/IRPS.2013.6532126

SECTION III: THE FITZHUGH-NAGUMO MODEL, THE PROPAGATION OF WAVES AND CHAOTIC NEURONS

5 NUCLEATION OF REACTION-DIFFUSION WAVES ON CURVED SURFACES

Frederike Kneer[1,2], Eckehard Schöll[1] and Markus A Dahlem[3]

[1] Institut für Theoretische Physik, Technische Universität Berlin, Hardenbergstrasse 36, D-10623 Berlin, Germany

[2]Institut für Softwaretechnik und Theoretische Informatik Technische Universität Berlin, Marchstr. 23, D-10587 Berlin, Germany

[3] Institut für Physik, Humboldt Universität zu Berlin, Robert-Koch-Platz 4, D-10115 Berlin, Berlin, Germany

ABSTRACT

We study reaction-diffusion waves on curved two-dimensional surfaces, and determine the influence of curvature upon the nucleation and propagation of spatially localized waves in an excitable medium modelled by the generic FitzHugh–Nagumo model. We show that the stability of propagating wave segments depends crucially on the curvature of the surface. As they propagate, they may shrink to the uniform steady state, or expand, depending on whether they are smaller or larger, respectively, than a critical nucleus. This critical nucleus for wave propagation is modified by the curvature acting like an effective space-dependent local spatial coupling, similar to diffuson, thus extending the regime of propagating excitation waves beyond the excitation threshold of flat surfaces. In particular, a negative gradient of Gaussian curvature Γ, as on the outside of a torus surface (positive Γ), when the wave segment symmetrically extends into the inside (negative Γ), allows for stable propagation of localized wave segments remaining unchanged in size and shape, or oscillating periodically in size.

KEYWORDS: reaction-diffusion, pattern formation, localized structures, Gaussian curvature

INTRODUCTION

Wave propagation in excitable extended media described by nonlinear reaction-diffusion equations has widespread applications in chemistry, biology, and medicine. A particularly important example are neuronal systems where spreading depression (SD) waves are associated with a pathological dysfunction of brain activity that occurs, for instance, during migraine or stroke. Understanding the effect of internal control mechanisms in neuronal wave dynamics is of great relevance not only for comprising the functionality of the human brain [1], but also for developing novel future therapies of pathological states which are connected with cortical SD waves in migraine aura or stroke [2, 3]. There is clinical and experimental evidence [4, 5] that spatially localized wave segments play a dominant role in these phenomena. In general, spatially localized wave segments, as they propagate, might shrink, expand, or remain unchanged in size and shape, in which case they are called particle-like waves or dissipative solitons. Spatially localized wave segments also represent critical structures which can be stabilized by global feedback [6–8]. They play an important role for the nucleation of propagating waves and wave segments in two-dimensional (2D) spatial domains. Generally, waves can be controlled by feedback or closed loop control. This is a robust and versatile concept which uses the internal dynamics of the system to generate a control signal which directs the system towards a desired dynamics. A plethora of examples are provided by global or nonlocal and in some cases time-delayed feedback control of wave propagation in reaction-diffusion systems [9–15]. On the other hand, the curvature of the medium itself also provides a means of internal control of the stability, as we will show in this paper.

Most previous studies have focussed on wave propagation in planar spatial domains, yet there is also a considerable body of work on reaction-diffusion waves in curved surfaces mostly on spirals and ring waves [16–24] but not to the best of our knowlege on nucleation. The cortex, however, represents a strongly curved surface. It is the purpose of this paper to study nucleation and propagation of wave segments on curved 2D surfaces. We will demonstrate that positive or negative Gaussian curvature of the spatial domain has a dramatically different effect upon the wave dynamics.

The paper is organized as follows. In section 2 we present the model. In section 3 we discuss wave solutions on a torus, which represents a curved surface on which locally both positive and negative Gaussian curvature occurs. We consider ring waves, wave segments, and dissipative solitons (critical nuclei) stabilized by feedback control. Specifically, we study ring wave break-up, curvature-induced changes of stability, and curvature-induced stabilization of wave-segments. In section 4 we draw conclusions.

MODEL

In this work we use the FitzHugh–Nagumo (FHN) system as a generic model for excitable systems [25], and study it in two spatial dimensions. It is a generalization of the van der Pol oscillator [26] and also known as Bonhoeffer-van der Pol oscillator [27, 28]. The model was first suggested by FitzHugh in 1961 [29], and independently by Nagumo et al in 1962 [30]. As a two-variable simplification of the four-variable Hodgkin–Huxley model, which describes the propagation of action potentials along the giant squid axon [31], it describes the response of an excitable nerve membrane to external current stimulation:

$$\frac{\partial u}{\partial t} = 3u - u^3 - v + D\nabla^2 u, \qquad (1)$$

$$\frac{\partial v}{\partial t} = \varepsilon\left(u + \beta\right). \qquad (2)$$

The original interpretation of the FHN equations (1), (2) is based on a single neuron: the variable umodels fast changes of the electrical potential across the membrane of a nerve cell axon (occurring as spikes in the time series), and v is the recovery variable related to the gating mechanism of the membrane channels [29]. The small parameter $\varepsilon \ll 1$ represents the time scale ratio of the two variables. The generic dynamical mechanism described by equations (1), (2) was also suggested to describe a fundamentally different ionic excitability of neuronal tissue that originates from bistable ion homeostasis and is the bais of SD [32, 33].

In general, the fast variable u is called the activator variable, whereas the slow variable v is referred to as inhibitor variable. The diffusion constant of the activator is D, which is chosen as 0.12 throughout this paper (and can simply be interpreted as a scaling of space), and inhibitor diffusion is assumed to be slow, and hence negligible. The threshold parameter β determines whether the systems is excitable ($\beta > 1$) or exhibits self-sustained perodic oscillations ($\beta < 1$). In the following we consider β in the excitable regime.

Figure 1 shows a schematic phase portrait of a spatially homogeneous system in the excitable regime ($\beta > 1$) with the cubic activator nullcline and the vertical inhibitor nullcline (solid red and green lines). The system has a single fixed point (u_s, v_s), which is stable for $\beta > 1$ and located on the left branch of the cubic nullcline. At $\beta = 1$ a supercritical Hopf bifurcation of a limit cycle occurs, and the fixed point becomes unstable and is shifted to the middle branch of the nullcline for $\beta < 1$.

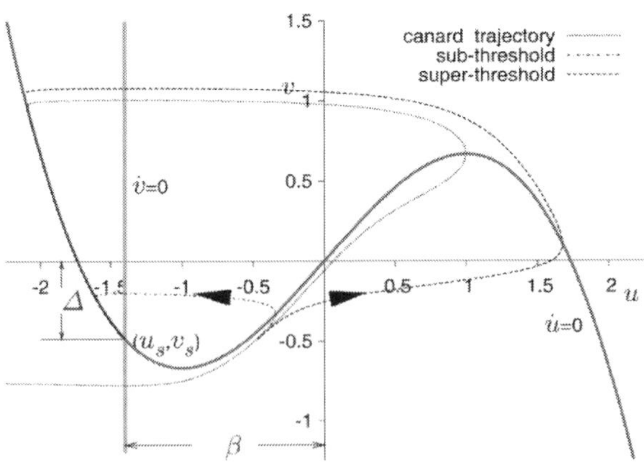

Figure 1. The nullclines $\dot{u} = 0$ (solid red) and $\dot{v} = 0$ (solid green) in the phase space of the homogeneous FHN system with $\beta = 1.4$. Their intersection at (u_s,

v_s) is a stable fixed point. Three trajectories are drawn for $\varepsilon = 0.04$: one canard-like trajectory (dotted), passing through the maximum of the nullcline $\dot{u} = 0$, and two trajectories starting at $v = v_s$ nearby but on opposite sides of the canard trajectory. They diverge sharply, producing threshold behaviour: the dashed and the dash-dotted trajectories represent super-threshold and subthreshold stimulation, respectively.

The excitable behaviour of the system is crucially determined by the cubic nonlinearity of the activator equation and the separation of time-scales between the two variables: when the system is perturbed by a sufficiently large (super-threshold) external stimulus, which can be regarded as setting the initial condition, the system undergoes a large excursion in phase space (spiking). Starting from its initial condition, the system performs, due to the strong timescale separation $\varepsilon \ll 1$, a fast transition to the stable right branch of the activator nullcline. After that, it travels slowly upwards approximately along this nullcline, until the phase points jumps back to the left branch, and returns, along the left branch of the nullcline, slowly downwards to the fixed point (recovery phase). Without further external stimulation the system remains in the stable fixed point (rest state).

The threshold-like behaviour of the FHN system is associated with the canard-like trajectory (dotted black line in figure 1), which is the trajectory passing through the local maximum of the cubic nullcline, and which is often referred to as the threshold of the FHN system. The region around the canard-like trajectory is extremely sensitive to initial conditions: for initial conditions only slightly below the canard-like trajectory the systems will perform a large excursion in phase space, whereas for initial conditions only slightly above the canard-like trajectory the excursion will be small (subthreshold excitation). In principle, the transition from small to large amplitude excitation is continuous; in fact, however, phase space excursions of intermediate amplitude are very rare. Correspondingly, small subthreshold stimulations (dash-dotted black line in figure 1) will result in fast relaxation, while super-threshold stimulations (dashed black line in figure 1) induce a full excursion in phase space, corresponding to a characteristic spike in the time evolution of the u-variable.

In the following we consider spatially inhomogeneous solutions, i.e., waves or wave segments in 2D. Depending upon the set of parameters (ε, β) there exist different wave solutions [5, 34]. Here we focus on localized wave segments which may either shrink or expand, as they propagate, or in the limit case, remain unchanged in size and shape, in which case they are called particle-like waves or dissipative solitons.

For each set of parameters (ε, β) with $\beta < \beta_{\partial R_\infty}$ (or $\varepsilon < \varepsilon_{\partial R_\infty}$, respectively) there exists a localized wave solution (wave segment), which represents a critical spatio-temporal structure, i.e. a particle-like wave, or dissipative soliton.

As we are interested in stationary propagating wave segments, equations (1)–(2) can be written as

$$c\frac{\partial u}{\partial \xi} = 3u - u^3 - v + D\nabla_\xi^2 u, \tag{3}$$

$$c\frac{\partial v}{\partial \xi} = \varepsilon\left(u + \beta\right), \tag{4}$$

where $\xi = x + ct$ is the co-moving coordinate with propagation velocity c. In a co-moving frame, the localized critical structure is related to a saddle-point with a single unstable eigenvector (one-dimensional (1D) unstable manifold) in phase space. The curve representing this solution in a parameter plane of the bifurcation diagram is called the rotor boundary ∂R. In phase space, the stable manifold of states on ∂R separates the attractor of a spiral wave (spatially nonconfined) and the stable, spatially uniform steady state. Thus, when the form of this dissipative soliton is perturbed, it either grows to a spiral wave, or shrinks and disappears. Perturbations above the critical size of the dissipative soliton grow, while perturbations below that critical nucleus shrink to the stable uniform state, i.e., the wave segments retracts. The internal cortical control of such dissipative solitons may be viewed as a strategy of the cortex to avoid re-entrant spiral waves, e.g., in migraine. Changing the nucleation size of this critical structure changes the susceptibility to pathological conditions such as SD.

In figure 2, we show the rotor boundary ∂R (black dashed) in a schematic bifurcation diagram of wavesize S as a function of the threshold parameter β. It separates the weakly excitable parameter regime (perturbations grow to a spiral wave) from the subexcitable parameter regime (wave segments in 2D below the critical size shrink in length, while in spatially 1D systems wave propagation is stable). There exists another boundary ∂R_∞ (dash-dotted vertical line), independent of size S, to the right of which all wave segments retract (corresponding to infinitely large critical size). Furthermore, the propagation boundary d ∂P_{1D} is shown, which separates the subexcitable parameter regime from the nonexcitable regime. In the nonexcitable parameter regime, perturbations shrink also in width and wave segments collapse, i.e., even in spatially 1D systems no wave propagation is possible.

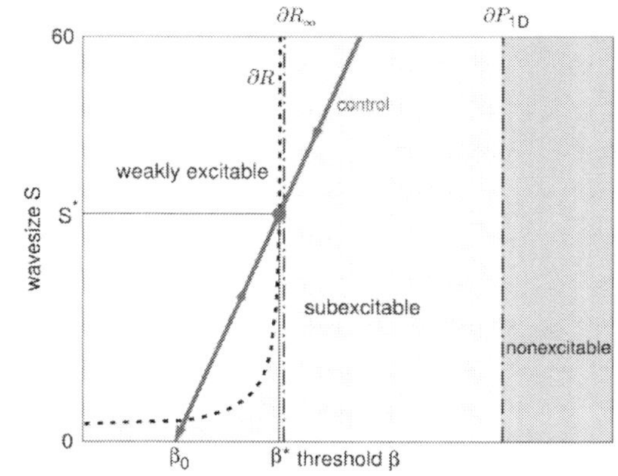

Figure 2. Schematic phase diagram of different regimes of the FitzHugh–Nagumo model in the plane of wavesize S and threshold β for fixed ε: weakly excitable (perturbations grow to spiral waves); subexcitable (perturbations shrink in length); nonexcitable (perturbations shrink in width, no propagation in one dimension). The respective boundaries are marked by ∂R_∞ and ∂P_{1D}. ∂R denotes the boundary of the critical nucleus of size S below which perturbations shrink. The red solid line marks the control loop, which stabilizes the critical nucleus (S^*, β^*) indicated by a red dot.

At this point we would like to remark that the critical nucleus (dissipative soliton), which has the dynamic signature of a saddle-point, can be stabilized by an internal feedback control loop which controls the excitation threshold β in equation (2) that is,

$$\beta(t) = \beta_0 + KS(t) \tag{5}$$

where K is the control strength, and the size of the wave segment S represents a measure of the active area occupied by the wave segment.

Equation (5) defines a control line (red solid line with arrows) in the (β, S) phase diagram in figure 2. As the temporal evolution of a perturbation in a controlled system is confined to the control line, it asymptotically approaches a stable wave segment (β^*, S^*) (if perturbed with convenient initial conditions). This follows from figure 2 since wave segments above ∂R, i.e., $\beta < \beta^*$, grow in size, while wave segments below ∂R, i.e., $\beta > \beta^*$, shrink.

The size of the wave segment S can be calculated via an integral over the excited area, e.g.,

$$S(t) = \int_{\mathbb{R}^2} u(\mathbf{r}, t) \mathrm{d}^2\mathbf{r}. \tag{6}$$

If the stabilized solution (β^*, S^*) is reached, S(t) becomes stationary. Thus, although the control is invasive, it does not produce new solution branches, and the stabilized critical nuclei of the controlled system are unstable solutions of the system without control at $\beta = \beta\ (t \longrightarrow \infty)$. Hence, it is adequate to define the size of the wave segment as the area where the activator concentration u (r, t) is larger than zero

$$S(t) = \int_{\mathbb{R}^2} \Theta(u(\mathbf{r}, t)) \mathrm{d}^2\mathbf{r} \tag{7}$$

with the Heaviside function Θ.

In the following simulations we will apply an internal feedback mechanism as in equation (5) with the wavesize S(t) from equation (7) to stabilize the critical nucleus. Furthermore, we will study the influence of the curvature of the excitable medium on the stability of localized waves. Hence the Laplace operator ∇^2 must be replaced with the Laplace–Beltrami operator Δ_{LB} [35] for surfaces given in curvilinear coordinates α^i, i = 1,2,:

$$\Delta_{\mathrm{LB}} = \sum_{i,j=1}^{2} g^{-\frac{1}{2}} \frac{\partial}{\partial \alpha^i} \left(g^{\frac{1}{2}} g^{ij} \frac{\partial}{\partial \alpha^j} \right) \tag{8}$$

where g = Det G and G with the matrix elements g^{ij} is the metric tensor of the parametrization, see appendix. The surface of a torus in the Euclidian space R^3 can be described by the parametrization (θ, φ) of the position vectors

$$(\theta, \varphi) \mapsto \begin{pmatrix} (R + r\cos\theta)\cos\varphi \\ (R + r\cos\theta)\sin\varphi \\ r\sin\theta \end{pmatrix} = \begin{pmatrix} x \\ y \\ z \end{pmatrix}. \tag{9}$$

The geometrical meaning of the major curvature radius R and the minor curvature radius r as well as the angles θ and φ is visualized in figure 3. The Laplace–Beltrami operator in torus coordinates reads

$$\Delta_{\text{LB}} = -\frac{\sin\theta}{r(R + r\cos\theta)}\frac{\partial}{\partial\theta} + \frac{1}{r^2}\frac{\partial^2}{\partial\theta^2} + \frac{1}{(R + r\cos\theta)^2}\frac{\partial^2}{\partial\varphi^2}. \tag{10}$$

We investigate sections of a torus with von Neumann boundary conditions (no flux boundary) on the equatorial section (at $\theta = 0$ and $\theta = \pi$) and periodic boundary conditions in the direction of the azimuthal angle φ. This restricts all traveling wave solutions as they have to obey the symmetries defined by these boundary conditions, i.e., the centre of mass of the critical nucleus is pinned either on the outside or inside of the torus.

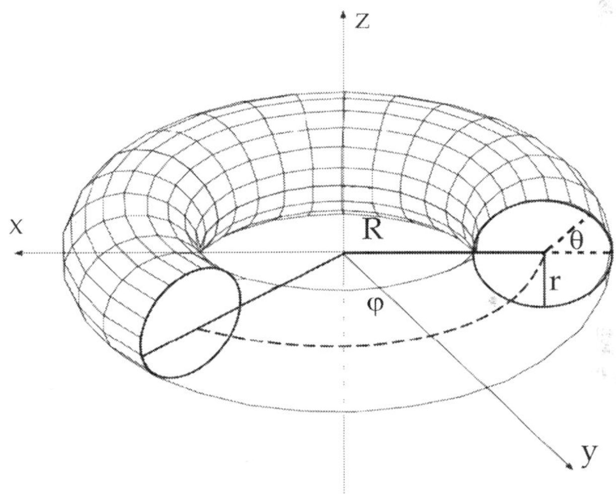

Figure 3. Parametrization of a torus by coordinates (θ, φ).

The FHN model with diffusion and internal feedback control equation (5) including the wavesize S(t) calculated with equation (7)

$$\frac{\partial u}{\partial t} = 3u - u^3 - v + D\Delta_{\text{LB}}u, \tag{11}$$

$$\frac{\partial v}{\partial t} = \varepsilon\left(u + \beta_0 + KS(t)\right) \tag{12}$$

is numerically solved using the explicit Euler method. With the discretization of space and time

$$x_j: \; = \; x_0 + j\delta x, \, j = 0, \, 1 \, ,\ldots, J$$

$$t_n: \; = \; t_0 + n\delta t, \, n = 0, \, 1 \, ,\ldots, N,$$

where x stands for θ or φ, respectively, the time derivative is calculated as

$$u_j^{n+1} = u_j^n + \left(3u_j^n - \left(u_j^n\right)^3 - v_j^n + D\Delta_{\mathrm{LB}} u_j^n\right)\delta t, \tag{13}$$

$$v_j^{n+1} = v_j^n + \varepsilon\left(u_j^n + \beta_0 + KS\left(t\right)\right)\delta t. \tag{14}$$

with $u_j^n := u\left(t_n, x_j\right)$. The Laplace–Beltrami operator equation (8) contains derivatives of first and second order with respect to θ and φ. The derivative of first order is solved using a forward–backward Euler algorithm

$$\frac{\partial u}{\partial \theta} \approx \frac{u_{j+1}^n - u_{j-1}^n}{2\delta\theta}. \tag{15}$$

The derivatives of second order are solved using the Euler method, first backward followed by forward

$$\frac{\partial^2}{\partial \theta^2} \approx \frac{u_{j+1}^n - 2u_j^n + u_{j-1}}{(\delta\theta)^2}. \tag{16}$$

As initial condition, the activator concentration u is set equal to u_s + 2 (which corresponds to a supra-threshold excitation in figure 1) in a rectangular area, and the inhibitor concentration v is set equal to v_s + 1.5 in a rectangular area shifted relative to the activator stimulus, in order to determine the propagation direction of the wave segment. Outside the rectangle, the initial condition is $u = u_s, \, v = v_s$, with the activator and inhibitor fixed point values u_s, v_s. In order to numerically obtain critical nuclei of smaller or larger size, the size of the rectangular area is varied.

RESULTS

Overview of Wave Solutions on a Torus

The main results are twofold. First, by investigating excitation waves on a torus, we show that the Gaussian curvature of the excitable medium changes the nucleation threshold in a systematic way. Second, and more surprisingly, we observe that a curved medium can even induce a change of stability. Unstable critical nuclei are transformed into stable propagating localized wave segments.

We analyze the nucleation of excitation waves in reaction-diffusion media on curved 2D surfaces, specifically on tori. A torus has positive and negative Gaussian curvature on the outside (θ=0) and the inside (θ = π), respectively, and a continuous transition in between with vanishing Gaussian curvature on the top $(\theta = \frac{\pi}{2})$ and bottom $(\theta = \frac{3\pi}{2})$, see figure 3. In general, a torus has, in contrast to a sphere, not only locally varying and even negative Gaussian curvature, but a torus also admits a global isothermal coordinate system, called toroidal coordinates $((\theta_i, \tilde{\varphi})$, see appendix), that is, coordinates where the metric is locally conformal to the Euclidean metric. Therefore an intuitive understanding of some of our results can be based on the particularly simple form of the Laplace–Beltrami operator, the 'diffusion operator', in these coordinates.

On tori, a stable solution besides the spatially homogeneous steady state are ring-shaped localized traveling wave solutions. Ring-shaped traveling waves have been analyzed, and in particular their critical properties have been discussed, namely, wave fronts with sufficiently large geodesic curvature break up on the torus inside [36]. We reconsider these travelling waves in order to compare them with the dynamics of critical nuclei. The stable manifold of a critical nucleus separates the attractor of a ring-shaped traveling wave and the spatially uniform steady state.

We restrict our study to nucleation of waves propagating strictly in the direction of the azimuthal angle φ (see section 2) and, furthermore, the centre of mass of the nucleation is pinned either on the outside or inside of the torus, i.e., the locations where the extreme values of the Gaussian curvature occur. In the following, we will simply refer to these solutions as inside or outside critical nuclei or, if stabilized, inside or outside traveling wave segments. These solutions are the symmetric solutions with respect to the equatorial section of the torus. Note that there exist also asymmetric solutions on tori that we have not studied. Furthermore, note that these two symmetric critical nuclei on the outside and inside assume the shape of a localized wave segment where the open ends extend in the direction of θ (perpendicular to the propagation direction), that is, into regions of decreasing and increasing Gaussian curvature, respectively. Our results are mainly explained by this gradient in the Gaussian curvature and not by the absolute value of the Gaussian curvature.

The results are displayed in two bifurcation diagrams. First, the same bifurcation diagram as already introduced in section 2 to define the regimes of excitability (weakly, sub-, and nonexcitable, see figure 2 in section 2) is shown in figure 4. The size S of the critical nucleus, see equation (7), is plotted versus the threshold parameter β of the local dynamics of the FHN system equations (1), (2). The reference branch of the critical nucleus from simulation on a flat medium is now labeled ‹flat› in figure 4 (black dashed). It separates the weakly excitable regime (to the left, decreasing β) from the subexcitable regime (to the right, increasing β), which ends at $\beta = \partial P_{1D}$, where the nonexcitable regime is reached.

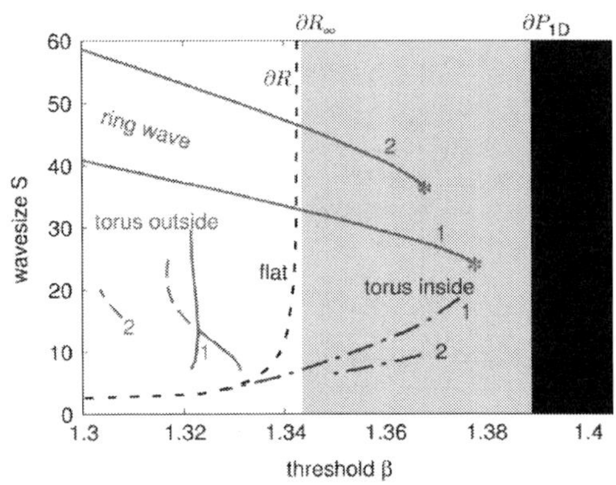

Figure 4. Bifurcation diagram of wavesize S as a function of threshold parameter β computed from equations (1), (2) with D = 0.12 and ε = 0.36; critical nucleus on a flat surface (black dashed line); 1: solutions on a torus with minor curvature radius $r = \frac{20}{2\pi}$ and major curvature radius $R = \frac{80}{2\pi}$; 2: solutions on a torus with minor curvature radius $r = \frac{20}{2\pi}$ and major curvature radius $R = \frac{40}{2\pi}$; stable ring wave solutions (red solid lines) with points of excitation block, i.e. propagation suppression (red asterisks); unstable inside critical nucleus (blue dash-dotted lines); unstable outside critical nucleus (green dashed lines); stable stationary and stable oscillating localized wave segment on the torus outside (green solid lines). Feedback equation (5) is applied to stabilize the states on the dashed and dash-dotted curves.

In figure 4, we show further solution branches simulated on two different tori. The torus labeled 1 has lower absolute values of Gaussian curvature than the torus labeled 2, since the latter torus has a two times smaller value of its major curvature radius R. For each torus, we have a branch of the ring-shaped traveling wave solution (red solid). Furthermore, for each torus, we have one branch of the inside (blue dash-dotted) and outside (green dashed) critical nucleus. The states on the curves of the critical nucleus (dashed or dash-dotted) are stabilized by applying an appropriate global feedback equation (5) with suitably chosen β_0 and K such that the respective state (β^*, S^*) is at the intersection with the line given by equation (5). In addition, on the torus outside, we find stable wave segments and stable oscillating waves (green solid), see section 3.4 below.

Second, figure 5 is a bifurcation diagram, where the propagation velocity c, see equation (3), (4), is plotted versus the threshold parameter β. The reference branches are on the one hand the propagation velocities of the stable fast and the unstable slow wave solution in spatially 1D systems with diffusion in one spatial direction only, [37], (grey dashed), and, on the other hand, a critical velocity c_{min} (black solid), below which stable wave propagation cannot be obtained [36].

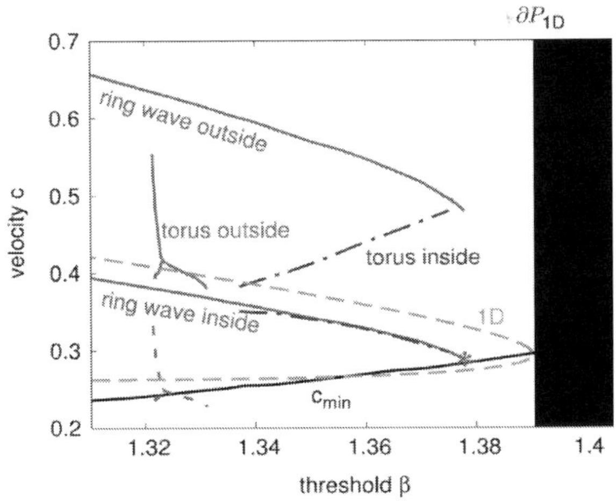

Figure 5. Bifurcation diagram of propagation velocity c as a function of threshold β computed from equations (1), (2) with D = 0.12 and ε = 0.36 on a torus with minor curvature radius $r = \frac{20}{2\pi}$ and major curvature radius $R = \frac{80}{2\pi}$; propagation velocity of stable ring wave solution (red solid lines) on the to-

rus inside (lower line) and torus outside (upper line), point of excitation block (red asterisk); propagation velocity of inside critical nucleus (blue dash-dotted lines) at centre of mass (lower line) and open ends (upper line); propagation velocity of stable stationary and stable oscillating localized outside wave segment (green solid lines) at centre of mass (upper line) and at open ends (lower line); hypothetical propagation velocity of outside stable wave segments (green dashed line) on torus inside; minimum propagation velocity (black solid line) calculated from equation (17) with c_{cr} and ε_{cr} computed from equations (1), (2) in one spatial dimension; propagation velocity of the stable and unstable wave solution in 1D (grey dashed). Feedback equation (5) is applied to stabilize the states on the blue dash-dotted curves.

In figure 5, we show further solution branches simulated on the less curved torus (torus 1). Two branches show the propagation velocity c in azimuthal (φ) direction of the ring-shaped traveling wave solution (red solid), the lower one is the propagation velocity on the torus inside, the upper one the propagation velocity on the torus outside. Furthermore, for the inside critical nucleus (blue dash-dotted), we display the propagation velocity in azimuthal (φ) direction at the centre of mass (lower line) and at the open ends (upper line), where the open ends are defined as the most distant lateral location where the activator concentration u equals zero.

For the outside stable wave segments, we show the propagation velocity c at the centre of mass (green solid) and, after the bifurcation into two branches with decreasing threshold parameter β, the maximum and minimum propagation velocity of the stable oscillating wave segment. In addition, for the outside critical nucleus (with propagation velocity c_o), we plot a 'hypothetical' branch (green dashed) that shows the propagation velocity $c_i = c_o\frac{R-r}{R+r}$ which a point of this wave segment would have on the torus inside if it existed there.

Ring Wave Break-Up at Saddle-Node Bifurcation

First we focus on the break-up of ring-shaped traveling waves on tori [36]. The ring-shaped traveling wave solution is a stable solution of equations (1)–(2) shown in figure 6. Thus the ring waves can be conceived as homoclinic solutions of the related ordinary differential equations (3), (4) in the co-moving frame. Ring waves have negative geodesic curvature on the torus inside and positive geodesic curvature on the torus outside, see figure 6(a). Thus, compared to 1D pulses (or infinitely extended wavefronts on a flat surface, respectively), ring waves propagate more slowly on the torus inside and faster on the torus outside, see figure 5 (red solid). If the propagation velocity falls below a critical value c_{min}, the ring wave breaks up on the torus inside [36], see figure 6(b). This excitation block is marked by an asterisk in figures 4 and 5. Below the minimal velocity c_{min}, stable wave propagation cannot be obtained.

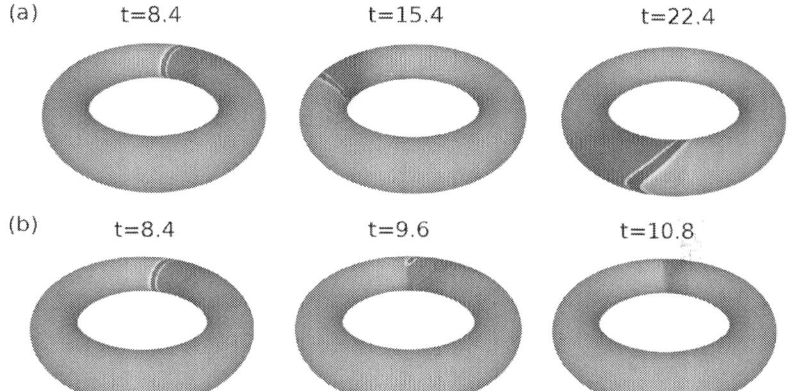

Figure 6. Snapshots of ring waves propagating counter-clockwise on a torus with minor curvature radius $r = \frac{20}{2\pi}$ and major curvature radius $R = \frac{80}{2\pi}$ computed from equations (1), (2) with D = 0.12 and ε = 0.36. (a) Stable ring wave, β = 1.378. (b) Ring-wave break-up, β = 1.379 See movies 1 and 2 in the supplementary data, available from stacks.iop.org/njp/16/053010/mmedia.

For 1D waves, it is known that the propagation failure is due to the coalescence of the homoclinic orbits of the fast and the slow wave (pulse) solution of the ordinary differential equations (ODE) problem equation (3), (4) [37]. In the related partial differential equations (PDE) problem equations (1), (2), the propagation boundary is a saddle-node bifurcation point of the stable fast wave branch and the unstable slow wave branch.

In figure 5, the fast wave branch and the slow wave branch of 1D traveling wave solutions are shown (upper and lower dashed grey line). At the propagation boundary ∂P_{1D}, they meet in a saddle-node bifurcation. Also in curved 2D media, the propagation failure is due to a saddle-node bifurcation, where the fast wave branch collides with the slow wave branch in a saddle-node bifurcation. In the 1D limit $R \longrightarrow \infty$, the threshold β, at which the ring waves break up, converges to the propagation boundary ∂P_{1D}. This is shown in figure 7, where the lines show the propagation failure in the (R, β) parameter space on two different tori; the upper line (dashed blue) is computed on a less curved torus with lower absolute values of Gaussian curvature compared to the lower line (dash-dotted green). The minimum velocity c_{min}, below which stable wave propagation is not possible, can be calculated as (see appendix)

$$c_{\min} = \frac{\varepsilon}{\varepsilon_{cr}} c_{cr}, \tag{17}$$

where ε_{cr} is the critical time separation parameter, where the homoclinic orbits of the 1D fast and slow wave (pulse) solution of the ODE equation (3), (4) coincide, and c_{cr} is the corresponding critical velocity. For the line c_{min} shown in figure 5, ε_{cr} and c_{cr} are computed with AUTO from equations (1), (2) in one spatial dimension. AUTO is a software tool for continuation and bifurcation problems in ordinary differential equations. Here, homoclinic solutions of the FHN model in a co-moving frame equations (3), (4) in one spatial dimension are continued in the threshold parameter β. The propagation velocity c of ring waves is affected by both the parameters of the local dynamics (ε and β) and the Gaussian curvature Γ of the torus. An increase of β or ε causes a deceleration of the ring wave on the torus inside. Also an increase of the Gaussian curvature Γ of a torus causes an increase of the absolute values of the geodesic curvature of the ring wave, which results in a deceleration of the ring wave on the torus inside. Thus, on more strongly curved tori (smaller R), the ring wave breaks up at smaller threshold β, see figure 7.

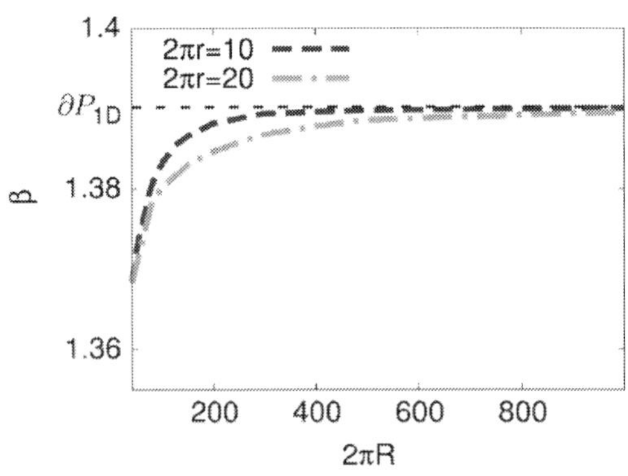

Figure 7. Excitation block (propagation failure) of ring-shaped traveling waves on tori in the (R, β) parameter space. No propagation is possible above the critical curves for different r; ∂P_{1D} denotes the 1D propagation boundary.

Curvature-Induced Changes of Nucleation

Next, we analyze the nucleation of excitation waves on the torus inside and outside, respectively.

Torus inside. The inside branches of the critical nucleus (blue dashed) in figure 4 are to the right of the reference curve (rotor boundary ∂R on flat surfaces). The larger the size S of the critical nucleus is, the stronger is the shift towards larger threshold β. On the more strongly curved torus (torus 2), the branch of the critical nucleus is shifted further. Thus, on the torus inside, there exist critical nuclei in the subexcitable parameter regime, where on flat surfaces all wave segments retract and vanish, cf figure 2. A qualitative explanation of this behaviour can be given by the relation of the Gaussian curvature Γ at the centre of mass of the critical nucleus ($\theta = \pi$) and at the open ends of the critical nucleus. Mathematically, the Gaussian curvature is described by the Laplace–Beltrami operator equation (8) in torus coordinates [35]. A torus admits a global isothermal orthogonal coordinate system, so-called toroidal coordinates $(\theta_i, \tilde{\varphi})$, that is, coordinates, where the metric is locally conformal to the Euclidean metric. The Laplace–Beltrami operator equation (8) given in toroidal coordinates is

$$\Delta_{\mathrm{LB}} = \frac{(\cosh \eta - \cos \theta_i)^2}{a^2}\left(\frac{\partial^2 u}{\partial \theta_i^2} + \frac{\partial^2 u}{\partial \tilde{\varphi}^2}\right). \tag{18}$$

where $a = (R^2 - r^2)^{\frac{1}{2}}$ is a measure for the scaling of the space, $\eta = \operatorname{arcoth}[R/(R^2 - r^2)^{\frac{1}{2}}]$ is a measure for the relation between the major radius R and the minor radius r and $\tilde{\varphi} = \varphi \sinh \eta$. The derivation can be found in the appendix. Introducing an effective coupling strength $C(\theta_i) = (\cosh \eta - \cos \theta)^2/a^2$, a torus can mathematically be interpreted as a flat medium with a spatial coupling being a function only of the location θ (θ_i can be expressed in terms of θ, see appendix), i.e. $\tilde{D}(\theta) = DC(\theta)$. This is similar to effective geometric potentials arising from curved surfaces in hard condensed-matter physics [40]. The coupling strength C (θ) is strictly monotonically increasing from the torus outside ($\theta = 0$) to the torus inside ($\theta = \pi$), see figure 8. For more strongly curved tori, the gradient of C (θ) is larger. The effective coupling strength C (θ) of the inside critical nucleus is larger at the centre of mass than at the open ends. Thus, the resultant diffusion perpendicular to the propagation direction is directed towards the open ends. This counteracts the retraction of the open ends in the parameter regime which is subexcitable in a flat medium. Larger critical nuclei reach over a region of larger difference in effective coupling strength, thus the shift towards larger threshold β is stronger. The propagation velocity c of the centre of mass of the inside critical nucleus is similar to the propagation velocity of the ring-shaped traveling waves at the torus inside, see figure 5.

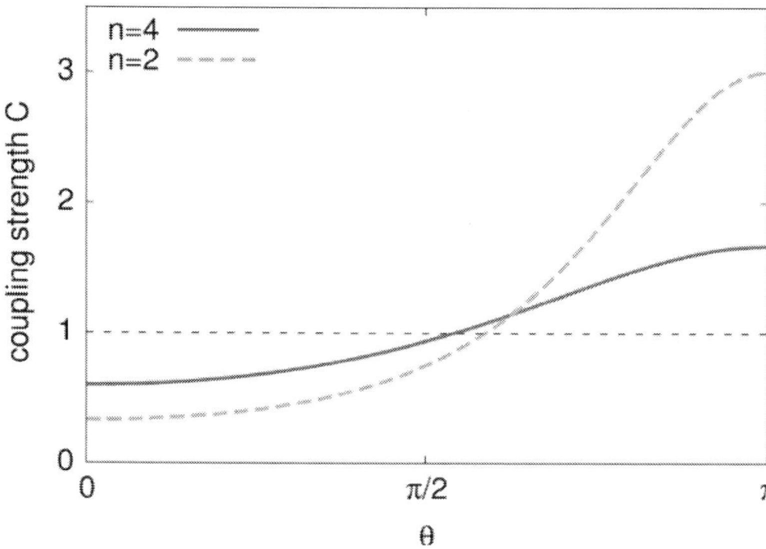

Figure 8. Coupling strength C as a function of common toroidal angular variable θ for two tori with different Gaussian curvature Γ with $n = \frac{R}{r}$.

Torus outside. On the torus outside, we find that, under certain conditions, unstable critical nuclei bifurcate into stable propagating wave segments (green solid line in figure 4). Furthermore, we find stable oscillating wave segments, whose size oscillates periodically in a self-sustained way. This striking bifurcation pattern will be explained in section 3.4. The outside branches (green) in figure 4are to the left of the flat reference branch ∂R (black dashed). On more strongly curved tori, the branch of the critical nucleus is further shifted.

The coupling strength relation between the torus inside and outside also explains this behaviour. As the coupling strength C (θ) at the centre of mass of the outside critical nucleus is smaller than at the open ends, the resultant diffusion perpendicular to the propagation direction is directed towards the centre of mass, which enhances the retraction of the open ends.

On the torus outside, critical nuclei with increasing size S are found at decreasing threshold β. This is distinct from the inside nucleation branch and the flat reference branch: the larger the size S of the critical nucleus is, the larger is the difference between the coupling strength at the centre of mass and the coupling strength at the open ends. Thus, larger critical nuclei at the torus outside are shifted to smaller threshold β, whereas on the torus inside larger critical nuclei are shifted to larger threshold β.

Critical nuclei with small size S extend over an area with almost constant coupling strength C, see figure 8. This supports the assumption that the branches of the inside and outside critical nucleus with small wavesize S in figure 4 lie close to the flat reference branch. This implies that the outside nucleation branch (green solid in figure 4) at small wavesize S terminates in a saddle-node bifurcation, and the unstable outside nucleation branch coalesces with the flat reference branch; this could, however, not be resolved numerically.

Curvature-Induced Stabilization

Depending upon the excitation parameter β, different space-time patterns occur (figure 9). Localized wave segments may either grow to become stable ring waves (figure 9(a)), or they may shrink and vanish (figure 9(d)). Additionally, as an effect of the curved surface, on the torus outside, we find stable propagating localized wave segments, see figure 9(c). Furthermore, we find stable oscillating wave segments, whose size oscillates periodically, see figure 9(b). In figure 10, we show the activator profile of a stable wave segment propagating with a stationary profile, and in figures 11(a) and (b) we show snapshots of the activator profile of an oscillating wave segment at its minimum size S and at its maximum size S, respectively.

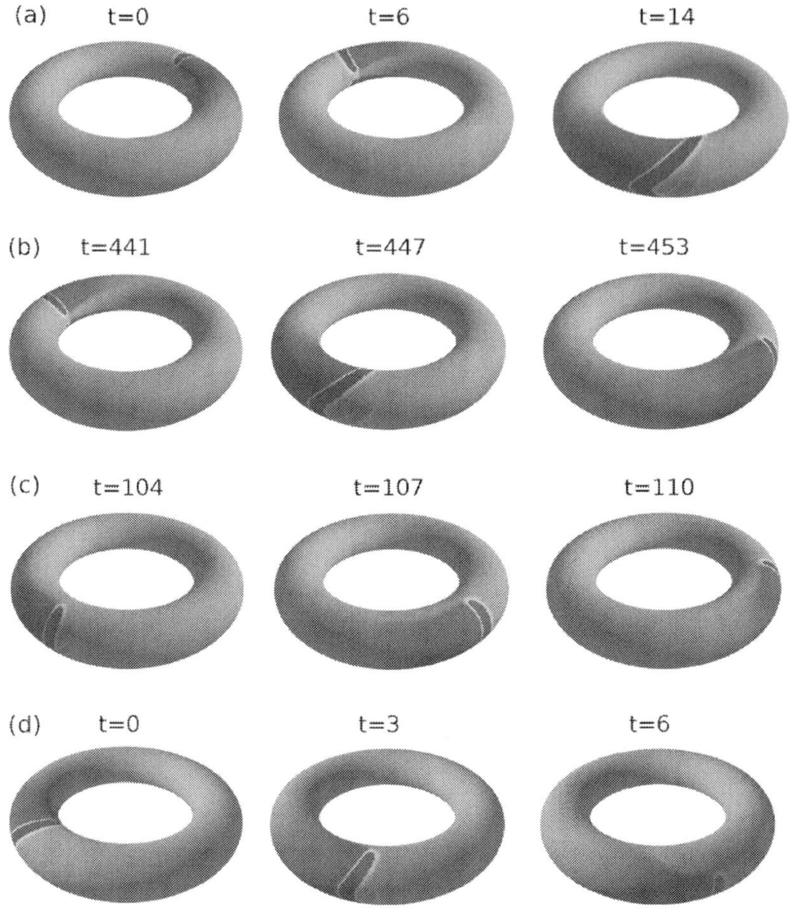

Figure 9. Snapshots of wave segments propagating counter-clockwise on a torus with minor curvature radius $r = \frac{20}{2\pi}$ and major curvature radius $R = \frac{80}{2\pi}$ computed from equations (1), (2) with D = 0.12 and ε = 0.36. (a) Wave segment growing towards ring-shaped traveling wave, β = 1.315. (b) Oscillating wave segment, β = 1.321476. (c) Stable propagating wave segment, β = 1.325. (d) Wave segment shrinking towards homogeneous steady state, β = 1.33. See movies 3–6 in the supplementary data, available from stacks.iop.org/njp/16/053010/mmedia.

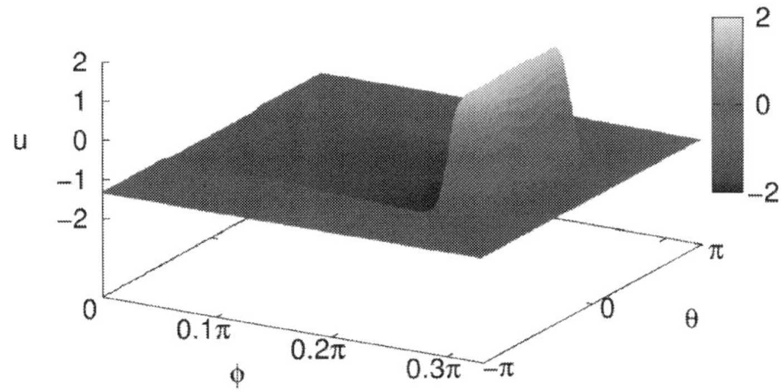

Figure 10. Snapshot of a stable propagating wave segment on the torus outside: activator concentration u (φ, θ) on a torus with minor curvature radius $r = \frac{20}{2\pi}$ and major curvature radius $R = \frac{80}{2\pi}$ computed from equations (1), (2) with β = 1.324, D = 0.12 and ε = 0.36.

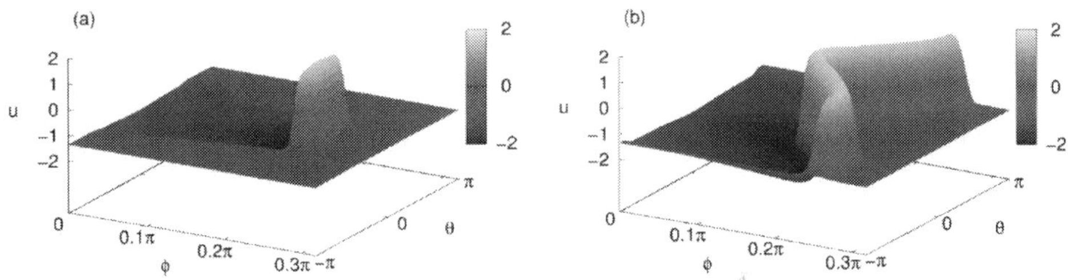

Figure 11. Snapshots of oscillating wave segment on the torus outside: same plots as in figure 10 with β = 1.321476, D = 0.12 and ε = 0.36. (a) t = 478.5 (minimum size), (b) t = 492.5 (maximum size).

The existence of stable wave segments on surfaces with positive Gaussian curvature can be qualitatively explained with the help of the space-dependent effective coupling strength C, as discussed in section 3.3. The open ends of a stable wave segment on the torus outside lie in an area of the torus where the coupling strength C is larger than the coupling strength at θ = 0, where the centre of mass of the wave segment is located. Thus the resultant effective diffusion perpendicular to the propagation direction caused by curvature is directed towards the centre of mass of the wave segment. The larger the size S of the perturbation is, the stronger is this effect. At the same time, in the excitable parameter regime (see figure 2), small perturbations grow in length. If these two effects are balanced, we find stable propagating wave segments. In figure 4, we show the branch of stable wave segments (green solid). Perturbations with size S larger than the stable wave segments (and smaller than ring waves) shrink, as the difference in effective coupling strength between the centre of mass and the open ends is large. Perturbations with size S smaller than the stable wave segments and larger than the small outside critical nucleus (which is not shown in figure 4 but supposed to lie close to the flat reference branch, see section 3.3) grow, as the difference in coupling strength between the centre of mass and the open ends is small.

In figure 5, we show the branch of the propagation velocity c at the centre of mass of the stable wave segments and the stable oscillating wave segments (green solid). Furthermore, we show a hypothetical branch, the related propagation velocity at the torus inside (green dashed).

It is impossible that the stable wave segments grow to ring-shaped traveling waves (without enlarging their geodesic curvature), as the hypothetical propagation velocity at the torus inside is smaller than the minimal velocity c_{min} (black solid line in figure 5).

For decreasing threshold β, the hypothetical propagation velocity at the torus inside of the stable outside wave solution (green dashed) accelerates, whereas the minimum velocity c_{min} (black solid) slows down. At the intersection point of these two branches, the stable wave solution bifurcates into a stable oscillating wave segment and an unstable critical nucleus.

The stable oscillating wave segments grow in length, until the propagation velocity c of the open ends falls below the minimum velocity c_{min}. The open ends become unstable and decrease in width. Although they continue growing in length, after the minimum velocity c_{min} is reached, the open ends asymptotically vanish.

CONCLUSIONS

SD is a pathological dysfunction of brain activity that occurs, e.g. during migraine or stroke. It appears as spatially localized wave segments propagating in the cortex [5]. Despite substantial progress in the understanding of SD, the inherent processes are still incompletely known. Here, we study the influence of the curvature of the cortex on SD.

For this purpose, nucleation and propagation of spatially localized reaction-diffusion waves are investigated on the surface of a torus. These unstable structures, which represent critical nuclei, are stabilized by an internal feedback control equation (5). We confine attention to the case that the centre of mass of the critical nuclei is pinned on the torus inside or outside, respectively.

We have shown that negative Gaussian curvature (torus inside) causes a shift of the nucleation branch to larger threshold β, i.e. there exist critical nuclei in a parameter regime that is subexcitable on flat surfaces. In view of SD waves in the cortex, this might indicate that SD is more likely to be initiated in areas with negative Gaussian curvature.

In addition, we have made the surprising discovery that curvature can induce a change of stability, i.e. on the torus outside we have found stable propagating localized wave segments, as well as wave segments periodically oscillating in size.

These findings are qualitatively explained by an effective coupling strength, which can be found as an equivalent mathematical description of the Gaussian curvature on surfaces that admit isothermal coordinates.

Furthermore, we have reviewed the behaviour of ring-shaped wave solutions (autowaves), which were first described by V A Davydov in 2003 [36]. In the context of critical propagation effects, we have confirmed that the propagation boundary of ring-shaped waves, constituting break-up on the torus inside, is caused by a saddle-node bifurcation, where the fast wave branch collides with the slow wave branch. Thereby, the propagation velocity of ring-shaped waves is compared to a semianalytically calculated minimum velocity c_{min} equation (17).

APPENDIX

Minimum Propagation Velocity

It is well known that wave propagation becomes impossible if the propagation velocity of a traveling wave falls below a critical value c_{min} [38]. Here, we consider the case of a traveling wave on a curved surface, which leads to slowing down of the wave by negative geodesic curvature of the wave front. The geodesic curvature of a wavefront is the curvature of the front projected in its tangential plane. This should not be confused with the Gaussian curvature Γ of a surface, which is the product of the two principal curvatures, i.e.

$$\Gamma = \frac{\cos\theta}{r\left(R + r\cos\theta\right)} \tag{A.1}$$

in common torus coordinates (θ, φ).

The FHN model (1), (2) in one spatial dimension for $1 < \beta < \sqrt{3}$ and sufficiently small ε, has a stable fast wave solution and an unstable slow wave solution corresponding to homoclinic orbits of the related ODE problem equation (3), (4), see [37]. There exists a critical line in the (ε, β) space, at which the fast wave branch is connected to the slow wave branch. For values of β or ε, respectively, above this critical line, propagation of traveling waves is not possible. The values of ε on this critical line are denoted by ε_{cr}, the corresponding critical propagation velocity is c_{cr}.

Here we show the derivation of the analytical dependency between the minimum velocity c_{min} and the critical time scale separation ε_{cr} and the critical propagation velocity c_{cr}.

As described in [38], the propagation of slightly curved fronts (curvature radius R ≪ rising front width L) can be approximated by

$$c\frac{\partial u}{\partial \xi} = 3u - u^3 - v + D\frac{\partial^2 u}{\partial^2 \xi} + \frac{D}{R}\frac{\partial u}{\partial \xi}, \tag{A.2}$$

$$c\frac{\partial v}{\partial \xi} = \varepsilon\left(u + \beta\right), \tag{A.3}$$

where the FHN model has been transformed to the co-moving coordinate ξ = x + ct, with propagation velocity c.

This can be written as

$$\left(c - \frac{D}{R}\right)\frac{\partial u}{\partial \xi} = 3u - u^3 - v + D\frac{\partial^2 u}{\partial^2 \xi}, \tag{A.4}$$

$$c\frac{\partial v}{\partial \xi} = \varepsilon\left(u + \beta\right). \tag{A.5}$$

Introducing rescaled parameters c* and ε*

$$c^* = c - \frac{D}{R}, \tag{A.6}$$

$$\varepsilon^* = \varepsilon\frac{c^*}{c}, \tag{A.7}$$

yields

$$c^*\frac{\partial u}{\partial \xi} = 3u - u^3 - v + D\frac{\partial^2 u}{\partial^2 \xi}, \tag{A.8}$$

$$c^*\frac{\partial v}{\partial \xi} = \varepsilon^*(u + \beta), \tag{A.9}$$

which has the same form as the FHN model in one spatial dimension.

Thus, for $c^* = c_{cr}$ and $\varepsilon^* = \varepsilon_{cr}$, the homoclinic solution of (A.8), (A.9) corresponds to the connection between the fast wave branch and the slow wave branch. Hence, the minimal velocity c_{min} of curved fronts can be derived from equation (A.7) by setting $c^* = c_{cr}$ and $\varepsilon^* = \varepsilon_{cr}$. This yields

$$c_{\min} = \frac{\varepsilon}{\varepsilon_{cr}} c_{cr}. \tag{A.10}$$

Appendix. Toroidal coordinates

A parametrization $f: \{\alpha^i\} \mapsto \{x^j\}$ gives the Laplace–Beltrami operator in curvilinear coordinates

$$\Delta_{LB} = \sum_{i,k} \frac{1}{\sqrt{g}} \frac{\partial}{\partial \alpha^i} \left(g^{ik} \sqrt{g} \frac{\partial}{\partial \alpha^k} \right), \tag{A.11}$$

where G is the metric tensor with matrix elements g_{ik}, which is the product of the transposed Jacobian matrix of f multiplied with the Jacobian matrix of f, and g = Det G, see [35]. The single components of the metric tensor thus are the scalar product

$$g_{ik} = \sum_j \frac{\partial f_j}{\partial \alpha^i} \frac{\partial f_j}{\partial \alpha^k} =: \left\langle \frac{\partial f}{\partial \alpha^i} \middle| \frac{\partial f}{\partial \alpha^k} \right\rangle. \tag{A.12}$$

A parametrization f is isothermal, if the derived coordinate system is orthogonal and conformal. In two spatial dimensions, a parametrization

$$f: (\alpha^1, \alpha^2) \mapsto \begin{pmatrix} x \\ y \\ z \end{pmatrix} \tag{A.13}$$

is orthogonal, if the scalar product of the basis vectors for $i \neq k$ equals zero,

$$\left\langle \frac{\partial f}{\partial \alpha^i} \middle| \frac{\partial f}{\partial \alpha^k} \right\rangle = 0. \tag{A.14}$$

The condition for conformal mapping is

$$\left\langle \frac{\partial f}{\partial \alpha^i} \middle| \frac{\partial f}{\partial \alpha^i} \right\rangle = \left\langle \frac{\partial f}{\partial \alpha^k} \middle| \frac{\partial f}{\partial \alpha^k} \right\rangle, \tag{A.15}$$

see [35]. This yields the following form of the Laplace–Beltrami operator

$$\Delta_{\mathrm{LB}} = \sum_{i,k} \frac{1}{\sqrt{g}} \frac{\partial}{\partial \alpha^i} \frac{\partial}{\partial \alpha^k} \delta^{ik} = \frac{1}{\sqrt{g}} \nabla^2. \tag{A.16}$$

To derive a global isothermal coordinate system for the surface of a torus, we start from the parametrization [39]

$$(\theta_i, \varphi) \mapsto \begin{pmatrix} \dfrac{a \sinh \eta \cos \varphi}{\cosh \eta - \cos \theta_i} \\ \dfrac{a \sinh \eta \sin \varphi}{\cosh \eta - \cos \theta_i} \\ \dfrac{a \sin \theta_i}{\cosh \eta - \cos \theta_i} \end{pmatrix} = \begin{pmatrix} x \\ y \\ z \end{pmatrix}, \tag{A.17}$$

where $a > 0$ is a scaling factor of space and $\eta > 0$ is a measure for the ratio of major curvature radius R and minor curvature radius r. As can easily be proved, these are orthogonal coordinates. As

$$g_{\theta_i \theta_i} = \frac{a^2}{(\cosh \eta - \cos \theta_i)^2}$$

and

$$g_{\varphi\varphi} = \frac{a^2 \sinh^2 \eta}{(\cosh \eta - \cos \theta_i)^2},$$

$g_{\theta_i \theta_i} \neq g_{\varphi\varphi}$, thus the parametrization (A.17) is not conformal. Introducing the variable

$$\tilde{\varphi} = \varphi \sinh \eta$$

yields

$$g_{\tilde{\varphi}\tilde{\varphi}} = g_{\theta_i \theta_i} = \sqrt{g} = \frac{a^2}{(\cosh \eta - \cos \theta_i)^2}. \tag{A.18}$$

Thus the Laplace–Beltrami operator in isothermal torus coordinates reads

$$\Delta_{\mathrm{LB}} = \frac{(\cosh \eta - \cos \theta_i)^2}{a^2} \left(\frac{\partial^2 u}{\partial \theta_i^2} + \frac{\partial^2 u}{\partial \tilde{\varphi}^2} \right). \tag{A.19}$$

To obtain the dependencies of a and η upon the major curvature radius R and the minor curvature radius r, which are parameters of the common parametrization

$$(\theta, \varphi) \mapsto \begin{pmatrix} (R + r\cos\theta)\cos\varphi \\ (R + r\cos\theta)\sin\varphi \\ r\sin\theta \end{pmatrix} = \begin{pmatrix} x \\ y \\ z \end{pmatrix}, \tag{A.20}$$

one needs to compare equation (A.20) with the isothermal parametrization

$$(\theta_i, \tilde{\varphi}) \mapsto \begin{pmatrix} \dfrac{a\sinh\eta\cos\left(\dfrac{\tilde{\varphi}}{\sinh\eta}\right)}{\cosh\eta - \cos\theta_i} \\ \dfrac{a\sinh\eta\sin\left(\dfrac{\tilde{\varphi}}{\sinh\eta}\right)}{\cosh\eta - \cos\theta_i} \\ \dfrac{a\sin\theta_i}{\cosh\eta - \cos\theta_i} \end{pmatrix} = \begin{pmatrix} x \\ y \\ z \end{pmatrix}. \tag{A.21}$$

A necessary and sufficient condition that a point from the domain of definition of the parametrization equation (A.20) lies on the 2D surface of a torus in the Euclidian R^3 is

$$\left(\sqrt{x^2 + y^2} - R\right)^2 + z^2 - r^2 = 0. \tag{A.22}$$

In toroidal coordinates equation (A.21) yields

$$x^2 + y^2 + z^2 - 2a\frac{\cosh\eta}{\sinh\eta}\sqrt{x^2 + y^2} + a^2 = 0. \tag{A.23}$$

By comparing the coefficients one obtains from equation (A.23) and equation (A.22)

$$R = a\frac{\cosh\eta}{\sinh\eta} = a\coth\eta, \tag{A.24}$$

$$r = a\frac{1}{\sinh\eta}, \tag{A.25}$$

and the inverse relations

$$a = \sqrt{R^2 - r^2}, \tag{A.26}$$

$$\eta = \operatorname{arcoth}\frac{R}{\sqrt{R^2 - r^2}} = \operatorname{arcoth}\frac{n}{\sqrt{n^2 - 1}}, \tag{A.27}$$

where $n = \frac{R}{r}$. As can be seen from the parametrizations (A.20) and (A.21), the transformation $\tilde{\varphi}(\varphi)$ is

$$\tilde{\varphi}(\varphi) = \varphi \sinh \eta. \tag{A.28}$$

To derive the dependency between θ_i and θ, the expressions $\sqrt{x^2 + y^2} - R$ of both coordinate systems are compared. This yields

$$r \cos \theta = a \frac{\sinh \eta}{\cosh \eta - \cos \theta_i} - R. \tag{A.29}$$

Replacing R and r with equations (A.24) and (A.25), this yields

$$\theta = \arccos\left(\frac{\cosh \eta \cos\theta_i - 1}{\cosh \eta - \cos\theta_i}\right) \cdot \begin{cases} +1 & \theta_i \geqslant 0 \\ -1 & \theta_i < 0 \end{cases} \tag{A.30}$$

The inverse function is

$$\theta_i = \arccos\left(\frac{R}{r} - \frac{R^2 - r^2}{r(R + r\cos\theta)}\right) \cdot \begin{cases} +1 & \theta \geqslant 0 \\ -1 & \theta < 0 \end{cases} \tag{A.31}$$

REFERENCES

1. Dahlem M A and Tusch J 2012 Predicted selective increase of cortical magnification due to cortical folding. J. Math. Neurosci 2 14
2. Karatas H, Erdener S E, Gursoy-Ozdemir Y, Lule S, Eren-Kocak E, Sen Z D and Dalkara T 2013 Spreading depression triggers headache by activating neuronal Panx1 channels Science 339 1092–5
3. Dreier J P 2011 The role of spreading depression, spreading depolarization and spreading ischemia in neurological disease Nat. Med. 17 439–47
4. Dahlem M A and Hadjikhani N 2009 Migraine aura: retracting particle-like waves in weakly susceptible cortex PLoS ONE 4 e5007
5. Dahlem M A, Graf R, Strong A J, Dreier J P, Dahlem Y A, Sieber M, Hanke W, Podoll K and Schöll E 2010 Two-dimensional wave patterns of spreading depolarization: retracting, re-entrant, and stationary waves Physica D 239 889–903
6. Krischer K and Mikhailov A S 1994 Bifurcation to traveling spots in reaction-diffusion systems Phys. Rev. Lett. 73 3165–8
7. Bode M and Purwins H G 1995 Pattern formation in reaction-diffusion systems-dissipative solitons in physical systems Physica D 86 53

8. Schenk C P, Or-Guil M, Bode M and Purwins H G 1997 Interacting pulses in three-component reactiondiffusion systems on two-dimensional domains Phys. Rev. Lett. 78 3781
9. Kim M, Bertram M, Pollmann M, von Oertzen A, Mikhailov A S, Rotermund H H and Ertl G 2001 Controlling chemical turbulence by global delayed feedback: pattern formation in catalytic CO oxidation on Pt(110) Science 292 1357
10. Zykov V S, Bordiougov G, Brandtstädter H, Gerdes I and Engel H 2004 Global control of spiral wave dynamics in an excitable domain of circular and elliptical shape Phys. Rev. Lett. 92 018304
11. Schlesner J, Zykov V, Engel H and Schöll E 2006 Stabilization of unstable rigid rotation of spiral waves in excitable media Phys. Rev. E 74 046215
12. Mikhailov A S and Showalter K 2006 Control of waves, patterns and turbulence in chemical systems Phys. Rep. 425 79–194
13. Dahlem M A, Schneider F M and Schöll E 2008 Failure of feedback as a putative common mechanism of spreading depolarizations in migraine and stroke Chaos 18 026110
14. Schneider F M, Schöll E and Dahlem M A 2009 Controlling the onset of traveling pulses in excitable media by nonlocal spatial coupling and time delayed feedback Chaos 19 015110
15. Kyrychko Y N, Blyuss K B, Hogan S J and Schöll E 2009 Control of spatio-temporal patterns in the Gray–Scott model Chaos 19 043126
16. Maselko J and Showalter K 1989 Chemical waves on spherical surfaces Nature 339 609–11
17. Abramychev A Yu, Davydov V A and Zykov V S 1990 Drift of spiral waves on nonuniformly curved surfaces J. Exp. Theor. Phys. 70 666
18. Davydov V A and Zykov V S 1991 Kinematics of spiral waves on nonuniformly curved surfaces Physica. D 49 71–74
19. Davydov V A and Zykov V S 1993 Spiral autowaves in a round excitable medium J. Exp. Theor. Phys. 76 414–9
20. Mikhailov A S, Davydov V A and Zykov V S 1994 Complex dynamics of spiral waves and motion of curves Physica D 70 1–39
21. Davydov V A, Manz N, Steinbock O, Zykov V S and Müller S C

2000 Excitation fronts on a periodically modulated curved surface Phys. Rev. Lett. 85 868–71

22. Davydov V A, Morozov V G and Davydov N V 2000 Ring-shaped autowaves on curved surfaces Phys. Lett. A 267 326–30
23. Davydov V A, Manz N, Steinbock O and Müller S C 2002 Critical properties of excitation waves on curved surfaces: curvature-dependent loss of excitability Europhys. Lett. 59 344–50
24. Manz N, Davydov V A, C Müller S and Bär M 2003 Dependence of the spiral rotation frequency on the surface curvature of reaction-diffusion systems Phys. Lett. A 316 311–6
25. Lindner B, García-Ojalvo J, Neiman A and Schimansky-Geier L 2004 Effects of noise in excitable systems Phys. Rep. 392 321–424
26. van der Pol B 1926 On relaxation oscillations Phil. Mag. 2 978–92
27. Bonhoeffer K F 1948 Activation of passive iron as a model for the excitation of nerve J. Gen. Physiol 32 69–91
28. Bonhoeffer K F 1953 Modelle der Nervenerregung Naturwissenschaften 40 301–11
29. FitzHugh R 1961 Impulses and physiological states in theoretical models of nerve membrane Biophys. J. 1 445–66
30. Nagumo J, Arimoto S and Yoshizawa S 1962 An active pulse transmission line simulating nerve axon Proc. IRE 50 2061–70
31. Hodgkin A L and Huxley A F 1952 A quantitative description of membrane current and its application to conduction and excitation in nerve J. Physiol. 117 500
32. Dahlem M A and Müller S C 2004 Reaction-diffusion waves in neuronal tissue and the window of cortical excitability Ann. Phys. 13 442–9
33. Hübel N, Schöll E and Dahlem M A 2014 Bistable dynamics underlying excitability of ion homeostasis in ion-based neuron models PLoS Comp. Biol. at press (arXiv:1310.1941v1).
34. Winfree A T 1991 Varieties of spiral wave behaviour: an experimentalist's approach to the theory of excitable media Chaos 1 303–34

35. Kühnel W 2005 Differentialgeometrie (Wiesbaden: Vieweg)
36. Davydov V A, Morozov V G and Davydov N V 2003 Critical properties of autowaves propagating on deformed cylindrical surfaces Phys. Lett. A 307 265–8
37. Krupa M, Sandstede B and Szmolyan P 1997 Fast and slow waves in the FitzHugh–Nagumo equation J. Differ. Equ. 133 49–97
38. Zykov V S 1984 Simulation of Wave Processes in Excitable Media (Moscow: Wiley) (Engl. transl. from 1992)
39. Moon P and Spencer D E 1971 Field Theory Handbook 2nd edn (Berlin: Springer)
40. Kosugi T 2013 Electronic properties and persistent spin currents of nanospring under static magnetic field J. Phys. Soc. Japan 531 173–99

6 IDENTIFYING CHAOTIC FITZHUGH–NAGUMO NEURONS USING COMPRESSIVE SENSING

Ri-Qi Su [1], Ying-Cheng Lai [1,2,] and Xiao Wang [3]

[1] School of Electrical, Computer, and Energy Engineering, Arizona State University, Tempe, AZ 85287, USA

[2] Department of Physics, Arizona State University, Tempe, AZ 85287, USA

[3] School of Biological and Health Systems Engineering, Arizona State University, Tempe, AZ 85287, USA

ABSTRACT

We develop a completely data-driven approach to reconstructing coupled neuronal networks that contain a small subset of chaotic neurons. Such chaotic elements can be the result of parameter shift in their individual dynamical systems and may lead to abnormal functions of the network. To accurately identify the chaotic neurons may thus be necessary and important, for example, applying appropriate controls to bring the network to a normal state. However, due to couplings among the nodes, the measured time series, even from non-chaotic neurons, would

appear random, rendering inapplicable traditional nonlinear time-series analysis, such as the delay-coordinate embedding method, which yields information about the global dynamics of the entire network. Our method is based on compressive sensing. In particular, we demonstrate that identifying chaotic elements can be formulated as a general problem of reconstructing the nodal dynamical systems, network connections and all coupling functions, as well as their weights. The working and efficiency of the method are illustrated by using networks of non-identical FitzHugh–Nagumo neurons with randomly-distributed coupling weights.

KEYWORDSL: compressive sensing; nonlinear system identification; neuronal networks; chaos; random networks

INTRODUCTION

In this paper, we address the problem of the data-based identification of a subset of chaotic elements embedded in a network of nonlinear oscillators. In particular, given such a network, we assume that time series can be measured from each oscillator. The oscillators, when isolated, are not identical in that their parameters are different, so dynamically, they can be in distinct regimes. For example, all oscillators can be described by differential equations of the same mathematical form, but with different parameters. Consider the situation where only a small subset of the oscillators are chaotic and the remaining oscillators are in dynamical regimes of regular oscillations. Due to mutual couplings among the oscillators, the measured time series from most oscillators would appear random. The challenge is to identify the small subset of originally ("truly") chaotic oscillators.

The problem of identifying chaotic elements from a network of coupled oscillators arises in biological systems and biomedical applications. For example, consider a network of coupled neurons that exhibit regular oscillations in a normal state. In such a state, the parameters of each isolated neuron are in the regular regime. Under external perturbations or slow environmental influences, the parameters of some neurons can drift into the chaotic regime. When this occurs, the whole network would appear to behave chaotically, which may correspond to a certain disease. The virtue of nonlinearity stipulates that the irregular oscillations at the

network level can emerge even if only a few oscillators have gone "bad". It is thus desirable to be able to pin down the origin of the ill-behaved oscillators—the few chaotic neurons among a large number of healthy ones.

One might attempt to use the traditional approach of time-delayed coordinate embedding to reconstruct the phase space of the underlying dynamical system [1–3] and then to compute the Lyapunov exponents [4,5]. However, since we are dealing with a network of nonlinear oscillators, the phase-space dimension is high and an estimate of the largest Lyapunov exponent would only indicate if the whole coupled system is chaotic or nonchaotic, depending on the sign of the estimated exponent. In principle, using time series from any specific oscillator(s) would give qualitatively the same result. Thus, the traditional approach cannot give an answer as to which oscillators are chaotic when isolated.

There were previous efforts in nonlinear systems identification and parameter estimation for coupled oscillators and spatiotemporal systems, such as the auto-synchronization method [6]. There were also works on revealing the connection patterns of networks. For example, a methodology was proposed to estimate the network topology controlled by feedback or delayed feedback [7–9]. Network connectivity can be reconstructed from the collective dynamical trajectories using response dynamics, as well [10,11]. In addition, the approach of random phase resetting was introduced to reconstruct the details of the network structure [12]. For neuronal systems, there was a statistical method to track the structural changes [13,14]. While many of these previous methods require complete or partial information about the dynamical equations of the isolated nodes and their coupling functions, completely data-driven and model-free methods exist. For example, the network structure can be obtained by calculating the causal influences among the time series based on, for example, the Granger causality method [15,16], the transfer-entropy method [17] or the method of inner composition alignment [18]. However, such causality-based methods are unable to reveal information about the dynamical equations of the isolated nodes. There were regression-based methods [19] for systems identification based on, for example, the least-squares approximation through the Kronecker-product representation [20], which would require large amounts of data. (Due to the L_1 nature of compressive sensing [21–25], the data requirement in our method can

be significantly relaxed.) The unique features of our method are: (1) it is completely data driven; (2) it can give an accurate estimate of all system parameters; (3) it can lead to faithful reconstruction of the full network structure, even for large networks; and (4) it requires a minimal data amount. While some of these features are shared by previous methods, no single previous method possesses all of these features.

Here, we develop a method to address the problem of identifying a subset of ill-behaved chaotic elements from a network of nonlinear oscillators, the majority of them being regular. The basic mathematical framework underlying our method is compressive sensing (CS), a paradigm for high-fidelity signal reconstruction using only sparse data [21–25]. The CS paradigm was originally developed to solve the problem of transmitting extremely large data sets, such as those collected from large-scale sensor arrays. Because of the extremely high dimensionality, direct transmission of such data sets would require a very broad bandwidth. However, there are many applications in which the data sets are sparse. To be concrete, say a data set of N points is represented by an $N \times 1$ vector, X, where N is a very large integer. Then, X being sparse means that most of its entries are zero and only a small number of k entries are non-zero, where $k \ll N$. One can use a random matrix $\mathbf{\Psi}$ of dimension $M \times N$ to obtain an $M \times 1$ vector Y: $Y = \mathbf{\Psi} \cdot X$, where $M \sim k$. Because the dimension of Y is much lower than that of the original vector X, transmitting Y would require a much smaller bandwidth, provided that **X**can be reconstructed at the other end of the communication channel. Under the constraint that the vector to be reconstructed is sparse, the feasibility of faithful reconstruction is guaranteed mathematically by the CS paradigm [21–25]. In the past decade, CS has been exploited in a large variety of applications, ranging from optical image processing [26] and reconstruction of nonlinear dynamical and complex systems [27,28] to quantum measurements [29].

It has been shown in a series of recent papers [27,28,30–33] that the detailed equations and parameters of nonlinear dynamical systems and complex networks can be accurately reconstructed from short time series using the CS paradigm. Here, we extend this approach to a network of coupled, mixed nonchaotic and chaotic neurons. We demonstrate that, by

formulating the reconstruction task as a CS problem, the system equations and coupling functions, as well as all of the parameters can be obtained accurately from sparse time series. Using the reconstructed system equations and parameters for each and every neuron in the network and setting all of the coupling parameters to zero, a routine calculation of the largest Lyapunov exponent can unequivocally distinguish the chaotic neurons from the nonchaotic ones.

We remark on the generality of our compressive sensing-based method. Insofar as time series from all dynamical variables of the system are available and a suitable mathematical base can be found in which the nodal and coupling functions can be expanded in terms of a sparse number of terms, the whole system, including all individual nodal dynamics, can be accurately reconstructed. With the reconstructed individual nodal equations, chaotic neurons can be identified through routine calculation of the largest Lyapunov exponent.

METHODS

Figure 1a shows schematically a representative coupled neuronal network. Consider a pair of neurons, one chaotic and another nonchaotic when isolated (say 1 and 10, respectively). When they are placed in a network, due to coupling, the time series collected from both will appear random and qualitatively similar, as shown in Figure 1b,c. It is visually quite difficult to distinguish the time series and to ascertain which node is originally chaotic and which is regular. The difficulty is compounded by the fact that the detailed coupling scheme is not known a priori. Say that the chaotic behavior leads to the undesirable function of the network and is to be suppressed. A viable and efficient method is to apply small pinning controls [34–37] to the relatively few chaotic neurons to drive them into some regular regime. (Here, we assume the network is such that, when all neurons are regular, the collective dynamics is regular. That is, we exclude the uncommon, but not unlikely, situation that a network system of coupled regular oscillators would exhibit chaotic behaviors.) Accurate identification of the chaotic neurons is thus key to implementing the pinning control strategy.

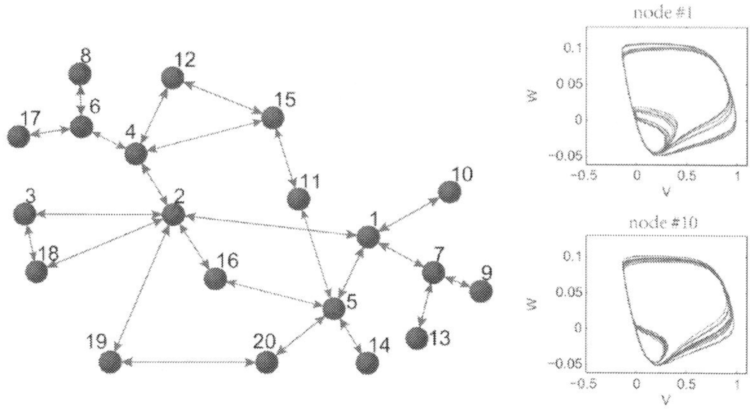

Figure 1. (**a**) Schematic illustration of a small neuronal network, where the dynamics of each neuron is mathematically described by the FitzHugh–Nagumo (FHN) equations. (b, c) Dynamical trajectories of two neurons from the coupled system, one being chaotic when isolated and another regular, respectively. The trajectories give little hint as to which one is originally chaotic and which one is regular, due to the coupling. Specifically, Neuron 1 is originally chaotic (by setting parameter a = 0.42 in the FHN equation), while all other neurons are regular (their values of the corresponding parameter in the FHN equation are chosen uniformly from the interval [0.43, 0.45]).

Given a neuronal network, our aim is thus to locate all neurons that are originally chaotic and neurons that are potentially likely to enter into a chaotic regime when they are isolated from the other neurons or when the couplings among the neurons are weakened. Our approach consists of two steps. Firstly, we employ the CS framework to estimate, from measured time series only, the parameters in the FHN equation for each neuron, as well as the network topology and various coupling functions and weights. As will be shown below, this can be done by expanding the nodal dynamical equations and the coupling functions into some suitable mathematical base, as determined by the specific knowledge of the actual neuronal dynamical system, and then casting the problem into that of determining the sparse coefficients associated with various terms in the expansion. The nonlinear system identification problem can then be solved using some standard CS algorithm. Secondly, we set all coupling parameters to zero and analyze the dynamical behaviors of each and every individual neuron by calculating the Lyapunov exponents. Those with positive largest exponent are identified as chaotic.

A typical time series from a neuronal network consists of a sequence of spikes in the time evolution of the cell membrane potential. We demonstrate that our CS-based reconstruction method works well even for such spiky time series. We also analyze the dependence of the reconstruction accuracy on the data amount and show that only limited data are required to achieve high accuracy in reconstruction.

The FitzHugh–Nagumo (FHN) Model of Neuronal Dynamics

The FHN model, a simplified version of the biophysically detailed Hodgkin–Huxley model [38], is a mathematical paradigm for gaining significant insights into a variety of dynamical behaviors in real neuronal systems [39,40]. For a single, isolated neuron, the corresponding dynamical system is described by the following two-dimensional, nonlinear ordinary differential equations:

$$\begin{aligned} \frac{dV}{dt} &= \frac{1}{\delta}[V(V-a)(1-V)-W], \\ \frac{dW}{dt} &= V-W-b+S(t), \end{aligned} \qquad (1)$$

where V is the membrane potential, W is the recover variable, S(t) is the driving signal (e.g., periodic signal) and a, b and $\delta$are parameters. The parameter δ is chosen to be infinitesimal, so that V (t) and W (t) are "fast" and "slow" variables, respectively. Because of the explicitly time-dependent driving signal S(t), Equation (1) is effectively a three-dimensional dynamical system, in which chaos can arise [41]. For a network of FHN neurons, the equations are:

$$\begin{aligned} \frac{dV_i}{dt} &= \frac{1}{\delta}[V_i(V_i-a)(1-V_i)-W_i]+\sum_{i=1}^{N} c_{ij}(V_j-V_i) \\ \frac{dW_i}{dt} &= V_i-W_i-b+S(t), \end{aligned} \qquad (2)$$

where c_{ij} is the coupling strength (weight) between the i-th and the j-th neurons (nodes). For $c_{ij} = c_{ji}$, the interactions between any pair of neurons are symmetric, leading to a symmetric adjacency matrix for the network. For $c_{ij} \neq c_{ji}$, the network is asymmetrically weighted.

Formulation of Reconstruction Problem in the CS Framework

The problem of CS can be stated as follows. Given a low-dimensional measurement vector $Y \in R^M$, one seeks to reconstruct the much higher-dimensional vector $X \in R^N$ according to:

$$Y=\Psi \cdot X, \tag{3}$$

where $N \gg M$ and $\boldsymbol{\Psi}$ is an $M \times N$ projection matrix. A sufficiently sparse vector X can be reconstructed by solving the following convex optimization problem [21–25]:

$$\min \|X\|_1 \text{ s.t. } Y=\Psi \cdot X, \tag{4}$$

where the $\|\mathbf{X}\|_1 = \sum_{i=1}^{N} |\mathbf{X}_i|$ is the l_1 norm of vector X.

Equation (2) is a non-autonomous coupling network of N oscillators, driven by external signal S(t). In general, for an isolated oscillator, the dynamics can be written as:

$$\dot{\mathbf{x}}_i(t) = \mathbf{F}_i[\mathbf{x}_i(t)] + \mathbf{S}_i(t), \tag{5}$$

where $x_i \in R^m$ is an m-dimensional dynamical variable and $S_i(t)$ denotes the external driving. The network equations can then be written as:

$$\dot{\mathbf{x}}_i = \mathbf{F}_i(\mathbf{x}_i) + \sum_{j=1, j\neq i}^{N} \mathbf{W}_{ij} \cdot [\mathbf{H}(\mathbf{x}_j) - \mathbf{H}(\mathbf{x}_i)] + \mathbf{S}_i(t), \tag{6}$$

where $W_{ij} \in R^{m\times m}$ is the weighted coupling matrix between node i and node j and H is the coupling function. Our goal is to reconstruct the nodal velocity field F_i and all of the coupling matrices W using time series x(t) and the given driving signal S(t). First, we group all terms directly associated with node i into $F'_i(x_i)$, by defining:

$$\mathbf{F}'_i(\mathbf{x}_i) \equiv \mathbf{F}_i(\mathbf{x}_i) - \mathbf{H}(\mathbf{x}_i) \cdot \sum_{j=1, j\neq i}^{N} \mathbf{W}_{ij}.$$

$$\dot{\mathbf{x}}_i = \mathbf{F}'_i(\mathbf{x}_i) + \sum_{j=1, j\neq i}^{N} \mathbf{W}_{ij}\mathbf{H}(\mathbf{x}_j) + \mathbf{S}_i(t). \tag{7}$$

Then, we choose a suitable base and expand $F'(x_i)$ into the following form:

$$\mathbf{F}'_i(\mathbf{x}_i) = \sum_{\gamma} \tilde{\mathbf{a}}_i^{(\gamma)} \cdot \tilde{\mathbf{g}}_i^{(\gamma)}(\mathbf{x}_i), \tag{8}$$

where $\tilde{g}_i^{(\gamma)}(x_i)$ are a set of orthogonal and complete base functions, which are chosen such that the coefficients $\tilde{a}_i^{(\gamma)}$ are sparse. While the coupling function $H(x_i)$, if it is nonlinear, can be expanded in a similar manner, for notational convenience, we assume that they are linear: $H(x_i)= x_i$. We then have:

$$\dot{\mathbf{x}}_i = \sum_{\gamma} \tilde{\mathbf{a}}_i^{(\gamma)} \cdot \tilde{\mathbf{g}}_i^{(\gamma)}(\mathbf{x}_i) + \sum_{j=1, j\neq i}^{N} \mathbf{W}_{ij} \cdot \mathbf{x}_j + \mathbf{S}_i(t), \tag{9}$$

where all of the coefficients $\tilde{a}_i^{(\gamma)}$ and W_{ij} are to be determined from time series x_i via CS. Specifically, the coefficient vector $\tilde{a}_i^{(\gamma)}$ determines the nodal dynamics, and the weighted matrices W_{ij} give the full topology and coupling strengths of the entire network.

Suppose we have measurements of all state variables $x_i(t)$ at M different values of t and assume further that for each tvalue, the values of the state variables at a slightly later time, $t + \delta t$, are also available, where $\delta t \ll \Delta t$, so that the derivative vector $\dot{x}_i$ can be estimated at each time instant. Equation (9) for all M time instants can then be written in the following matrix form:

$$\mathbf{G}_i = \begin{pmatrix} \tilde{\mathbf{g}}_i(t_1) & \mathbf{x}_1(t_1) & \cdots & \mathbf{x}_k(t_1) & \cdots & \mathbf{x}_N(t_1) \\ \tilde{\mathbf{g}}_i(t_2) & \mathbf{x}_1(t_2) & \cdots & \mathbf{x}_k(t_2) & \cdots & \mathbf{x}_N(t_2) \\ \vdots & \vdots & \cdots & \vdots & \cdots & \vdots \\ \tilde{\mathbf{g}}_i(t_M) & \mathbf{x}_1(t_M) & \cdots & \mathbf{x}_k(t_M) & \cdots & \mathbf{x}_N(t_M) \end{pmatrix}, \tag{10}$$

where the index k in $x_k(t)$ runs from one to N, $k \neq i$, and each row of the matrix is determined by the available time series at one instant of time. The derivatives at different times can be written in a vector form as: $X_i =[\dot{x}_i(t_1), \cdots, \dot{x}_i(t_M)]^T$. The coefficients from the functional expansion and the weights associated with all links in the network, which are to be determined, can be combined concisely into a vector a_i, as follows:

$$\mathbf{a}_i = [\tilde{\mathbf{a}}_i, \mathbf{W}_{1i}, \cdots, \mathbf{W}_{i-1,i}, \mathbf{W}_{i+1,i}, \cdots, \mathbf{W}_{N,i}]^T. \tag{11}$$

where $[\cdot]^T$ denotes the transpose. For a properly chosen expansion base and a general complex network whose connections are typically sparse, the vector a_i to be determined is sparse, as well. Finally, Equation (9) can be written in the standard CS form as:

$$\mathbf{X}_i = \mathbf{G}_i \cdot \mathbf{a}_i + \mathbf{S}_i(t), \tag{12}$$

a linear equation in which the dimension of the unknown coefficient vector $\mathbf{a}_i$ can be much larger than that of X_i, and the measurement matrix G_i will have many more columns than rows. In a conventional sense, this equation is ill defined, but since a_i is sparse, insofar as its number of non-zero coefficients is smaller than the dimension of X_i, the vector a_i can be uniquely and efficiently determined by CS [21–25].

RESULTS

We consider the FHN model with sinusoidal driving: $S(t)= r \sin \omega_0 t$. The model parameters are $r = 0.32$, $\omega_0 = 15.0$, $\delta = 0.005$ and $b = 0.15$. For $a = 0.42$, an individual neuron exhibits chaos. The time series are generated by the fourth-order Runge–Kutta method with step size $h = 10^{-4}$. We sample three consecutive measurements at time interval $\tau = 0.05$ apart and then use a standard two-point formula to calculate the derivative. Representative chaotic time series and the corresponding dynamical trajectory are shown in Figure 2.

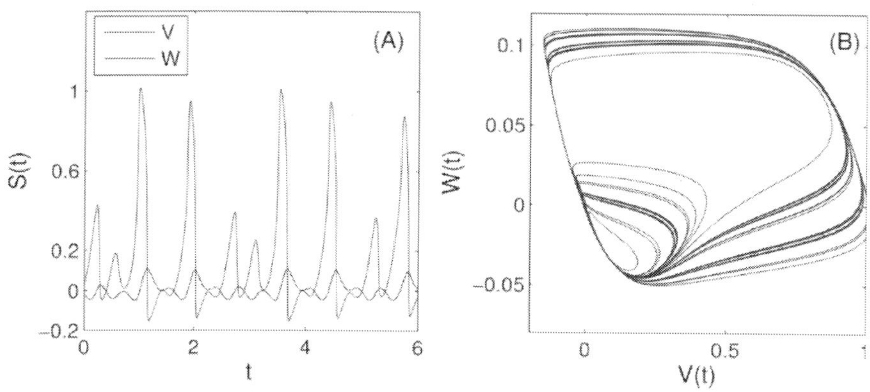

Figure 2. (a) Chaotic time series of the membrane potential V and recovery variable Wfrom a single neuron for a = 0.42; and (b) the corresponding dynamical trajectory.

We first present the reconstruction result for an isolated neuron, by setting to zero all coupling terms in Equation (10). The vector a_i to be determined then contains the unknown parameters associated with a single neuron only. We choose power series of order four as the expansion base, so that there are 17 unknown coefficients to be determined. We use 12 data points generated from a random starting point. The results of the reconstruction are shown in Figure 3a,b for variables V and W, respectively. The last two coefficients associated with each variable represent the strength of the driving signal. Since only the variable W receives sinusoidal input, the last coefficient in W is nonzero. By comparing the positions of nonzero terms and our previously assumed vector form, $g_i(t)$, we can fully reconstruct the dynamical equations of any isolated neuron. In particular, we see from Figure 3a,b that all estimated coefficients agree with their respective true values. Figure 3c shows how the estimated coefficients converge to the true values as the number of data points is increased. We see that, for over 10 data points, we can already reconstruct faithfully all of the parameters.

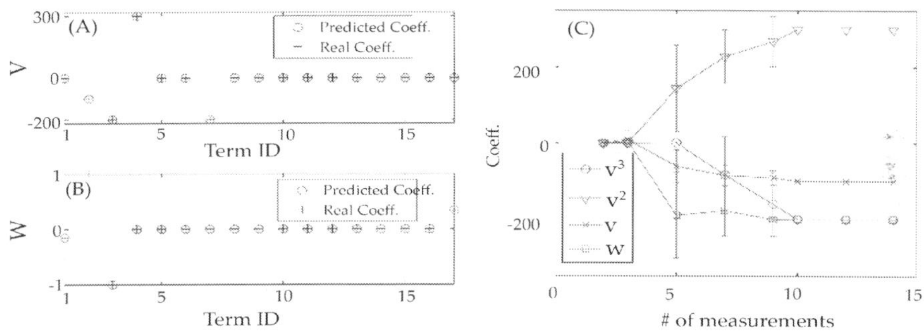

Figure 3. (a, b) Predicted coefficients from compressive sensing (CS) and a comparison with the actual parameter values in the dynamical equations of variables V and W. The number of data points used is 12. (c) Predicted parameters for a single neuron as the number of data points is increased. The sampling interval is $\Delta t = 0.05$. All results are averaged over 10 independent time series.

Next, we consider the network of coupled FHN neurons as schematically shown in Figure 1a, where the coupling weights among various pairs of nodes are uniformly distributed in the interval [0.3, 0.4]. The network is random with connection probability $p = 0.04$. From time series, we construct the CS matrix for each variable of all nodes. Since the couplings occur among the variables V of different neurons, the strengths of all incoming links can be found in the unknown coefficients associated with different V variables. Extracting all coupling terms from the estimated coefficients, we obtain all off-diagonal terms in the weighted adjacency matrix.

To assess the reconstruction accuracy, we define E_{nz} as the average normalized difference between the non-zero terms in the estimated coefficients and the real values:

$$E_{nz} = \frac{1}{M_{nz}} \sum_{k=1}^{M_{nz}} \| \frac{c'_k - c_k}{c_k} \|,$$

where M_{nz} is the number of non-zero terms in the actual coefficients, c'_k and c_k are the k-th nonzero terms in the estimated coefficients and the true one, respectively. For convenience, we define R_m as the relative number of data points normalized by the number of total unknown coefficients. Figure 4 shows the reconstructed adjacency matrix as compared with the real one for $R_m = 0.7$. We see that our method can predict all links correctly, in spite of the small errors in the predicted weight values. The errors are mainly due to the fact that there are large coefficients in the system equations, but the coupling weights are small.

Using the weighted adjacency matrix, we can identify the coupling terms in the vector function $F'_i(x_i)$, so as to extract the terms associated with isolated nodal velocity field F_i. We can then determine the value of parameter a and calculate the largest Lyapunov exponent for each individual neuron. The results are shown in Figure 5a,b. We see that, for this example, Neuron 1 has a positive largest exponent, while the largest exponents for all others are negative, so 1 is identified as the only chaotic neuron among all neurons in the network.

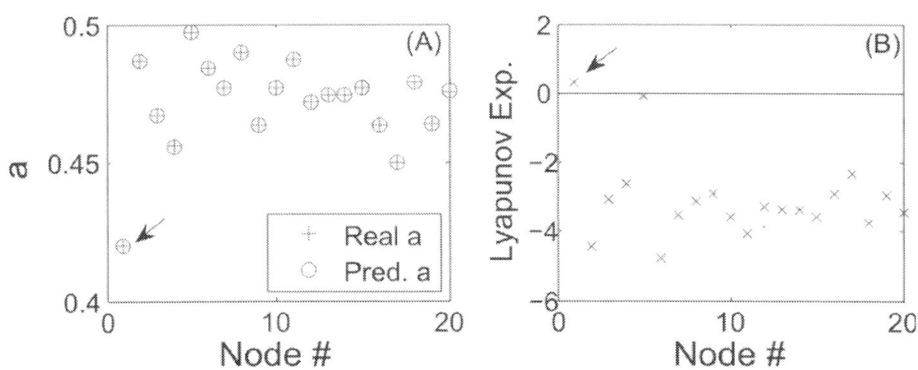

Figure 5. (**a**) Estimated values of parameter a for different neurons (red circles), as compared with the actual values (black crosses). The random network size is N = 20 with connection probability p = 0.04. The normalized data amount used in reconstruction is $R_m = 0.7$. (**b**) The largest Lyapunov exponents calculated from the reconstructed system equations. The reference line denotes a null value.

Next, we discuss the relationship between reconstruction error and data requirement. As shown in Figure 6, for different network sizes N, the reconstruction error decreases with R_m. For R_m larger than a threshold, the normalized error E_{nz} is small. For N = 40, the threshold is about 0.6, and it is 0.5 for N = 60. That is because, for fixed connection probability, a larger network will have more sparse connections, requiring a smaller value of R_m for accurate reconstruction.

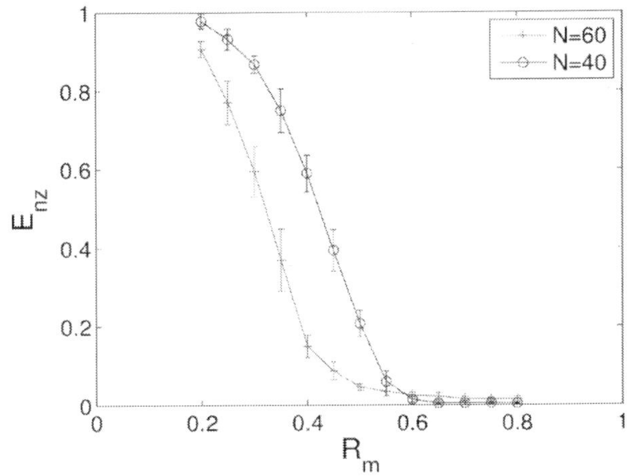

Figure 6. For neural networks of size N = 40 and N = 60, the normalized error associated with non-zero terms, E_{nz}, as a function of normalized data amount R_m. The results are averaged over 10 independent measurement realizations.

Finally, we study the performance of our method with respect to systematically varying of network size and edge density. As with any method, larger networks require more computation. We study networks of a size up to N = 100 nodes. Figure 7a shows the normalized error associated with the nonzero terms, E_{nz}, for different network sizes N and normalized data amount R_m. For a given network size, similar to Figure 6, E_{nz} gradually decreases to a certain low level as the relative data amount R_m is increased. We further observe that smaller values of R_m are required to reconstruct larger networks of the same connecting probability P. Note that R_m is the relative data amount defined with respect to the number of unknown coefficients, so for larger networks, the absolute data amount required actually increases. In Figure 7b, we show the contour plot of values of E_{nz} in the parameter plane (R_m, P) for a fixed network size (N = 60). We see that, for a fixed value of R_m, as P is increased, the error E_{nz} also increases, which is anticipated, as denser networks lead to a denser projection matrix in compressive sensing.

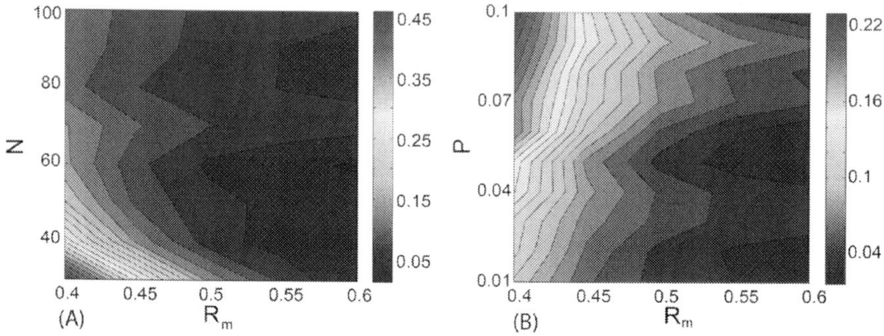

Figure 7. (**a**) For a random network of fixed connecting probability p = 0.04, a contour plot of the normalized error associated with nonzero terms, E_{nz}, in the parameter plane (R_m, N). (**b**) For a random network of fixed size N = 60, a contour plot of E_{nz} in the parameter plane (R_m, P). All results are obtained from 10 independent network realizations. See the text for explanations.

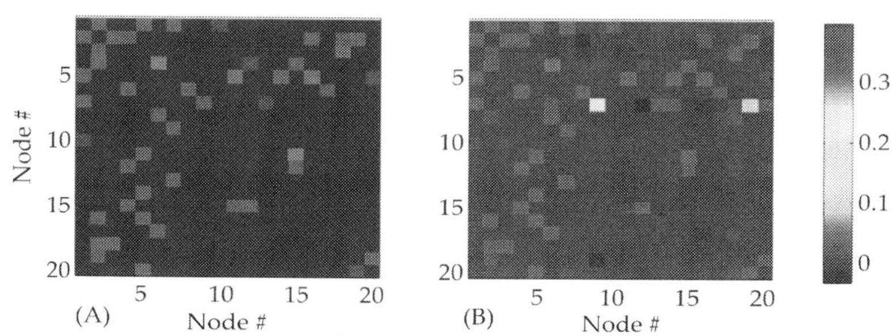

Figure 8. For the network in Figure 1a, (a) the actual and (b) estimated weighted adjacency matrix. The normalized data amount used in the reconstruction is R_m = 0.7.

CONCLUSIONS

We develop a completely data-driven method to detect chaotic elements embedded in a network of nonlinear oscillators, where such elements are assumed to be relatively few. From a biomedical perspective, the chaotic elements can be the source of certain diseases, and their accurate identification is desirable. In spite of being only a few, the chaotic oscillators can cause the time series from other, originally regular oscillators to appear random, due to the network interactions among the oscillators. The standard method in nonlinear time series, the method of delay-coordinate embedding, cannot be used to identify the local chaotic elements, because the method can give information only about the global dynamics. For example, one can attempt to estimate the largest Lyapunov exponent by using time series, either from a chaotic oscillator or from an originally regular oscillator, and the embedding method would yield qualitatively or even quantitatively similar results. Our compressive sensing-based method, however, overcomes such difficulties by generating an accurate estimate of all system equations, which include the local dynamical equations of each individual node and all coupling functions. Isolating the coupling functions from the local velocity fields, we can obtain the original dynamical equations for each individual oscillator, enabling efficient calculation of the Lyapunov exponents for all oscillators and, consequently, accurate identification of the chaotic oscillators. We illustrate this methodology by using model networks of FHN neurons. One key virtue of compressive sensing, namely the low data requirement, enables us to accomplish the task of identifying chaos with short time series. Our method is generally applicable to any nonlinear dynamical networks, insofar as time series from the oscillators are available.

Comparing with our previous works on compressive sensing-based nonlinear system identification and reverse engineering of complex networks [27,28,30–33], the new technical features of the present work are the following. Firstly, we demonstrate that the compressive sensing-based system identification is effective for spiky time series that are typical of neuronal networks. Secondly, local velocity fields and non-uniform weights of node-to-node interactions can be reconstructed accurately for neuronal networks with both fast and slow variables in the presence of external driving. Thirdly, the method works regardless of the ratio between the number of originally chaotic and nonchaotic oscillators. The great flexibility, the extreme low data requirement and high accuracy make our method appealing for various problems arising from nonlinear system identification, especially in biology and biomedicine.

There are a number of limitations to our method. For example, for any accessible node in the network, time series of all dynamical variables are required. If information from one node or some of the nodes in the network is inaccessible, or "hidden" from the outside world, it is not feasible to recover the nodal dynamical system of such nodes and their neighbors [31,33]. The "hidden dimensions" problem, in which some dynamical variables are not given, is another obstacle to realistic applications. Our compressive sensing-based method also requires reasonable knowledge about the underlying complex system, so that a suitable mathematical base can be identified for expansions of the various nodal and coupling functions. Further efforts are certainly needed.

Acknowledgments

This work was supported by AFOSR (Air Force Office of Scientific Research) under Grant No. FA9550-10-1-0083 and by NSF (National Science Foundation) under Grant No. CDI-1026710.

Author Contributions

Ri-Qi Su, Ying-Cheng Lai, Xiao Wang designed research project. Ri-Qi Su did computations. Ri-Qi Su and Ying-Cheng Lai analyzed results. Ying-Cheng Lai, Ri-Qi Su and Xiao Wang wrote the paper. All authors have read and approved the final manuscript.

Conflicts of Interest

The authors declare no conflict of interest.

REFERENCES

1. Takens, F. Detecting strange attractors in fluid turbulence. In Dynamical Systems and Turbulence; Rand, D., Young, L.S., Eds.; Springer-Verlag: Berlin, Germany, 1981; Volume 898, pp. 366–381.
2. Packard, N.H.; Crutchfield, J.P.; Farmer, J.D.; Shaw, R.S. Geometry from a time series. Phys. Rev. Lett **1980**, 45, 712–716.
3. Kantz, H.; Schreiber, T. Nonlinear Time Series Analysis, 1st ed; Cambridge University Press: Cambridge, UK, 1997.
4. Wolf, A.; Swift, J.B.; Swinney, H.L.; Vastano, J.A. Determining Lyapunov exponents from a time series. Physica D**1985**, 16, 285–317.

5. Eckmann, J.P.; Kamphorst, S.O.; Ruelle, D.; Ciliberto, S. Liapunov exponents from time series. Phys. Rev. A **1986**, 34, 4971–4979.
6. Parlitz, U. Estimating model parameters from time series by autosynchronization. Phys. Rev. Lett **1996**, 76, 1232.
7. Yu, D.; Righero, M.; Kocarev, L. Estimating topology of networks. Phys. Rev. Lett **2006**, 97, 188701.
8. Tang, W.K.S.; Yu, M.; Kocarev, L. Identification and monitoring of biological neural network. Proceedings of the IEEE International Symposium on Circuits and Systems (IEEE ISCAS 2007), New Orleans, LA, USA, 27–30 May 2007; pp. 2646–2649.
9. Yu, D.; Parlitz, U. Inferring network connectivity by delayed feedback control. PLoS One **2011**, 6, e24333.
10. Timme, M. Revealing network connectivity from response dynamics. Phys. Rev. Lett **2007**, 98, 224101.
11. Shandilya, S.G.; Timme, M. Inferring network topology from complex dynamics. New J. Phys **2011**, 13, 013004.
12. Levnajić, Z.; Pikovsky, A. Network reconstruction from random phase resetting. Phys. Rev. Lett **2011**, 107, 034101.
13. Berry, T.; Hamilton, F.; Peixoto, N.; Sauer, T. Detecting connectivity changes in neuronal networks. J. Neurosci. Methods **2012**, 209, 388–397.
14. Hamilton, F.; Berry, T.; Peixoto, N.; Sauer, T. Real-time tracking of neuronal network structure using data assimilation. Phys. Rev. E **2013**, 88, 052715.
15. Brovelli, A.; Ding, M.; Ledberg, A.; Chen, Y.; Nakamura, R.; Bressler, S.L. Beta oscillations in a large-scale sensorimotor cortical network: directional influences revealed by Granger causality. Proc. Nat. Acad. Sci. USA **2004**, 101, 9849–9854.
16. Zhou, D.; Xiao, Y.; Zhang, Y.; Xu, Z.; Cai, D. Causal and structural connectivity of pulse-coupled nonlinear networks. Phys. Rev. Lett **2013**, 111, 054102.
17. Stetter, O.; Battaglia, D.; Soriano, J.; Geisel, T. Model-free reconstruction of excitatory neuronal connectivity from calcium imaging signals. PLoS Comput. Biol **2012**, 8, e1002653.
18. Hempel, S.; Koseska, A.; Kurths, J.; Nikoloski, Z. Inner

composition alignment for inferring directed networks from short time series. Phys. Rev. Lett **2011**, 107, 054101.

19. Makarov, V.A.; Panetsos, F.; Feo, O. A method for determining neural connectivity and inferring the underlying network dynamics using extracellular spike recordings. J. Neurosci. Methods **2005**, 144, 265–279.
20. Yao, C.; Bollt, E.M. Modeling and nonlinear parameter estimation with Kronecker product representation for coupled oscillators and spatiotemporal systems. Physica D **2007**, 227, 78–99.
21. Candès, E.; Romberg, J.; Tao, T. Stable signal recovery from incomplete and inaccurate measurements. Commun. Pure Appl. Math **2005**, 59, 1207–1223.
22. Candès, E.; Romberg, J.; Tao, T. Robust uncertainty principles: exact signal reconstruction from highly incomplete frequency information. IEEE Trans. Inf. Theory **2006**, 52, 489–509.
23. Donoho, D. Compressed sensing. IEEE Trans. Inf. Theory **2006**, 52, 1289–1306.
24. Baraniuk, R.G. Compressed sensing. IEEE Signal Process. Mag **2007**, 24, 118–121.
25. Candeś, E.; Wakin, M. An introduction to compressive sampling. IEEE Signal Process. Mag **2008**, 25, 21–30.
26. Willett, R.; Marcia, R.; Nichols, J. Compressed sensing for practical optical imaging systems: A tutorial. Opt. Eng **2011**, 50, 072601.
27. Wang, W.X.; Yang, R.; Lai, Y.C.; Kovanis, V.; Grebogi, C. Predicting catastrophes in nonlinear dynamical systems by compressive sensing. Phys. Rev. Lett **2011**, 106, 154101.
28. Wang, W.X.; Yang, R.; Lai, Y.C.; Kovanis, V.; Harrison, M.A.F. Time-series based prediction of complex oscillator networks via compressive sensing. EuroPhys. Lett **2011**, 94, 48006.
29. Shabani, A.; Kosut, R.L.; Mohseni, M.; Rabitz, H.; Broome, M.A.; Almeida, M.P.; Fedrizzi, A.; White, A.G. Efficient measurement of quantum dynamics via compressive sensing. Phys. Rev. Lett **2011**, 106, 100401.

30. Wang, W.X.; Lai, Y.C.; Grebogi, C.; Ye, J.P. Network reconstruction based on evolutionary-game data. Phys. Rev. X**2011**, 1, 021021.
31. Su, R.Q.; Wang, W.X.; Lai, Y.C. Detecting hidden nodes in complex networks from time series. Phys. Rev. E **2012**, 85, 065201.
32. Su, R.Q.; Ni, X.; Wang, W.X.; Lai, Y.C. Forecasting synchronizability of complex networks from data. Phys. Rev. E**2012**, 85, 056220.
33. Su, R.Q.; Lai, Y.C.; Wang, X.; Do, Y.H. Uncovering hidden nodes in complex networks in the presence of noise. Sci. Rep **2014**, 4, 3944.
34. Wang, X.F.; Chen, G. Pinning control of scale-free dynamical networks. Physica A **2002**, 310, 521–531.
35. Li, X.; Wang, X.F.; Chen, G. Pinning a complex dynamical network to its equilibrium. IEEE Trans. Circ. Syst. I **2004**, 51, 2074–2087.
36. Sorrentino, F.; di Bernardo, M.; Garofalo, F.; Chen, G. Controllability of complex networks via pinning. Phys. Rev. E**2007**, 75, 046103.
37. Yu, W.; Chen, G.; Lü, J. On pinning synchronization of complex dynamical networks. Automatica **2009**, 45, 429–435.
38. Hodgkin, A.L.; Huxley, A.F. A quantitative description of membrane current and its application to conduction and excitation in nerve. J. Physiol **1952**, 17, 500–544.
39. FitzHugh, R. Impulses and physiological states in theoretical models of nerve membrane. Biophys. J **1961**, 1, 445–466.
40. Nagumo, J.; Arimoto, S.; Yoshizawa, S. An active pulse transmission line simulating nerve axon. Proc. IRE **1962**, 50, 2061–2070.
41. Ott, E. Chaos in Dynamical Systems, 2nd ed; Cambridge University Press: Cambridge, UK, 2002.

SECTION IV: NEURONAL NETWORKS DYNAMICS

7 A DYNAMIC CAUSAL MODEL STUDY OF NEURONAL POPULATION DYNAMICS

André C. Marreiros, Stefan J. Kiebel, Karl J. Friston

The Wellcome Trust Centre for Neuroimaging, Institute of Neurology, University College London, UCL, 12 Queen Square, London, UK WC1N 3BG, UK

ABSTRACT

In this paper, we compare mean-field and neural-mass models of electrophysiological responses using Bayesian model comparison. In previous work, we presented a mean-field model of neuronal dynamics as observed with magnetoencephalography and electroencephalography. Unlike neural-mass models, which consider only the mean activity of neuronal populations, mean-field models track the distribution (e.g., mean and dispersion) of population activity. This can be important if the mean affects the dispersion or vice versa. Here, we introduce a dynamical causal model based on mean-field (i.e., population density) models of neuronal activity, and use it to assess the evidence for a coupling between the mean and dispersion of hidden neuronal states using observed electromagnetic responses. We used Bayesian model comparison to compare homologous mean-field and neural-mass models, asking whether empirical responses support a role for population variance in shaping neuronal dynamics. We used the mismatch negativity (MMN) and somatosensory evoked potentials (SEP) as representative neuronal responses in physiological and non-physiological paradigms respectively. Our main conclusion was that although neural-mass models may be sufficient for cognitive paradigms, there is clear evidence for an effect of dispersion at the high levels of depolarization evoked in SEP paradigms. This suggests that (i) the dispersion of neuronal states within populations generating evoked brain signals can be manifest in observed brain signals and that (ii) the evidence for their effects can be accessed with dynamic causal model comparison.

KEYWORDS: Neural-mass models, Nonlinear, Modelling, Laplace assumption, Mean-field, Neuronal, Bayesian

INTRODUCTION

Neuronal activity generated by coupled neuronal populations can be autonomous or elicited by sensorimotor or cognitive perturbation. Neural models have been in used to study this activity for many years (e.g., Wilson and Cowan 1972; Nunez 1974; Freeman 1975; Lopes da Silva et al., 1976; Jansen and Rit, 1995; Jirsa and Haken 1996; Valdes et al., 1999; David and Friston 2003). These models form the basis for understanding the genesis of neuroimaging signals (e.g., Riera et al., 2007). We continue this modelling initiative, by using the formulation of population dynamics described by Marreiros et al. (2009) as the basis of a dynamic causal model (DCM) of evoked electromagnetic responses. We try to establish its face validity using Bayesian model comparison and evaluate its ability to explain empirical data, in relation to simpler variants.

This is the third and final paper in a trilogy that tries to integrate the Fokker-Planck formalism and stochastic formulations of neuronal dynamics with DCM. The first (Marreiros et al. 2008) introduced the idea that the sigmoid activation could be regarded as accommodating dispersion of neuronal states (as opposed to dispersion of the threshold). The second (Marreiros et al. 2009) used this perspective to motivate a simple method-of-moments scheme to model density dynamics in terms of the mean and dispersion of neuronal states. This simplicity is important because the current paper uses this scheme in dynamic casual models of real data.

This work rests on a key distinction between models that summarize the activity of a neuronal population with a single state (e.g., its mean activity) and those that model the distribution of states in terms of probability densities (i.e., density dynamics). We follow the terminology established in Deco et al. (2008), which adopts a pragmatic approach: if the model can be cast as a set of ordinary differential equations describing the evolution of neuronal states, it is called a neural-mass model (NMM; e.g., Freeman 1975; Lopes da Silva et al., 1976; Rodrigues et al., 2006). This is motivated by treating the current state as a point 'mass' (i.e., delta function) approximation to the underlying density on the population›s states. Conversely, models based on stochastic differential equations that include random fluctuations are referred to as mean-field models (MFM; e.g., Knight 1972a,b; Sampolinsky and Zippelius, 1982; Nykamp and Tranchina 2000; Omurtag et al. 2000; Frank et al., 2001; Haskell et al., 2001). This nomenclature appeals to the use of the term 'mean-field' in statistical physics and machine learning. A mean-field assumption approximates a full density (on the states of a population of neurons or multiple populations) with a density that factorises into a series of simpler marginal densities; where the marginals influence each other through average or mean-field effects. These models necessary entail a stochastic or probabilistic representation of activity, which is usually encoded by the sufficient statistics of a probability density (like the mean and variance of a Gaussian distribution). Although we will not deal with them in this paper, models that are formulated as partial differential equations of space and time are referred to as neural field models (NFM; e.g., Wilson and Cowan 1972; Amari 1972; 1975; Robinson et al., 2005), because they model the spatiotemporal dynamics on spatially extensive (e.g., cortical) fields.

It should be noted that this operational taxonomy of models is insensitive to the deep history of their development or their original descriptions; for example, there is a fundamental difference between the work of Wilson and Cowan (1972) and that of Amari (1972; 1975) on 'neural fields' and that of Freeman (1975) and that of Lopes da Silva et al. (1976) on 'neural-masses': the former provides spatiotemporal descriptions of neuronal dynamics, without recourse to experimental evidence; whereas the latter are derived from observing empirical population responses and writing down equations that reproduce physiological responses. Both neural field and mass models are parsimonious models of mean activity (e.g., firing rate or membrane potential) and have been used to emulate a wide range of brain rhythms and dynamics. Neural-mass models are particularly appropriate for data that reflect the average behaviour of neuronal populations; such as the electroencephalogram (EEG) and magnetoencephalogram (MEG) (David et al., 2006; Marreiros et al., 2009).

In a previous paper (Marreiros et al., 2009) we formulated neural-mass models, currently used as generative models in dynamic causal modelling, as a limiting case of mean-field models; in which the variance of the activity in any one neuronal population is fixed. Unlike neural-mass models, mean-field models consider the full density on the hidden states of neuronal populations including the variance or dispersion. We derived a generic mean-field treatment of neuronal populations or ensembles based on a Laplace or Gaussian approximation to the population or ensemble density. The Laplace approximation (also known as a saddle-point approximation) approximates the integral of an exponential function using a second-order Taylor expansion. When the function is a probability density, the implicit assumption is that the density is approximately Gaussian. Because a Gaussian density can be specified in terms of its first two moments, the ensuing scheme is formally identical to the second-moment method described by Rodriguez and Tuckwell (1996). This scheme summarizes density dynamics with (ordinary differential) equations of motion for the sufficient statistics of the ensemble density. This reduces to a neural-mass model when the second-order statistics (i.e., variance) of neuronal states is assumed to be constant. The key behaviour we were interested in was the coupling between the mean and variance under the Laplace approximation, which is lost in neural-mass approximations. In this work, we use the mean-field density dynamics as the basis of a dynamic causal model (DCM) of observed data. The resulting framework allows one to adjudicate between models which include (or not) the high-order statistics of hidden neuronal states when predicting EEG/MEG time series. This adjudication is based on the relative evidence for different models of the same data that obtain from Bayesian inversion (i.e., fitting) of the models.

The aim of this work was to evaluate dynamic causal models based on density dynamics and compare them to established neural-mass models. DCM is a natural extension of the convolution models used in the standard analysis of biological time series (see David et al., 2006). DCM models neuronal dynamics in each source or region and interactions within and between distributed sources. Currently, DCM uses neural-mass models and implicitly considers only the mean neuronal state for each population; i.e., special cases of the more general population density formulation, in which we ignore all but the first-order statistics (i.e., the mean). Here, we replace the neural-mass model with second-order dynamics under the Laplace approximation using the Fokker-Planck formalism (Frank, 2004) to give a mean-field model (Marreiros et al., 2009). This allows us to model the interactions between mean neuronal states (e.g., firing rates) and their dispersion within each population and to compare homologous DCMs based on neural-mass and mean-field models in terms of their model evidence. This enabled us to ask whether including density dynamics is warranted; i.e., to see if there is any evidence for an effect of second-order statistics or dispersion.

The paper is composed of two sections. In the first, we summarize density dynamics under mean-field and Laplace assumptions and describe the resulting DCM framework. In the second section, we use two EEG data sets and Bayesian model comparison (BMC) to assess the relative evidence for neural-mass and mean-field models. In addition, we establish the face validity of neural-mass DCMs and their mean-field generalizations using synthetic data, generated using the conditional estimates of the network parameters, for each of the empirical examples.

THEORY

Neural-mass and field models can reproduce neuronal dynamics reminiscent of observed evoked responses. However, to emulate more complex dynamics we may need to take into account the high-order statistics of ensemble dynamics. In a previous paper (Marreiros et al., 2009) we derived a generic mean-field treatment of neuronal dynamics, based on a Laplace approximation to the ensemble density. This model is formulated in terms of equations of motion for the moments of the ensemble density, reducing to an NMM when the second-order moment (variance) is ignored. The interesting behaviour in these mean-field models arises from the coupling between the mean and variance of ensemble activity, which is ignored in neural-mass approximations. Here, we will use the Laplace and neural-mass approximations in DCMs of electrophysiological responses to sensory input. We start by reviewing briefly mean-field and neural-mass models (MFM and NMM) for M/EEG and then turn to Bayesian estimation, inference and model comparison in the context of DCM.

Modelling Neuronal Dynamics with Mean-Field Models

In DCM, neural-mass models are used to model the evolution of the mean response of neuronal populations to exogenous or experimental perturbations (David et al., 2006; Kiebel et al., 2008). Mean-field approximations go further and model the distribution of the population response. However, employing MFMs can be computationally expensive, because one has to consider the density at all points in neuronal state-space, as opposed to a single quantity (e.g., the mean). In other words, to represent the full probability distribution of states over a population, we would need to encode the probability of finding a neuron in every state at each point in time, which would require an infinite number of variables. Under the Laplace approximation, the population or ensemble density reduces to a Gaussian form, whose sufficient statistics is composed of only the conditional mean and covariance. This simplification allows one to model interactions between the first two moments (i.e., mean and variance) of neuronal states in an efficient and tractable way.

In brief, the dynamics of neurons in an ensemble can be described with the stochastic differential equation $dx = f(x,u)dt + \omega d\Gamma$, where $f(x,u)$ corresponds to the equations of motion describing neuronal processes and ω controls the amplitude of uncorrelated random fluctuations $d\Gamma$ (i.e., a Weiner process) in neuronal states $x(t) \in \Re^n$. Here, $u(t) \in \Re^m$ is a real valued vector of time-dependent exogenous or experimental inputs. To compute the associated density dynamics, one can use the Fokker-Planck formalism: $\dot{q} = -\nabla \cdot fq + \nabla \cdot D \nabla q$, where $q(x,t)$ is an approximate density on neuronal states (e.g., post synaptic potentials and conductances) and $D(\omega)$ is a diffusion tensor. The time-dependent density $q(x,t)$ is an ensemble density, which approximates the probability of finding a neuron from the ensemble in state x and time t. The high dimensionality and complexity of this Fokker-Planck formalism can be finessed with a mean-field approximation, $q(x,t) \approx \Pi\, q(x^{(i)},t)$, which describes the evolution of separable ensembles that are coupled by mean-field effects. Moreover, the Laplace assumption allows us to summarize the ensemble density with a Gaussian density, described by the mean and variance of the states of the i-th population. By parameterizing the densities in terms of these time-dependent sufficient statistics, we have; $q(x^{(i)},t) = N(\mu^{(i)},\Sigma^{(i)})$. Please see Marreiros et al. (2009) for a fuller explanation.

Crucially, the Laplace assumption allows the density dynamics (based on stochastic differential equations) to be expressed as an ordinary differential equation for its sufficient statistics (mean and variance) in a compact form (Marreiros et al., 2009)

$$\dot{\mu}_j^{(i)} = f_j^{(i)}(\mu,u) + \frac{1}{2}tr\left(\Sigma^{(i)}\partial_x^2 f_j^{(i)}\right)$$
$$\dot{\Sigma}^{(i)} = \partial_x f^{(i)}\Sigma^{(i)} + \Sigma^{(i)}\partial_x f^{(i)T} + D^{(i)} + D^{(i)T} \quad (1)$$

This allows one to describe the population dynamics knowing only the flow, its gradient and curvature at each point in state-space. The reason that there are no superscripts on the sufficient statistics $\mu := \{\mu^{(1)}, \mu^{(2)},\ldots,\Sigma^{(1)},\Sigma^{(2)},\ldots,\}$ is that the sufficient statistics of one population can influence the sufficient statistics of another through extrinsic and intrinsic connections. Eq. (1) shows explicitly how the first and second moments of the density depend on each other; the variance affects the mean when the curvature of the flow is nonzero (which is always the case when the equations of motion are nonlinear in the states). The effect of the mean on the variance depends on the gradients of the flow, which only change with the mean, when the curvature is nonzero. Therefore, the MFM formulation enables us to model interactions between the mean of neuronal states (e.g., postsynaptic potentials) and their variance over each population modelled (cf., Harrison et al., 2005).

A special case of the MFM is obtained if we assume $\Sigma^{(i)}$ is fixed for all populations. This is the NMM, where we ignore all but the first moment of the density (i.e., the mean or mode). Because the curvature $\partial_x^2 f_j^{(i)}$ is generally constant and negative, this induces a decay term, giving density dynamics of the form

$$\dot{\mu}_j^{(i)} = f_j^{(i)}(\mu, u) + \frac{1}{2} tr\left(\Sigma^{(i)} A_x^2 f_j^{(i)}\right) \qquad (2)$$

This corresponds to a neural-mass model and will be the NMM used in this paper. The value for $\Sigma^{(i)}$ we used was the stationary solution to Eq. (1), in the absence of input (i.e., the same steady-state value as in the MFM). Further assuming that $\Sigma^{(i)}$ is spherical (i.e., all off-diagonal terms are zero) means the decay terms disappears because the leading diagonal; $\partial_x{}^2 f_j^{(i)}$ is generally 0 for most neuronal models. This is because the self-decay terms are generally linear in the states (see Eq. (3) below for an example). In this case, we can ignore the second-order statistics completely and the dynamics reduce to the original equations of motion for a single neuron: $\dot{\mu}^{(i)} = f^{(i)}(\mu,u)$.

The equations of motion $f^{(i)}$ for each population considered in this paper are described in Marreiros et al. (2009) and conform to a simplified Morris and Lecar (1981) model, where the states are composed of transmembrane potential and three conductances: leaky, excitatory and inhibitory. This conductance-based model can be expressed with the equations of motion for the states of the i-th population

$$x^{(i)} = \begin{bmatrix} V^{(i)} \\ g_E^{(i)} \\ g_I^{(i)} \end{bmatrix}$$

$$f^{(i)} = \begin{bmatrix} \frac{1}{C}\left(g_L\left(V_L - V^{(i)}\right) + g_E^{(i)}\left(V_E - V^{(i)}\right) + g_I^{(i)}\left(V_I - V^{(i)}\right) + \gamma_i^u u\right) \\ \kappa_E\left(\varsigma_E^{(i)} - g_E^{(i)}\right) \\ \kappa_I\left(\varsigma_I^{(i)} - g_I^{(i)}\right) \end{bmatrix}$$

$$\varsigma_k^{(i)} = \sum_j \gamma_{ij}^k \sigma\left(\mu_V^{(j)} - V_R, \Sigma^{(j)}\right) \qquad (3)$$

The presynaptic input $\varsigma_k^{(i)}: k \in E,I$ scales with the expected firing rate in all populations, where the sigmoid function $\sigma(\cdot)$ is a Gaussian cumulative density on the depolarization (Marreiros et al., 2008) and the coupling parameters γ_{ij}^k specify the connectivity among populations. V_R is the threshold potential for neuronal firing. These parameters can be used to ensure that each population couples to one and only one conductance type (i.e., each population can only release one sort of neurotransmitter). Critically, in this work, the model also allows for conduction delays on the connections (not shown in the equations for simplicity). Given Eq. (3) we can write down the derivatives that are required to evaluate the density dynamics in Eq. (1) (cf. Eq. (18) in Marreiros et al. 2009), where, dropping superscripts for clarity:

$$\partial_x f = \begin{bmatrix} -\frac{1}{C}\sum_k g_k & \frac{1}{C}(V_E - V) & \frac{1}{C}(V_I - V) \\ 0 & -\kappa_E & 0 \\ 0 & 0 & -\kappa_I \end{bmatrix}$$

$$\partial_x^2 f_V = \begin{bmatrix} 0 & -\frac{1}{C} & -\frac{1}{C} \\ -\frac{1}{C} & 0 & 0 \\ -\frac{1}{C} & 0 & 0 \end{bmatrix}$$

$$\partial_x^2 f_g = \begin{bmatrix} 0 & 0 & 0 \\ 0 & 0 & 0 \\ 0 & 0 & 0 \end{bmatrix} \tag{4}$$

This particular form of neuronal model is among the simplest that are nonlinear in the states (note that the rate of change of voltage depends on conductance times voltage). This nonlinearity is critical in the present context because, as discussed above, in its absence there is no coupling between the mean and dispersion (i.e., the neural-mass and mean-field formulations would behave identically). We are not suggesting, in the choice of this model, that it is a sufficient or complete model of neuronal dynamics; this would be a wider question for model comparison. We are using this minimal model to ask whether mean-field formulations provide a better account of observed neuronal responses than their neural-mass counterparts.

Having established the form of the ordinary differential equations (Eqs. (1) and (2)) for the mean-field and neural-mass models of population activity, we now describe how they are embedded in a DCM to provide a spatiotemporal forward model of observed electromagnetic responses.

Dynamic Causal Modelling for EEG/MEG

Dynamic causal modelling provides a generative model for M/EEG responses (David et al., 2006; Kiebel et al., 2008). The idea behind this approach is that M/EEG data are the response of a distributed network of interacting neuronal sources to experimental inputs. Here every source contains different neuronal populations, each described by a NMM or a MFM. Each population has its own (intrinsic) dynamics governed by the neural-mass or the mean-field equations above, but also receives extrinsic input, either directly as sensory input or from other sources. The dynamics of these sources are specified fully by a set of first-order differential equations that are formally related to other neural-mass and mean-field models of M/EEG (e.g., Breakspear et al. 2006; Rodrigues et al. 2006).

DCM for event related potential models the activity of a source using three neural subpopulations, each assigned to one of three cortical layers; an excitatory subpopulation in the granular layer, an inhibitory subpopulation in the supra-granular layer and a population of deep pyramidal cells in the infra-granular layer. These are connected using the connectivity rules described in the study of Felleman and Van Essen (1991). See Fig. 1. Here, it is assumed that the depolarization of pyramidal cell populations gives rise to observed M/EEG data, which are expressed in the sensors through a conventional lead-field. The full spatiotemporal model takes the form of a nonlinear state-space model with hidden states modelling (unobserved) neuronal dynamics, while the observation (lead-field) equation is instantaneous and linear in the states. The ensuing DCM is specified in terms of its state-equation (Eqs. (1) or (2)) and an observer or output equation

$$h(\vartheta) = L(\theta) \begin{bmatrix} \alpha_1 \mu_V^{(1)} \\ \alpha_2 \mu_V^{(2)} \\ \vdots \end{bmatrix} \tag{5}$$

where $\mu_V^{(i)}$ is the mean depolarization of the i-th population and $h(\theta)$ is the predicted EEG or MEG signal. Here, $\vartheta \supset \{\theta, \alpha_i, \gamma_{ij}^k, \kappa_I, \kappa_E, \omega \ldots)$ are the unknown quantities that parameterize the state and observer equations (David et al., 2006). The parameters also control any unknown attributes of the stimulus function encoding exogenous input; we use Gaussian bump functions parameterized by their latencies and dispersions. We assume the MEG or EEG signal is a linear mixture of depolarizations in the pyramidal populations; where the columns of $L(\theta)$ are conventional lead-fields, which account for passive conduction of the electromagnetic field from the sources to the sensors (Mosher et al. 1999). The parameters of the lead-field, θ encode the location and orientation of the underlying sources and α scale their contribution to the dipole. Please see David et al. (2006) and Kiebel et al. (2008) for further background and Daunizeau et al. (2009) for details of the particular spatial model we used in this study.

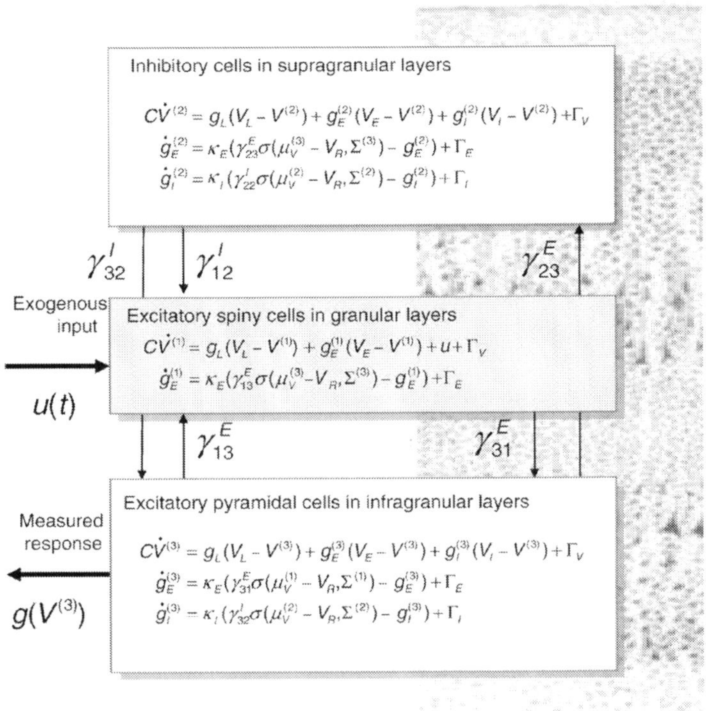

Figure 1. Neuronal state-equations for a source model with a layered architecture composed of three interconnected populations (spiny-stellate, interneurons, and pyramidal cells), each of which has three different states (voltage, excitatory and inhibitory conductances).

The predicted signal h(θ) corresponds to a generalized convolution of exogenous inputs (i.e., experimental stimulus functions). Under Gaussian assumptions about measurement noise, this generalized convolution gives a likelihood model for observed M/EEG data:

$$y = \mathrm{vec}(h(\vartheta) + X\beta) + \varepsilon \Rightarrow$$
$$p(y \mid \vartheta, \lambda) = \mathcal{N}(\mathrm{vec}(h(\vartheta) + X\beta), \mathrm{diag}(\lambda) \otimes V) \qquad (6)$$

Observation noise, is assumed to be zero-mean Gaussian and independent over channels, where λ is a vector of unknown channel-specific error variances and Vrepresents a temporal autocorrelation matrix. Low-frequency noise or drift components are modelled by confounding variables in the columns of the matrix, X(this was simply a constant term in this paper) with associated parameters $\beta \subset \vartheta$. For computational expediency, we reduce the dimensionality of the sensor data, while retaining the maximum amount of information. This is assured by projecting the data onto a subspace defined by its principal modes; computed using singular value decomposition; see Fastenrath et al. (2009). Having established how the mean-field and neural-mass models specify a likelihood model for observed signals, we now consider how the ensuing DCM is fitted or inverted.

Bayesian Estimation, Conditional Inference and Model Comparison

A DCM is fitted to data by tuning the free parameters to minimize the discrepancy between predicted and observed MEG/EEG time series, under complexity constraints. In addition to minimizing prediction error, the parameters are constrained by a prior specification of the range they are likely to lie in the study of Friston et al. (2003). These constraints, which take the form of a prior density $-p(\vartheta)$, are combined with the likelihood, $p(y|\vartheta)$, to form a posterior density $p(\vartheta|y) \propto p(y|\vartheta)p(\vartheta)$ according to Bayes› rule. The priors $p(\vartheta)$ are usually specified under log-normal assumptions to impose positivity constraints; and are therefore specified by the prior mean and variance of log-parameters. Table 1 lists the priors for the free parameters of the neuronal model and the values we used for its fixed parameters.

Table 1. Prior densities on the neuronal parameters.

Free parameters	
Extrinsic coupling parameters	$\gamma_{ij}^{E} = \begin{cases} \frac{1}{2}e^{\theta} & \text{forward} \\ \frac{1}{4}e^{\theta} & \text{backward} \\ \frac{1}{4}e^{\theta} & \text{lateral} \end{cases} \quad p(\theta) = N(0, 1)$

Intrinsic coupling parameters	$\gamma_{ij}^E = e^{\theta}\begin{bmatrix} 0 & 0 & \frac{1}{2} \\ 0 & 0 & 1 \\ \frac{1}{2} & 0 & 0 \end{bmatrix}$ $\gamma_{ij}^I = e^{\theta}\begin{bmatrix} 0 & \frac{1}{4} & 0 \\ 0 & 0 & 1 \\ 0 & 1 & 0 \end{bmatrix}$ $p(\theta) = N\left(0, \frac{1}{32}\right)$
Capacitance	$C = 8e^{\theta}$ $p(\theta) = N\left(0, \frac{1}{32}\right)$
Time constants	$\tau_e = \frac{1}{\kappa_e} = 4e^{\theta}\text{ms}$ $\tau_i = \frac{1}{\kappa_i} = 16\text{ms}$ $p(\theta) = N\left(0, \frac{1}{32}\right)$
Diffusion tensor	$D(\omega) = e^{\theta}\begin{bmatrix} \frac{1}{8} & 0 & 0 \\ 0 & 1 & 0 \\ 0 & 0 & 1 \end{bmatrix}$ $p(\theta) = N\left(0, \frac{1}{64}\right)$
Conduction delays	$\Delta_{ij} = \begin{cases} 2e^{\theta}\text{ms} & \text{intrinsic} \\ 16e^{\theta}\text{ms} & \text{extrinsic} \end{cases}$ $p(\theta) = N\left(0, \frac{1}{128}\right)$
Stimulus function parameters	$u(t) = exp\left(-\frac{(t-\tau_u e^{\theta latency})^2}{\tau_u e^{\theta duration}}\right) : \begin{cases} \tau_u = 128\text{ms : MMN} \\ \tau_u = 16\&32\text{ms : SEP} \end{cases}$ $p(\theta_i) = N(0, \frac{1}{16})$
Fixed parameters	
Potentials	$V_L = -70\mu V$ $V_E = 60\mu V$ $V_I = -90\mu V_1$ $V_R = -40\mu V$
Conductance	$g_L = 1$

For a given DCM, say model m, inversion corresponds to approximating the moments of the posterior or conditional distribution given by Bayes' rule

$$p(\vartheta | y, m) = \frac{p(y | \vartheta, m) p(\vartheta, m)}{p(y | m)} \tag{7}$$

The estimation procedure employed in DCM is described in the study of Friston et al. (2003). The posterior moments (mean and covariance) are updated iteratively using variational Bayes under a fixed-form Laplace (i.e., Gaussian) approximation to the conditional density $-q(\vartheta)$. This can be regarded as an expectation-maximization (EM) algorithm that employs a local linear approximation of the predicted responses (Eq. (6)) about the current conditional expectation. The E-step conforms to a Fisher-scoring scheme (Fahrmeir and Tutz 1994) that performs a descent on a variational free-energy F(q,λ,m), with respect to the conditional moments. In the M-step, the error variances λ are updated in exactly the same way to provide their maximum likelihood. The estimation scheme can be summarized as follows:

$$\begin{aligned} \text{E-step} &: q \leftarrow \min_q F(q,\lambda,m) \\ \text{M-step} &: \lambda \leftarrow \min_\lambda F(q,\lambda,m) \end{aligned}$$

$$\begin{aligned} F(q,\lambda,m) &= \langle \ln q(\vartheta) - \ln p(y|\vartheta,\lambda) - \ln p(\vartheta|m) \rangle_q \\ &= KL(q \| p(\vartheta|y,\lambda)) - \ln p(y|\lambda,m) \end{aligned} \tag{8}$$

The free-energy is simply a function of the log-likelihood, the log-prior and the approximation to the conditional density we seek. The free-energy is the Kullback-Leibler divergence between the real and approximate conditional density minus the log-likelihood. This means that when the free-energy is minimized, the discrepancy between the true and approximate conditional density is suppressed (because the divergence is non-negative). At this point the free-energy approximates the negative log-evidence $F \approx -\ln p(y|\lambda,m)$. This scheme is identical to that employed by DCM for fMRI and evoked responses (Friston et al., 2003; David et al., 2006).

The log-evidence is an important quantity because it allows one to compare different models (Penny et al. 2004). We can approximate the log-evidence for model m with $\ln p(y|m) \approx -F$. The most likely model is the one with the largest log-evidence. Model comparison rests on the likelihood ratio (i.e., Bayes-Factor) of the evidence or relative log-evidence for two models. For models i and j the Bayes-Factor is

$$B_{ij} = \frac{p(y|m_i)}{p(y|m_j)} \Rightarrow \ln B_{ij} = \ln p(y|m=i) - \ln p(y|m=j) \tag{9}$$

Strong evidence in favour of one model typically requires the difference in log-evidence to be three or more (Penny et al. 2004). Under flat priors on models this corresponds to a conditional confidence that the winning model is $\exp(3) \approx 20$ times more likely than the alternative. This indicates that the data provide 'strong' (10:1 to 30:1) evidence in favour of one model over the other. See http://en.wikipedia.org/wiki/Bayes_factor for the range of Bayes factors indicating 'very strong' (30:1 to 100:1) and 'decisive' (more than 100:1) evidence for a model. In the next section, we will use the free-energy bound on log-evidence to compare the different models elaborated above.

SIMULATIONS AND EMPIRICAL RESULTS

Our key question was: can we find evidence for coupling between the mean and dispersion of neuronal states in empirical data? However, we anticipated that the answer would be context sensitive; in the sense that some evoked responses may induce large fluctuations in dispersion, whereas others may not. This context sensitivity can be seen from the form of Eq. (1), where changes in the dispersion of neuronal states depend upon the systems Jacobian $\partial_x f^{(i)}$ or rate of change of flow with state. The Jacobian depends on depolarization and conductance (Eq. (4)), which depends on presynaptic input $\varsigma_k^{(i)}$. This implies that we would expect to see large fluctuations in dispersion and the ensuing effect on the mean under high levels of extrinsic presynaptic input. We therefore chose to perform our model comparison using two sorts of evoked responses. The first used a traditional 'cognitive' paradigm (a mismatch negativity paradigm) in which auditory stimuli can be regarded as delivering low amplitude physiological inputs to cortical sources. In contrast, the second paradigm was a somatosensory evoked potential (SEP) paradigm; in which neuronal sources are excited with a non-physiological electrical stimulus, eliciting transient but high amplitude presynaptic inputs. We predicted that if there was any evidence for the mean-field model, relative to the neural-mass model, then we would be more likely to see it in the SEP paradigm, relative to the mismatch negativity paradigm. In what follows, we describe these paradigms and the results of our model comparisons.

Mismatch negativity paradigm

In this section, we analyze data from a multi-subject mismatch negativity (MMN) study (Garrido et al. 2007a). This analysis is largely the same as we have presented previously when looking at plasticity using DCM. The only difference here is that we used the conductance-based neuronal model described in Marreiros et al. (2009) and compared the neural-mass and mean-field variants of this model. In brief, the MMN is the differential response to an unexpected (rare or oddball) auditory stimulus relative to an expected (standard) stimulus. The MMN has been studied extensively and is regarded as a marker for error detection, caused by a deviation from a learned regularity, or familiar auditory context. According to Näätänen et al. (2001) the MMN is caused by two underlying functional processes, a sensory memory mechanism and an automatic attention-switching process that might engage frontal generators (Giard et al. 1990). It has been shown that the temporal and frontal MMN sources have distinct behaviours over time (Rinne et al. 2000) and that these sources interact with each other (Jemel et al. 2002). Thus the MMN could be generated by a temporofrontal network (Doeller et al. 2003; Opitz et al. 2002; Escera et al. 2003), as revealed by M/EEG and fMRI studies. In a predictive coding framework, these findings can also be framed as adaptation and experience-dependent plasticity in an auditory network (Jääskeläinen et al., 2004; Friston, 2005; Garrido et al., 2007a,b).

Using DCM, we modelled the MMN generators with a temporofrontal network composed of bilateral sources over the primary and secondary auditory and frontal cortex. Following Garrido et al. (2007a), we used a five-source network with forward and backward extrinsic (between-source) connections. Exogenous or auditory input (modelled with $u(t) \in \Re$, a parameterized bump function of time; see Table 1) enters via subcortical structures into two bilateral sources in posterior auditory cortex (lA1 and rA1). These have forward connections to two bilateral sources in anterior auditory cortex; i.e., superior temporal gyri (lSTG and rSTG). These sources are laterally and reciprocally connected via the corpus callosum. The fifth source is located in the right inferior frontal gyrus (rIFG) and is connected to the rSTG with reciprocal unilateral connections. Using these sources and prior knowledge about the functional anatomy cited above, we specified the DCM network in Fig. 2. Here, we were interested in comparing the NMM and MFM formulations of this network, in terms of their negative free-energy. To simplify the analysis, we modelled only the responses evoked by standard stimuli (from 0 ms to 256 ms).

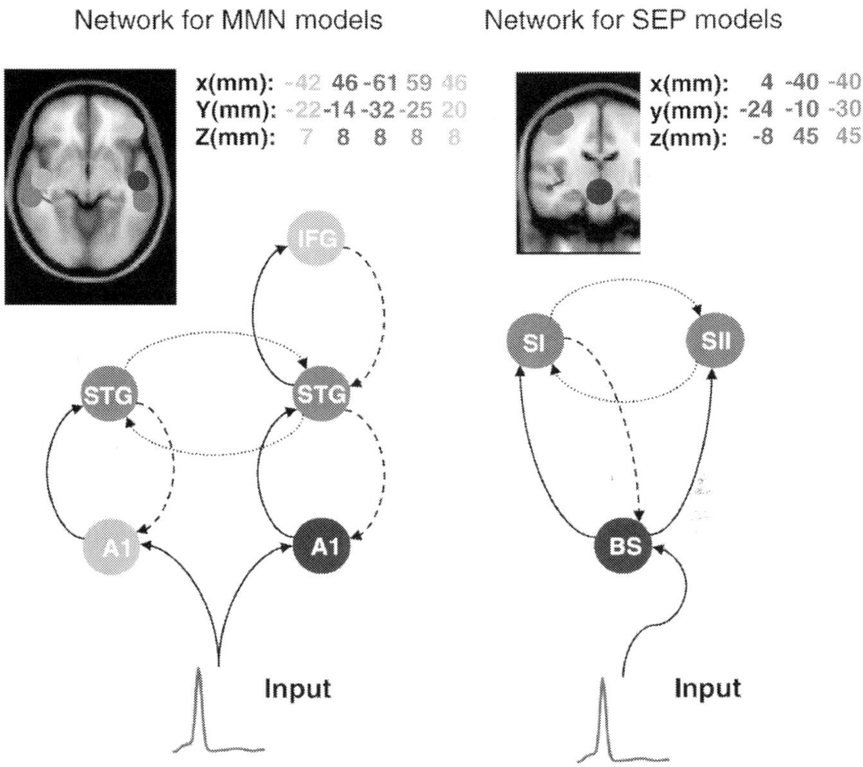

Figure 2. DCM networks used for the mismatch negativity (MMN) and somatosensory evoked potential (SEP) paradigms. Forward connections (solid lines), backward connections (dashed lines) and lateral connections (dotted lines) couple sources. The inserts shows the prior locations of sources; these are composed of vertices on a canonical cortical mesh within 16 mm of the source locations shown in the Talairach and Tournoux coordinates. See text and Garrido et al. (2007a) for a full model description.

Empirical Results

Two DCMs (NMM and MFM variants) were inverted for all twelve subjects and compared using their log-evidence. Fig. 3 shows the differences in log-evidences for each subject. For all but one subject, there was decisive evidence for the NMM over the MFM. The log-evidence at the group level (> 100), pooled over all subjects (given the data are conditionally independent over subjects) was similarly decisive. Although the relative log-evidence is quantitatively meaningful in its own right, one can also treat it as a log-odds ratio and use its distribution over subjects to compute a classical p-value (Stephan et al., 2010). In this instance, a one-sample t-test was extremely significant (T = 3.58, df = 11, p = 0.002). This means that we can reject the null hypothesis that the data are explained equally well by neural-mass and mean-field formulations of the same neuronal model.

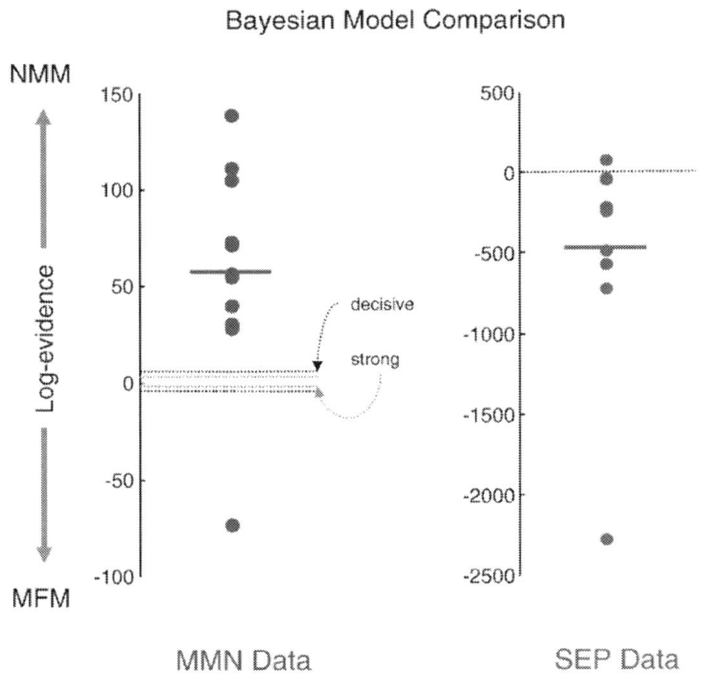

Figure 3. Bayesian model comparisons for NMM in relation to MFM. Right: relative log-evidence for the NMM for each subject using the network in Fig. 2. The NMM log-evidences are consistently better than the MFM log-evidences, with a pooled difference > 100 over subjects. Left: the same results for the SEP data; the group log-evidence difference was > 100 in favour of the MFM. The solid lines indicate the mean over subjects.

These results suggest the NMM is a better model for explaining evoked auditory responses. Note that the complexities of the NMM and MFM are the same; the MFM has more states but does not have more unknown parameters. This may seem counterintuitive because the dispersion in the MFM may appear to make it more complicated. However, the dispersion is a sufficient static of a density on hidden states and is not itself subject to random effects. This means, given the model parameters, it is a deterministic quantity and does not add to model complexity (i.e., it is specified by the same parameters as the neural-mass model). This is important because the results in Fig. 3 are remarkably consistent over subjects and cannot be explained by differences in model complexity. In short, the NMM provides a better prediction of the observed responses than the MFM, in this paradigm. Furthermore, the differences between the NMM and MFM predictions are fairly subtle (see Fig. 4). This suggests that the population variance is actually quite stable over peristimulus time, because the model selection clearly favours the predictions from the neural-mass model (Fig. 5).

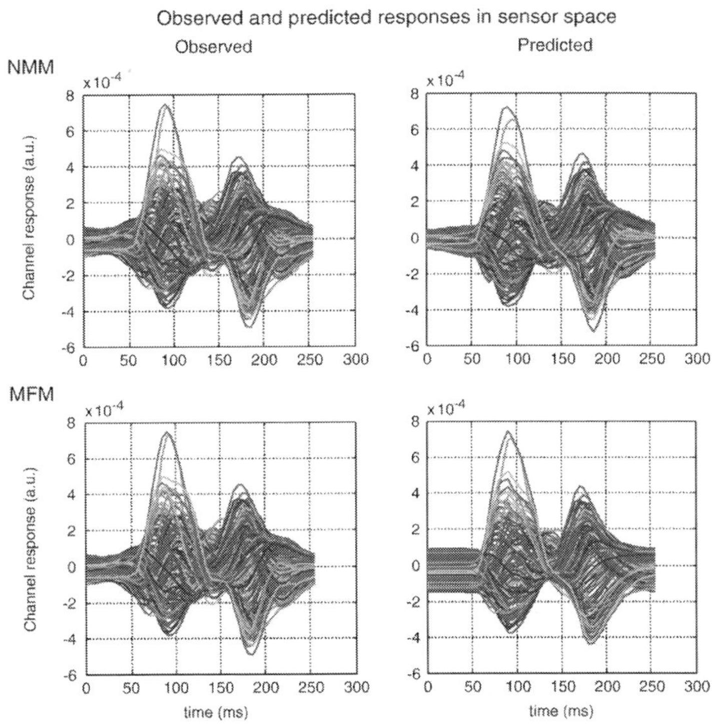

Figure 4. Upper panels: observed (left) and predicted (right) evoked responses over 128 channels and peristimulus time. Each coloured line corresponds to a different channel. These results are from the neural-mass DCM of the first subject from the MMN paradigm. Lower panels: the same but showing MFM predictions. The observed response is duplicated because it is adjusted for the confounding DC or constant term in our model (see Eq. (5)). This adjustment renders the observed data slightly different, depending on the model fit.

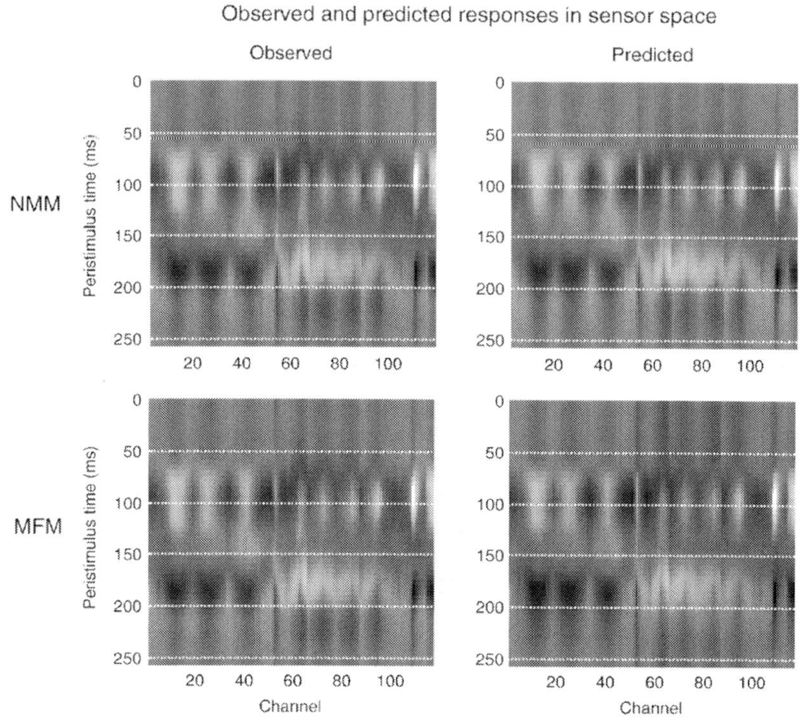

Figure 5. These are the same data as shown in Fig. 4 but in image format. Upper panels: observed (left) and predicted (right) evoked responses over 128 channels and peristimulus time (grey scale normalised to the maximum of each image). These results are from the NMM-DCM of the first subject from the MMN paradigm. Lower panels: the same but showing MFM predictions.

Simulations

We next performed some comparative evaluations and validations of DCM using neural-mass and mean-field models, using synthetic data based on the empirical results above. These are presented to show that the empirical model comparison above is sufficiently sensitive to disambiguate between neural-mass and mean-field variants of the same model. After generating data from a known model, we used model comparison to ask whether one can recover the correct model over its alternative. We integrated the NMM and MFM with known (true) model parameters derived from the real data above (the conditional means from a DCM of the grand average over subjects) and added random measurement noise with a standard deviation of 10% of the peak response in channel space. This was roughly the amplitude of noise in the real data—see Fig. 4. Finally, we used the synthetic data generated by both models to invert the neural-mass and mean-field DCMs. Table 2lists the resulting log-evidences. Each column contains the log-evidences for each data set. The maximum values are found on the diagonal; i.e., the true model had the greatest evidence and the relative evidence for the correct model was 'decisive'. These results confirm that these models can be disambiguated using DCM, under empirically realistic levels of noise.

Table 2. Log-evidences for neural-mass (NMM) and mean-field (MFM) models using synthetic data generated by a five-source MMN model (see Fig. 2) using NMM and MFM formulations. The diagonal values in bold show higher log-evidences for the true model.

Models	**Synthetic data**	
	NMM	MFM
NMM	**– 662.7**	– 952.9
MFM	– 844.2	**– 665.5**

Somatosensory Evoked Potential Paradigm

To explore the context sensitivity of these results, we analyzed data from a study of paired associative stimulation (PAS) (Litvak et al., 2007), which involves repetitive magnetic cortical stimulation timed to interact with median nerve stimulation-induced peripheral signals from the hand. The PAS paradigm has been shown to induce long-lasting changes in somatosensory evoked potentials (Wolters et al., 2005) as measured by single-channel recordings overlying somatosensory cortex. The SEP generators evoked by compound nerve stimulation have been studied extensively with both invasive and non-invasive methods in humans and in animal models (Allison et al., 1991). Litvak et al. (2007) characterised the topographical distribution of PAS-induced excitability changes as a function of the timing and composition of afferent somatosensory stimulation, with respect to a transcranial magnetic stimulation (TMS). The temporal response pattern of the SEP is composed of a P14 component generated subcortically and then an N20–P30 complex from the sensorimotor cortex, which is followed by a P25–N35 complex (Allison et al. 1991). The remainder of the SEP can be explained by a source originating from the hand representation in S1 (Litvak et al., 2007).

We chose these data as examples of fast sensory transients that might engage a more circumscribed network than the auditory stimuli in the MMN paradigm above. We anticipated that the density dynamics of neuronal populations that were stimulated electromagnetically, may disclose the effects of dispersion (see above). We analyzed the SEP data from eleven subjects following median nerve stimulation (i.e., in the absence of transcranial magnetic stimulation) as above. The network architecture was based on previous reports (Buchner et al., 1995; Ravazzani et al., 1995, Litvak et al., 2007): we modelled the somatosensory system with three sources, each comprising three neuronal populations. In this model (see Fig. 2, Litvak et al., 2007 and Marreiros et al. 2008) exogenous input was delivered to the brainstem source (BS), which accounts for early responses in the medulla. The input was a mixture of two parameterized bump functions with prior latencies based on known conduction delays (see Table 1). This region connects to two sources SI and SII in Brodmann area 3 (Marreiros et al., 2008). We inverted the resulting DCMs using the sensor data from 4 ms to 64 ms, following stimulation. We report the results form the first ten subjects because the DCM inversion failed to converge for the last subject.

Empirical Results

Fig. 3 (right) shows the log-evidence differences. In stark contrast to the MMN results, there was 'decisive' evidence in all but one subject for the MFM over the NMM. Moreover, the large group difference in log-evidence of (> 100) favours the MFM model; i.e., if we had to account for the data from all subjects with the same model, then the evidence for the MFM was decisive. The classical p-value was similarly significant ($T = 2.19$, $df = 9$, $p = 0.028$) but less significant than in the MMN analyses due to larger inter-subject variability. These results indicate that the MFM is, in this instance, a demonstrably better explanation for somatosensory evoked potentials. It is important to appreciate that exactly the same model was used for the MMN and SEP data, including the prior density on the free parameters (with the exception of the exogenous input). However, the results of model comparison are, as anticipated, completely the opposite and remarkably consistent over subjects for both paradigms.

Anecdotally, the superior performance of the mean-field model seemed to be its ability to fit both the early N20–P30 complex and later waveform components of the SEP (although no model was able to fit the P14 components convincingly). This contrasts with the neural-mass model that was unable to reproduce the fast initial transients but was able to model the slower components that followed. The results from the first subject in channel space illustrate this difference nicely (Fig. 6); showing that the predictions of both models are almost indistinguishable after 30 ms. Phenomenologically, this means that the dispersion of neuronal states in the MFM confers a greater range on the time constants of population dynamics, which allows the MFM to reproduce fast, large-amplitude responses in, we presume, relatively circumscribed neuronal populations responding synchronously to extrinsic afferents.

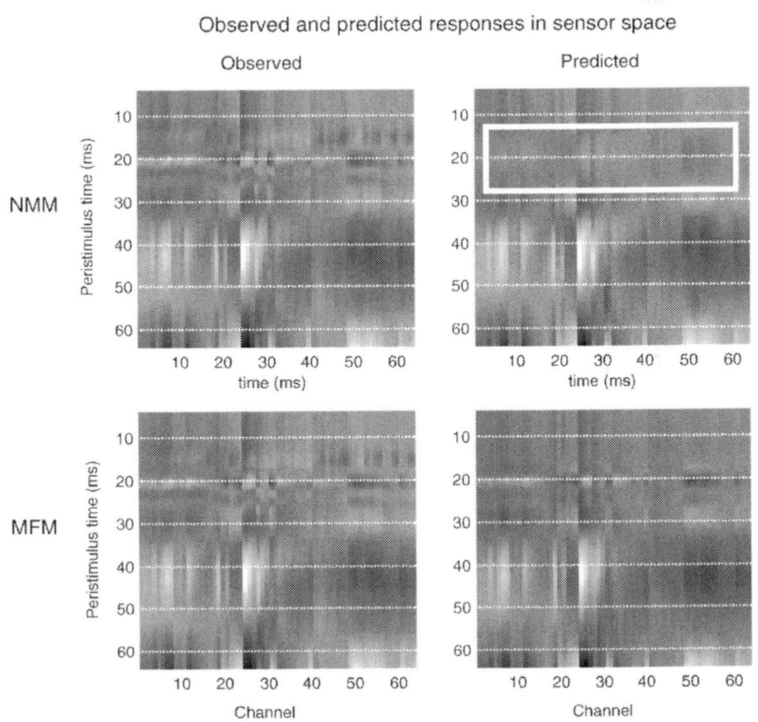

Figure 6. Observed and predicted responses from the SEP paradigm in the first subject. These images adopt the same format as Fig. 5 but showing responses over 64 channels and 4 ms to 64 ms of peristimulus time. The key thing to note here is the failure of the neural-mass DCM to model the early (N20) components of the SEP (white rectangle), relative to its mean-field homologue.

Simulations

To ensure the model comparison retained its sensitivity in this SEP setting, we again generated synthetic data using the conditional means of the parameters estimated from the empirical SEP data. We used an NMM and an MFM for generation and inversion and evaluated all combinations to ensure that model selection identified the correct model. For the integration of the forward models, we used the conditional means of parameters over subjects. We added random noise to these synthetic data, with a standard deviation that was 5% of the peak response in sensor space. We used the three-source model above (Litvak et al., 2007; Marreiros et al., 2008) to generate and model data. Table 3 presents the log-evidences for each of the four inversions. The highest evidences were obtained for the models that were used to generate the synthetic data: these correspond to the diagonal entries. Again, the results conform that model comparison can identify the correct model of these somatosensory responses, and do so decisively.

Table 3. Log-evidences for neural-mass (NMM) and mean-field (MFM) models using synthetic data generated by a three-source SEP model (see Fig. 2) using NMM and MFM formulations. The diagonal values in bold show higher log-evidences for the true model.

Models	**Synthetic data**	
	NMM	MFM
NMM	**− 185.0**	− 175.8
MFM	− 369.6	**− 102.1**

A quantitative Illustration of Density Dynamics

Fig. 7 (upper left panel), shows the sufficient statistics of population activity for source in the first SEP subject. These are the mean and covariance of neuronal states, in source space. These are obtained by integrating the ensemble dynamics in Eq. (1), using the equations of motion in Eq. (3) and Fig. 1 and the conditional parameter estimates. Generally, when the mean depolarization increases, the covariance decreases, getting close to zero when the mean approaches its maximum. This is seen here at about 30 ms. This concentration of neuronal states augments the decay of mean depolarization (see Eq. (1)). Note that at around 20 ms the N20 is modelled by polyphasic dynamics in the mean depolarization that rests on a coupling with dispersion. These coupling and ensuing dynamics are the ones missing in the neural-mass model. In the lower panel, we see the conditional response estimate, in sensor space, in terms of the observed (dotted lines) and predicted (solid lines) time series for all modes. These results are representative of DCM prediction accuracy.

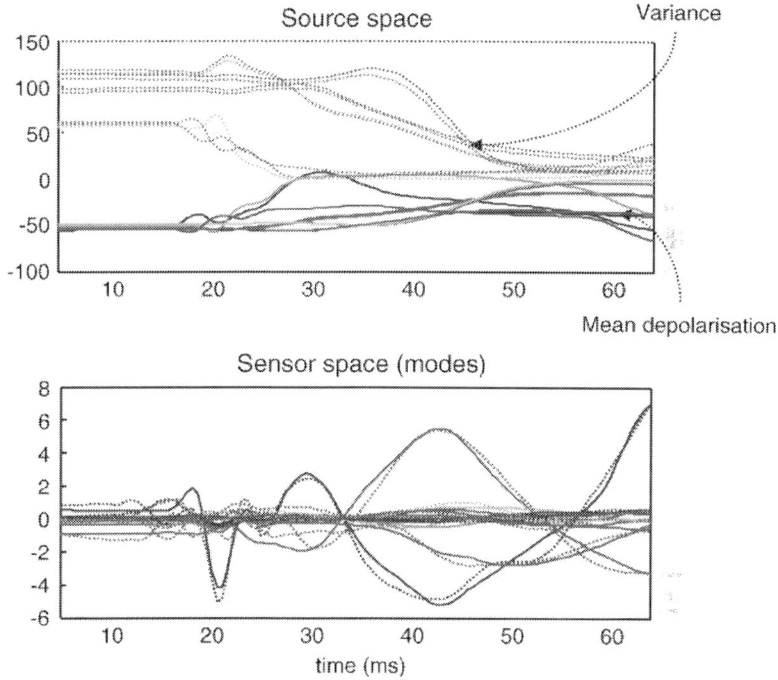

Figure 7. Standard DCM output for MFM of the SEP data. Upper panel: conditional estimates of the mean (solid lines) and covariance (dotted lines) of neuronal states, in source space (coloured lines correspond to different neuronal subpopulations). Lower panel: conditional estimates of responses, in sensor space (coloured lines correspond to different spatial modes; solid line: predicted; dotted line: observed).

DISCUSSION

We have introduced a mean-field DCM for M/EEG data, which approximates the full density dynamics of population activity in neuronal sources of observed electromagnetic responses with a Gaussian density. This work was motivated by the observation that neural-mass models, which consider only the first moment of the density of each neuronal population, can be seen as a limiting case of mean-field models (Marreiros et al., 2009). The mean-field model used physiological plausible priors with the hope of creating a reasonably realistic conductance-based model (cf., Rodrigues et al., 2006). We have shown, using model inversion and simulations that one can disambiguate between MFM and NMM models (Tables 2 and 3) and found that the NMM was the best model for explaining the MMN data. In contrast, we found that the MFM was a better model of the SEP data (Fig. 3), in the vast majority of subjects and at the group level. We deliberately chose these distinct data sets in the hope of disclosing this dissociation between neural-mass and mean-field models.

This difference in performance between the two models on the two data sets lies in the difference between Eqs. (1) and (2). Our results suggest that the MFM captures the faster SEP population dynamics better than the NMM. This may be because the SEP paradigm evokes larger presynaptic inputs to small circumscribed neuronal populations as compared to the MMN and related cognitive paradigms. It is this input that induces changes in conductance which synchronise hidden states and cause a subsequent suppression of mean activity. It can be seen from Eq. (1) that changes in covariance depend on the derivative of flow. Eq. (4) shows that this depends on depolarization and conductance. In support of this, the MFM solutions for the SEP data do indeed show a reciprocal coupling between mean depolarization and variance (see Fig. 7). Having said this, the appropriateness of a model for any particular data or paradigm data cannot necessarily be deduced analytically. The aim of this paper was to show that questions about density dynamics of this sort can be answered using Bayesian model comparison. Future studies with NMM and MFM may provide heuristics about the relative utility of these models. In particular, it will be interesting to use MFMs when more complex dynamics are induced by extreme perturbations from steady-state dynamics (e.g., transcranial magnetic stimulation).

Most DCMs in the literature are deterministic; in that they allow for observation noise on the sensors but do not consider random fluctuations on hidden states. Here, the hidden states in mean-field DCMs are sufficient statistics of a density, which accommodates random fluctuations on neuronal states. This is important because it means we can model systems specified in terms of stochastic differential equations (cf., Eq. (3)) with ordinary differential equations (Eq. (1)) through the Fokker-Planck formalism. A potential application of this approach, beyond finessing prediction of EEG signals, could be in the context of EEG-fMRI fusion; where the second-order statistics of neuronal activity (cf., power) may be an important predictor of BOLD signals.

It is important to appreciate that the scheme described in this paper is not tied to any particular model of neuronal dynamics. The same comparisons presented above could be repeated easily, using any model of neuronal dynamics that are entailed by their equations of motion. Indeed, we anticipate that people will want to compare neural-mass and mean-field implementations of different models (see software note). The only constraint on these comparisons is that the equations of motion should be nonlinear in the neuronal states. This is because linear models preclude a coupling of first and second-order moments and render the behaviour of the neural-mass and mean-field formulation identical.

CONCLUSION

We have shown that it is possible to implement a mean-field DCM, which considers the mean and variance of neuronal population activity. The modulation of second-order statistics may be a useful extension of DCM for evoked responses, as measured with magnetoencephalography and electroencephalography. Critically, the role of higher moments can be assessed empirically in a Bayesian model comparison framework. In this initial work, we conclude that, although conventional neural-mass models may be sufficient for modelling responses in conventional cognitive paradigms, it is easy to find evidence for coupling among the moments of neuronal ensemble densities in observed EEG data.

Software Note

Matlab demonstration and analysis routines referred to in this paper are available as academic freeware as part of the SPM software from http://www.fil.ion.ucl.ac.uk/spm.

Acknowledgments

The Portuguese Foundation for Science and Technology and the Wellcome Trust supported this work. We thank Marta Garrido for the MMN data and Vladimir Litvak and Joseph Classen for the SEP data.

REFERENCES

1. Allison, T., McGarthy, G., Wood, C.C., Jones, S.J., 1991. Potentials evoked in human and monkey cerebral cortex by stimulation of the median nerve. A review of scalp and intracranial recordings. Brain 114, 2465–2503.
2. Amari, S., 1972. Characteristics of random nets of analog neuron-like elements. IEEE Trans. Syst. Man Cybern. 2 (5), 643–657.
3. Amari, S., 1975. Homogeneous nets of neuron-like elements. Biol. Cybern. 17, 211–220. Breakspear, M., Roberts, J.A., Terry, J.R., Rodrigues, S., Mahant, N., Robinson, P.A., 2006. A unifying explanation of primary generalized seizures through nonlinear brain modeling and bifurcation analysis. Cereb. Cortex 16, 1296–1313.
4. Buchner, H., Adams, L., Muller, A., Ludwig, I., Knepper, A., Thron, A., Niemann, K., Scherg, M., 1995. Somatotopy of human hand somatosensory cortex revealed by dipole source analysis of early somatosensory evoked potentials and 3D-NMR tomography. Electroencephalogr. Clin. Neurophysiol. 96, 121–134.
5. David, O., Friston, K.J., 2003. A neural-mass model for MEG/EEG: coupling and neuronal dynamics. NeuroImage 20, 1743–1755.
6. David, O., Kiebel, S.J., Harrison, L.M., Mattout, J., Kilner, J.M., Friston, K.J., 2006. Dynamic causalmodeling of evoked responses inEEGandMEG.NeuroImage 30 (4), 1255–1272.
7. Daunizeau, J., Kiebel, S.J., Friston, K.J., 2009 Aug 15. Dynamic causal modelling of distributed electromagnetic responses. Neuroimage 47 (2), 590–601.
8. Deco, G., Jirsa, V.K., Robinson, P.A., Breakspear, M., Friston, K.J., 2008. The dynamic brain: from spiking neurons to neural-masses and cortical fields. PLoS Comput. Biol. 4 (8), e1000092.
9. Doeller, C.F., Opitz, B., Mecklinger, A., Krick, C., Reith, W., Schroger, E., 2003. Prefrontal cortex involvement in preattentive

auditory deviance detection: neuroimaging and electrophysiological evidence. Neuroimage 20, 1270–1282.

10. Escera, C., Yago, E., Corral, M.J., Corbera, S., Nunez, M.I., 2003. Attention capture by auditory significant stimuli: semantic analysis follows attention switching. Eur. J. Neurosci. 18, 2408–2412.
11. Fahrmeir, L., Tutz, G., 1994. Multivariate Statistical Modelling Based on Generalized Linear Models. Springer-Verlag, New York.
12. Fastenrath, M., Friston, K.J., Kiebel, S.J., 2009. Dynamical causal modelling for M/EEG: spatial and temporal symmetry constraints. NeuroImage 44 (1), 154–163.
13. Felleman, D.J., Van Essen, D.C., 1991. Distributed hierarchical processing in the primate cerebral cortex. Cereb. Cortex 1, 1–47.
14. Frank, T.D., 2004. Nonlinear Fokker-Planck Equations: Fundamentals and Applications. Springer Series in Synergetics. Springer, Berlin.
15. Frank, T.D., Daffertshofer, A., Beek, P.J., 2001. Multivariate Ornstein-Uhlenbeck processes with mean-field dependent coefficients: application to postural sway. Phys. Rev. E. Stat. Nonlin. Soft Matter Phys. 63 (011), 905.
16. Freeman, W.J., 1975. Mass Action in the Nervous System. Academic Press, New York.
17. Friston, K., 2005. Philos. Trans. R. Soc. Lond., B Biol. Sci. 360, 815–836.
18. Friston, K.J., Harrison, L.M., Penny, W.D., 2003. Dynamic causal modelling. NeuroImage 19 (4), 1273–1302.
19. Garrido, M.I., Kilner, J.M., Kiebel, S.J., Stephan, K.E., Friston, K.J., 2007a. Dynamic causal modelling of evoked potentials: a reproducibility study. NeuroImage 36, 571–580.
20. Garrido, M.I., Kilner, J.M., Kiebel, S.J., Friston, K.J., 2007b. Evoked brain responses are generated by feedback loops. PNAS 104 (52), 20961–20966.
21. Giard, M.H., Perrin, F., Pernier, J., Bouchet, P., 1990. Brain generators implicated in the processing of auditory stimulus deviance: a topographic event-related potential study. Psychophysiology 27, 627–640.

22. Harrison, L.M., David, O., Friston, K.J., 2005. Philos. Trans. R. Soc. Lond. B Biol. Sci. 360 (1457), 1075–1091.
23. Haskell, E., Nykamp, D.Q., Tranchina, D., 2001. Population density methods for largescale modelling of neuronal networks with realistic synaptic kinetics: cutting the dimension down to size. Network 12, 141–174.
24. Jääskeläinen, I.P., Ahveninen, J., Bonmassar, G., Dale, A.M., Ilmoniemi, R.J., Levänen, S., Lin, F.H., May, P., Melcher, J., Stufflebeam, S., Tiitinen, H., Belliveau, J.W., 2004. Human posterior auditory cortex gates novel sounds to consciousness. Proc. Natl. Acad. Sci. U. S. A. 101, 6809–6814.
25. Jansen, B.H., Rit, V.G., 1995. Electroencephalogram and visual evoked potential generation in a mathematical model of coupled cortical columns. Biol. Cybern. 73, 357–366.
26. Jemel, B., Achenbach, C., Muller, B.W., Ropcke, B., Oades, R.D., 2002. Mismatch negativity results from bilateral asymmetric dipole sources in the frontal and temporal lobes. Brain Topogr. 15, 13–27.
27. Jirsa, VK, Haken, H, 1996 Jul 29. Field theory of electromagnetic brain activity. Phys. Rev. Lett. 77 (5), 960–963.
28. Kiebel, S.J., Garrido, M.I., Moran, R.J., Friston, K.J., 2008. Dynamic causal modelling for EEG and MEG. Cogn. Neurodyn. 2 (2), 121–136.
29. Knight, B.W., 1972a. The relationship between the firing rate of a single neuron and the level of activity in a population of neurons. Experimental evidence for resonant enhancement in the population response. J. Gen. Physiol. 59, 767–778.
30. Knight, B.W., 1972b. Dynamics of encoding in a population of neurons. J. Gen. Physiol. 59, 734–766.
31. Litvak, V., Zeller, D., Oostenveld, R., Maris, E., Cohen, A., Schramm, A., Gentner, R., Zaaroor, M., Pratt, H., Classen, J., 2007. LTP-like changes induced by paired associative stimulation of the primary somatosensory cortex in humans: source analysis and associated changes in behaviour. Eur. J. Neurosci. 25 (9), 2862–2874.
32. Lopes da Silva, F.H, van Rotterdam, A., Barts, P., van Heusden, E., Burr, W., 1976. Models of neuronal populations: the basic mechanisms of rhythmicity. Prog. Brain Res. 45, 281–308.

33. Marreiros, A.C., Kiebel, S.J., Daunizeau, J., Friston, K.J., 2008. Population dynamics: variance and the sigmoid activation function. Neuroimage 42 (1), 147–157.
34. Marreiros, A.C., Kiebel, S.J., Daunizeau, J., Harrison, L.M., Friston, K.J., 2009. Population dynamics under the Laplace assumption. Neuroimage 44 (3), 701–714.
35. Morris, C., Lecar, H., 1981. Voltage oscillations in the barnacle giant muscle fiber. Biophys. J. 35, 193–213.
36. Mosher, J.C., Leahy, R.M., Lewis, P.S., 1999. EEG and MEG: forward solutions for inverse methods. IEEE Trans. Biomed. Eng. 46, 245–259.
37. Näätänen, R., Tervaniemi, M., Sussman, E., Paavilainen, P., Winkler, I., 2001. “Primitive intelligence” in the auditory cortex. Trends Neurosci. 24, 283–288.
38. Nunez, P.L., 1974. The brain wave equation: a model for the EEG. Math. Biosci. 21, 279–297.
39. Nykamp, D.Q., Tranchina, D., 2000. A population density approach that facilitates largescale modeling of neural networks: analysis and an application to orientation tuning. J. Comput. Neurosci. 8, 19–50.
40. Omurtag, A., Knight, B.W., Sirovich, L., 2000. On the simulation of large populations of neurons. J. Comput. Neurosci. 8, 51–63.
41. Opitz, B., Rinne, T., Mecklinger, A., von Cramon, D.Y., Schroger, E., 2002. Differential contribution of frontal and temporal cortices to auditory change detection: fMRI and ERP results. Neuroimage 15, 167–174.
42. Penny, W.D., Stephan, K.E., Mechelli, A., Friston, K.J., 2004. Comparing dynamic causal models. NeuroImage 22, 1157–1172.
43. Ravazzani, P., Tognola, G., Grandori, F., Budai, R., Locatelli, T., Cursi, M., Di Benedetto, G., Comi, G., 1995. Temporal segmentation and multiple source analysis of shortlatency median nerve SEPs. J. Med. Eng. Technol. 19, 70–76.
44. Riera, JJ, Jimenez, JC, Wan, X, Kawashima, R, Ozaki, T., 2007 Aprr. Nonlinear local electrovascular coupling. II: from data to neuronal masses. Hum. Brain Mapp. 28 (4), 335–354.
45. Rinne, T., Alho, K., Ilmoniemi, R.J., Virtanen, J., Näätänen, R.,

2000. Separate time behaviors of the temporal and frontal mismatch negativity sources. Neuroimage 12, 14–19.

46. Robinson, P.A., Rennie, C.J., Rowe, D.L., O›Connor, S.C., Gordon, E., 2005. Multiscale brain modelling. Phil. Trans. Roy. Soc. B 360, 1043–1050.
47. Rodriguez, R., Tuckwell, H.C., 1996. Statistical properties of stochastic nonlinear dynamical models of single spiking neurons and neural networks. Phys. Rev. E 54, 5585.
48. Rodrigues, S., Terry, J.R., Breakspear, M., 2006. On the genesis of spikewave oscillations in a mean-field model of human thalamic and corticothalamic dynamics. Phys. Lett. A 355, 352–357.
49. Sampolinsky, H., Zippelius, A., 1982. Relaxational dynamics of the Edwards-Anderson model of the mean-field theory of spin-glasses. Physical Review B 25 (11), 6860–6875.
50. Stephan, K.E., Penny, W.D., Moran, R.J., den Ouden, H.E., Daunizeau, J., Friston, K.J., 2010 Nov 12. Ten simple rules for dynamic causal modeling. Neuroimage 49 (4), 3099–3109.
51. Valdes, P.A., Jimenez, J.C., Riera, J., Biscay, R., Ozaki, T., 1999. Nonlinear EEG analysis based on a neural-mass model. Biol. Cybern. 81, 415–424.
52. Wilson, H.R., Cowan, J.D., 1972. Excitatory and inhibitory interactions in localized populations of model neurons. Biophys. J. 12, 1–24.

8 INTERACTION BETWEEN SYNAPTIC INHIBITION AND GLIAL-POTASSIUM DYNAMICS LEADS TO DIVERSE SEIZURE TRANSITION MODES IN BIOPHYSICAL MODELS OF HUMAN FOCAL SEIZURES

E. C. Y. Ho[1], Wilson Truccolo[1, 2]

[1] Department of Neuroscience & Institute for Brain Science, Brown University, Providence, USA

[2] U.S. Department of Veterans Affairs, Center for Neurorestoration and Neurotechnology, Providence, USA

ABSTRACT

How focal seizures initiate and evolve in human neocortex remains a fundamental problem in neuroscience. Here, we use biophysical neuronal network models of neocortical patches to study how the interaction between inhibition and extracellular potassium ($[K^+]_o$) dynamics may contribute to different types of focal seizures. Three main types of propagated focal seizures observed in recent intracortical microelectrode recordings in humans were modelled: seizures characterized by sustained

(~30−60 Hz) gamma local field potential (LFP) oscillations; seizures where the onset in the propagated site consisted of LFP spikes that later evolved into rhythmic (~2−3 Hz) spike-wave complexes (SWCs); and seizures where a brief stage of low-amplitude fast-oscillation (~10−20 Hz) LFPs preceded the SWC activity. Our findings are fourfold: (1) The interaction between elevated $[K^+]_o$ (due to abnormal potassium buffering by glial cells) and the strength of synaptic inhibition plays a predominant role in shaping these three types of seizures. (2) Strengthening of inhibition leads to the onset of sustained narrowband gamma seizures. (3) Transition into SWC seizures is obtained either by the weakening of inhibitory synapses, or by a transient strengthening followed by an inhibitory breakdown (e.g. GABA depletion). This reduction or breakdown of inhibition among fast-spiking (FS) inhibitory interneurons increases their spiking activity and leads them eventually into depolarization block. Ictal spike-wave discharges in the model are then sustained solely by pyramidal neurons. (4) FS cell dynamics are also critical for seizures where the evolution into SWC activity is preceded by low-amplitude fast oscillations. Different levels of elevated $[K^+]_o$ were important for transitions into and maintenance of sustained gamma oscillations and SWC discharges. Overall, our modelling study predicts that the interaction between inhibitory interneurons and $[K^+]_o$ glial buffering under abnormal conditions may explain different types of ictal transitions and dynamics during propagated seizures in human focal epilepsy.

KEYWORDS: Focal epilepsy, Seizure dynamics, Spike-wave discharges, Gamma oscillations

INTRODUCTION

Epilepsy is a neurological disorder characterized by recurrent seizures affecting an estimated 65 million people worldwide (Thurman et al. 2011). Its apparent simple characterization belies the often complex and varied cellular and synaptic basis underlying the abnormal neuronal activity during epileptic seizures. As a result, despite significant advances in understanding epilepsy over the past several decades, key knowledge gaps remain with respect to the disorder. One example is the mechanism(s) concerning the transition of neural dynamics from normal to ictal as focal seizures propagate to distal regions in the brain, especially in neocortex.

Seizures have been hypothesized to originate from a wide variety of cellular and network mechanisms (Richardson et al. 2008). It is possible that, even in the same patient, different seizures might be triggered by different factors. Facing the diversity of possible triggers of seizure transition, recent theoretical studies have proposed canonical models that focus on general unifying dynamical principles (Kramer et al. 2012; Jirsa et al. 2014; Sritharan and Sarma 2014; Wang et al. 2014; Wei et al. 2014b; Naze et al. 2015). Recent animal experiments (Trevelyan et al. 2006; Zhang et al. 2012; Cammarota et al. 2013; žiburkus et al. 2013) and human studies (Truccolo et al. 2011; Schevon et al. 2012; Ahmed et al. 2014) have also suggested that complex dynamics between different types of neurons can be in play as cortical networks transition into seizures. In particular, these studies have indicated the potential critical role of fast-spiking (FS) inhibitory interneurons (see Paz and Huguenard 2015 for a review) and their interactions with changes in extracellular potassium concentration $[K^+]_o$ (e.g. žiburkus et al. 2006). Elucidating such interplay between interneuron activity and ionic concentrations during seizure initiation and propagation can provide the much-needed clues for the development of seizure prediction, early detection and novel therapeutic approaches.

Furthermore, studies by Truccolo et al. (2011, 2014) and Wagner et al. (2015) have characterized spiking in ensembles of single neurons and local field potential (LFP) activity recorded via 96-microelectrode arrays in humans during focal seizures. Three main types of seizure activity were recorded in neocortical patches distal (~2–4 cm) to the identified seizure onset zones: seizures consisting of sustained (~30–60 Hz) gamma-band oscillations, seizures where the onset consisted of a direct transition into LFP spike discharges which evolved later into (~2–3 Hz) spike-wave complexes (SWCs), and seizures where the onset included a brief transient of low voltage fast oscillations (10–20 Hz) before the evolution into SWC discharges. (For a more comprehensive classification of seizures in terms of onset patterns and relationships to pathology in larger datasets, see Perucca et al. 2014). In the gamma seizures, despite the narrowband nature of the oscillations, neuronal spiking activity remained highly irregular and asynchronous (Truccolo et al. 2014). Classification of recorded single neurons into putative principal and interneuron cells suggested that inhibition was preserved throughout the various ictal stages in these gamma seizures (Truccolo et al. 2011). On the other hand, a recent investigation of SWC seizures in these datasets indicated that inhibitory activity seems to shut off just before the emergence of large amplitude spike-wave discharges (Ahmed et al. 2014). Ahmed et al. (2014) hypothesized the role of depolarization block in interneurons and examined their sensitivity to $[K^{+}]_o$ in a single cell model of a FS inhibitory interneuron.

Here, we examine how the interaction between FS inhibitory interneuron activity and extracellular potassium concentration in biophysical neuronal network models can capture the different types of propagated seizure dynamics observed in these human neocortical recordings. We used neuronal network models of neocortical patches that include pyramidal and FS interneurons (Wang and Buzsáki 1996), and $[K^{+}]_o$ diffusion coupled with glial cell activity. The neuronal network models spanned 9×9 minicolumns, each 25 μm wide and containing 12 pyramidal neurons and 4 FS interneurons (for a total of 9×9×16=1296 model neurons). Coupled $[K^{+}]_o$ and glial buffering dynamics were adapted from previous models in Fröhlich et al. (2010).

Several previous computational models (Kager at al. 2000; Bazhenov et al. 2004; Traub et al. 2005; Anderson et al. 2007; Cressman Jr. et al. 2009; Ullah et al. 2009; Fröhlich et al. 2010; Krishnan and Bazhenov, 2011; Krishnan et al. 2013, Wei et al. 2014a, 2014b) have explored the role of intracellular and extracellular concentrations of various types of ions (e.g. potassium, sodium, chloride, calcium, etc.), as well as inhibitory/excitatory conductances and metabolic factors, on seizure activity. Our model specification was inspired by several of these previous studies, including, for example, the formulation of glial-buffering processes (see Section 2). Nevertheless, we emphasize that, to our knowledge, the work presented here is the first to examine how the interaction between inhibitory activity and abnormal levels of extracellular potassium concentration can lead to three different types of seizures (gamma, SWC, low-voltage fast-oscillations followed by SWCs) observed in intracortical recordings from patients with focal epilepsy. Overall, we demonstrate that transition into and maintenance of both gamma-band and SWC seizures in the model can be explained by variations in inhibition strength (and therefore changes in the excitation/

inhibition balance) and $[K^+]_o$ levels. We consider this to be the main novel contribution reported in our study.

METHODS

Model Description and Parameter Choices

Single Neuron Model

We used a single-compartment conductance based model to represent both the fast-spiking inhibitory interneurons and pyramidal cells (Wang and Buzsáki 1996; Skinner 2006; Anderson et al. 2007). The differential equation for the membrane potential of a given neuron corresponds to:

$$C \times \frac{dV^l}{dt} = I^l_{int}(t) + I^l_{syn}(t) + I^l_{ext}(t), \quad (1)$$

for l=1,2,…,1296 neurons. I^l_{int} denotes the intrinsic current of the neuron in question, I^l_{syn} denotes synaptic currents and I^l_{ext} denotes external input. The specific capacitance C is assumed to be 1 μF/cm^2. I^l_{int} has the following general form:

$$\begin{aligned} I^l_{int} &= -I_{Na} - I_K - I_L - I_{KCa} \\ &= -g_{Na} \times m^3_\infty \times h \times (V - E_{Na}) \\ &\quad -g_K \times n^4 \times (V - E_K([K^+]_o)) \\ &\quad -g_L \times (V - E_L([K^+]_o)) - I_{KCa}, \end{aligned} \quad (2)$$

which represent sodium, delayed-rectifier potassium, leak and calcium-mediated potassium currents, respectively. The leak reversal potential is dependent on various ionic concentrations via the Goldman-Hodgkin-Katz voltage equation:

$$\begin{aligned} E_L([K^+]_o) &= 26.64\text{mV} \\ &\times \log_e \frac{1 \times [K^+]_o + 0.085 \times [Na^+]_o + 0.1 \times [Cl^-]_i}{1 \times [K^+]_i + 0.085 \times [Na^+]_i + 0.1 \times [Cl^-]_o}, \end{aligned} \quad (3)$$

where $[K^+]_o$,$[K^+]_i$,$[Na^+]_o$,$[Na^+]_i$,$[Cl^-]_o$,$[Cl^-]_i$ denote ionic concentrations, with subscript i and $o$representing intracellular and extracellular, respectively. The coefficient before each type of ionic concentration is the relative membrane permeability of that ion. The "fast" activation variable m_∞ of the sodium current is assumed to have reached a steady state and has the form:

$$m_\infty = \frac{\alpha_m}{\alpha_m + \beta_m},$$
$$\alpha_m(V) = \frac{-0.1\ \text{mV}^{-1} \times (V + 35\ \text{mV})}{\exp(-0.1\ \text{mV}^{-1} \times (V + 35\ \text{mV})) - 1},$$
$$\beta_m(V) = 4 \times \exp\left(-\frac{V + 60\ \text{mV}}{18\ \text{mV}}\right), \quad (4)$$

and the inactivation variable h of the sodium and the activation variable n of delayed-rectifier potassium current obey the following differential equation:

$$\frac{dX}{dt} = 5\ \text{ms}^{-1} \times (\alpha_X \times (1 - X) - \beta_X \times X), \quad (5)$$

where X can be h or n, with

$$\alpha_h(V) = 0.07 \times \exp\left(-\frac{V + 58\ \text{mV}}{20\ \text{mV}}\right),$$
$$\beta_h(V) = \frac{1}{\exp(-0.1\ \text{mV}^{-1} \times (V + 28\ \text{mV})) + 1},$$
$$\alpha_n(V) = -\frac{0.01\ \text{mV}^{-1} \times (V + 34\ \text{mV})}{\exp(-0.1\ \text{mV}^{-1} \times (V + 34\ \text{mV})) - 1},$$
$$\beta_n(V) = 0.125 \times \exp\left(-\frac{V + 44\ \text{mV}}{80\ \text{mV}}\right). \quad (6)$$

The purpose of the calcium-mediated potassium current (I_{KCa}) is to provide the spike-frequency adaptation of the model pyramidal cells. It has the following form:

$$I_{KCa} = g_{KCa} \times (V - E_K) \times \frac{[Ca^{2+}]_i}{1\text{mM} + [Ca^{2+}]_i}, \quad (7)$$

where g_{KCa} and $[Ca^{2+}]_i$ denote the conductance for calcium-mediated potassium currents and the intracellular calcium concentration, respectively. The $[Ca^{2+}]_i$ dynamics are given according to Cressman et al. (2009) and Ullah et al. (2009):

$$\frac{d[Ca^{2+}]_i}{dt} = -\frac{0.002\frac{\text{mol}}{\text{cm}\cdot\text{mC}} \times g_{Ca} \times (V - E_{Ca})}{1 + \exp\left(-\frac{(V+25\ \text{mV})}{2.5\ \text{mV}}\right)} - \frac{[Ca^{2+}]_i}{80\ \text{ms}}, \quad (8)$$

where g_{Ca} and E_{Ca} denote the conductance and reversal potential related to Ca^{2+}, respectively.

Network Structure and Connectivity

The structure of our network of model neurons is based on the arrangement of neurons in minicolumns in the neocortical area (Buxhoeveden and Casanova 2002). In this network model, each minicolumn consists of 16 single-compartment model neurons, of which 12 are excitatory pyramidal cells and 4 are inhibitory interneurons. The neurons are arranged into four layers within each minicolumn (Fig. 1). Each layer consists of four neurons in a square arrangement, with each vertex of the square being the location of one neuron. Pyramidal cells occupy the top three layer and the bottom layer is for the four inhibitory interneurons. (We do not attempt here to replicate the actual 6-layer neocortical structure). The entire cortical network consists of 9 ×9 such minicolumns for a total of 1296 model neurons. Details of the structure configuration of the cortical network model and various distance parameters between model neurons are given in Fig. 1 and Table 1. Unless otherwise stated, we chose a global random connectivity to avoid boundary effects associated with the small size of the simulated cortical network.

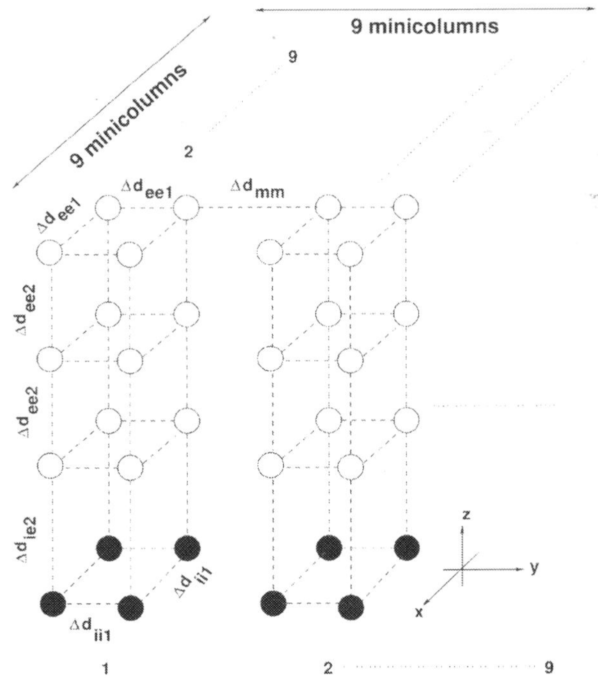

Figure 1. Model cortical network: spatial configuration. Open circles represent pyramidal cells and black circles represent inhibitory interneurons. Two mini-

columns are shown. Table 1specifies the values for the distance parameters in the figure

Table 1: Network geometry parameters. Please refer to Fig. 1 for the description of parameters

Parameter	**Units**	**Values**
Δd_{ee1}	cm	5×10^{-4}
Δd_{ee2}	cm	5×10^{-4}
Δd_{ie2}	cm	30×10^{-4}
Δd_{ii1}	cm	5×10^{-4}
Δd_{mm}	cm	15×10^{-4}

Glial and Potassium Ionic Dynamics

Each neuron in the network is endowed with a potassium pump. An associated potassium uptake dynamics related to nearby glial cells are included in the model to regulate the extracellular potassium concentration $[K^+]_o$. The potassium pump is represented by:

$$I_{Kpump} = \frac{I_{Kmax}}{\left(1 + \frac{[K^+]_{o(eq)}}{[K^+]_o}\right)^2}, \tag{9}$$

where I_{Kmax} and $[K^+]_{o(eq)}$ are defined in Table 3. The glial potassium uptake mechanism is modelled in terms of a reversible binding process between extracellular potassium ions and the glial buffer (Kager et al. 2000; Bazhenov et al. 2004; Park and Durand 2006; Fröhlich et al. 2008, 2010). The binding/unbinding process can be represented as

$$K^+ + B \underset{k_b}{\overset{k_f}{\rightleftharpoons}} KB, \tag{10}$$

where K^+ and B denote the extracellular potassium ions and the free buffer in the glia, respectively, and k_b and k_f correspond to the backward and forward glial binding rate, respectively (Tables 3 and 4). K^+ and B combine reversibly to form the bound form of the buffer KB. We fixed the maximal amount of free glial buffer $[B]_{max}$ to 500mM (Table 3, see also Kager et al. 2000).

The above process (10) can be written as a set of differential equations:

$$\frac{\partial[K^+]_o}{\partial t}\Big|_{glia} = k_b \times \frac{([B]_{max} - [B])}{1.1} - k_f \times [K^+]_o \times [B],$$

$$\frac{d[B]}{dt} = k_b \times ([B]_{max} - [B]) - k_f \times [K^+]_o \times [B],$$

$$k_f = \frac{k_b/1\text{mM}}{1 + \exp\left(\frac{[K^+]_o - [K^+]_{o(th)}}{\theta}\right)}, \tag{11}$$

where $[K^+]_{o(th)}$ and θ are specified in Table 3.

We vary $[K^+]_{o(th)}$ and θ to simulate a range of physiological and pathological states for the potassium glial-buffering system. $[K^+]_{o(th)}$ corresponds to the equilibrium level (set point) for the extracellular potassium concentration being targeted by the glial buffering system. How tightly the glial buffering system tracks this equilibrium set-point depends on the parameter θ. For example, the lower the absolute values of θ, the closer the glial buffer system attempts to keep the $[K^+]_o$ concentration to the target $[K^+]_{o(th)}$ level. Under this condition, the system may fail if there is a big enough difference between $[K^+]_o$ and $[K^+]_{o(th)}$, leading to bi-stability and abnormal functioning. In more physiological conditions, the absolute value of θ is kept at higher values so that the glial buffering system tracks $[K^+]_{o(th)}$ in a softer fashion, allowing a broader range of $[K^+]_o$ levels and more effectively avoiding bi-stability. We note that while we vary $[K^+]_{o(th)}$ and θ in order to change the glial buffering dynamics and to induce abnormal states, we do not yet have a hypothesis about specific causes that would lead to these changes in first place.

The diffusion of extracellular potassium ions across the network was modelled according to the diffusion equation:

$$\frac{\partial[K^+]_o}{\partial t}\Big|_{diff} = D \times \nabla^2[K^+]_o, \tag{12}$$

where ∇^2 is the Laplace operator $\frac{\partial^2}{\partial x^2} + \frac{\partial^2}{\partial y^2} + \frac{\partial^2}{\partial z^2}$ and D is the diffusion constant. In the simulations, ∇^2 is approximated by a set of finite difference equations, since the model neurons are discretely located (Fig. 1). We impose the "no-flux" or Neumann boundary condition (i.e.

$\frac{\partial [K^+]_o}{\partial x} = 0$ on the y–z planes at both the x boundaries and so on) in the network simulations.

Therefore, the total time derivative of $[K^+]_o$ of a particular model neuron, $\frac{d[K^+]_o}{dt}$ (unit in mM $\cdot ms^{-1}$), is the sum of all the potassium, pump, glia and diffusion currents

$$\begin{aligned}\frac{d[K^+]_o}{dt} &= \frac{50}{96489}\frac{\text{mol}}{\text{cm}\cdot\text{mC}} \times (I_{KCa} + I_K - I_{Kpump} \\ &+ g_L|_{[K^+]} \times (V - E_K([K^+]_o))) \\ &+ \left.\frac{\partial [K^+]_o}{\partial t}\right|_{\text{glia}} + \left.\frac{\partial [K^+]_o}{\partial t}\right|_{\text{diff}},\end{aligned} \tag{13}$$

where $g_L|_{[K+]}$ (≥ 0) is the value of the leak conductance attributable to potassium. The value can be estimated by fitting the equation,

$$\begin{aligned}g_L \times (V - E_L([K^+]_o)) = g_L|_{[K^+]} &\times (V - E_K([K^+]_o)) \\ +g_L|_{[Cl^-]} &\times (V - E_{Cl}) \\ +g_L|_{[Na^+]} &\times (V - E_{Na}),\end{aligned}$$

with the constraint that

$$g_L = g_L|_{[K^+]} + g_L|_{[Cl^-]} + g_L|_{[Na^+]}, \tag{14}$$

where $g_L|_{[Cl-]}$ (≥ 0) and $g_L|_{[Na+]}$ (≥ 0) are the values of leak conductance due to Cl^- and Na^+respectively, while E_{Cl} and E_{Na} are their respective reversal potentials as shown in equations (16) and (17). $g_L|_{[K+]}$ values at ~ 6–13 % of total leak give reasonable fits over a wide range of physiological $[K^+]_o$ and membrane potential values. (Various fitting scenarios indicate that >70 % of the total leak conductance value is due to chloride). Based on the above procedure, we assume a $g_L|_{[K+]}$ value at 6 % of the g_L value.

The potassium reversal potential E_K is affected by $[K^+]_o$ via the Nernst equation:

$$E_K = 26.64\text{mV} \times \log_e\left(\frac{[K^+]_o}{[K^+]_i}\right), \tag{15}$$

where the intracellular potassium concentration $[K^+]_i$ is set at 133 mM (Table 3). Similarly, E_{Cl} is given by

$$E_{Cl} = 26.64\text{mV} \times \log_e\left(\frac{[Cl^-]_i}{[Cl^-]_o}\right) \sim -74\text{mV}, \tag{16}$$

and E_{Na} by

$$E_{Na} = 26.64\text{mV} \times \log_e \left(\frac{[Na^+]_o}{[Na^+]_i} \right) \sim 54\text{mV}. \tag{17}$$

See also Table 3 for values of $[Na^+]_o$, $[Na^+]_i$, $[Cl^-]_o$ and $[Cl^-]_i$.

Synaptic Model

Unless otherwise stated, the synaptic current I^l_{syn} corresponds to

$$I^l_{syn} = \sum_{m \neq l} \Gamma_{m \to l} \times g^{m \to l}_{syn} \times s_m(t - \tau_{delay}) \times (E_{\{e,i\}} - V^l),$$

with $s(t)$ representing the synaptic gating variable of each model neuron, such that

$$\begin{aligned} s_l &\to s_l + s_{max} \text{ (when } V^l \text{ crosses 0 from below)}, \\ \frac{ds_l}{dt} &= -\frac{s_l}{\tau_{\{e,i\}}} \text{ (otherwise)}. \end{aligned} \tag{18}$$

In the above, $\Gamma_{m \to l}$ denotes the binary connectivity matrix element in which a value of 1 represents the existence of a synaptic connection in neuron *l* from neuron *m*, and a value 0 represents the absence of such connection. Table 5 lists the probabilities of connection between pyramidal neurons ($P(e \to e)$), from pyramidal neurons to interneurons ($P(e \to i)$), from interneurons to pyramidal neurons ($P(i \to e)$) and between interneurons ($P(i \to i)$). The term $E_{\{e,i\}}$ denotes the excitatory or inhibitory reversal potential value, respectively. (Whether the reversal potential is excitatory or inhibitory depends on the particular neuron–i.e. neuron *m*–from which neuron *l* receives synaptic current).

As done in Ho et al. (2012) for computational efficiency, we use a discontinuous model to represent the opening of the synapses (equation set (18)) whenever there is a spike. Parameters $_{\{e,i\}}$ and s_{max} (for s_l) denote the excitatory (inhibitory) synaptic decay time constant and the value of the maximal opening of the synaptic gates per spike, respectively (whether the decay time constant should be excitatory or inhibitory depends on whether neuron *l* itself is a pyramidal cell or an inhibitory interneuron). Table 5 provides more details on the terms of the above equations.

Background Synaptic Activity

Unless otherwise stated, each model neuron is driven by an external input I_{ext} with the form (Rudolph et al. 2004) to mimic the activities of the fluctuating synaptic background:

$$I_{ext}^{l} = g_e^l(t) \times (E_e - V^l) + g_i^l(t) \times (E_i - V^l),$$

$$\frac{dg_{\{e,i\}}^{l}(t)}{dt} = \frac{g_{\{e,i\}0} - g_{\{e,i\}}^{l}(t)}{\tau_{\{e,i\}}} + \sqrt{\frac{2 \times \sigma_{\{e,i\}}^2}{\tau_{\{e,i\}}}} \times \chi_{\{e,i\}}^{l}(t), \qquad (19)$$

where $E_{\{e,i\}}$ denotes the excitatory/inhibitory reversal potential and $\tau_{\{e,i\}}$ is the excitatory/inhibitory time constant. The excitatory/inhibitory conductance values $g_{\{e,i\}}$ are stochastic variables following an Ornstein-Uhlenbeck process (equation set (19)), with mean $g_{\{e,i\}0}$ and SD $\sigma_{\{e,i\}}$ (excitatory/inhibitory fluctuations). The stochastic term $\chi_{\{e,i\}}$ is such that $\int_0^{\epsilon} \chi_{\{e,i\}}(s) \times ds$ is a Gaussian distributed random variable with mean zero and variance $\epsilon>0$. Further details on the numerical implementation of the stochastic elements of equation set (19) can be found in Ho et al. (2012).

Table 5 provides more details on the terms in the above equation.

Implementing the Simulations

The code for the model cortical network simulation was written in C++ (gcc, version 4.4.7, http://gcc.gnu.org). We used open MPI (version 1.8, http://www.open-mpi.org) for parallel processing of each simulation. Each typical 1296-neuron simulation used eight CPUs. A 10-hour simulation on the Brown University computer cluster (https://www.ccv.brown.edu/) usually yielded ~ 2-3 minutes of data. Parameter scanning was automated via the use of custom Perl (version 5.18.4) and Linux shell scripts. Post simulation analysis of data was carried out using Matlab (version R2013b, http://www.mathworks.com), Octave (version 3.8.1, http://www.gnu.org/software/octave/) and customized Perl scripts. Single neuron simulations (Fig. 2) were performed using xppaut (Ermentrout 2002). Graphical rendering of data was performed by gnuplot (version 4.6.3, http://www.gnuplot.info), xfig (version 3.2) and gimp (version 2.8.10, http://www.gimp.org). The parula palette of the time-frequency plots is from https://github.com/Gnuplotting/gnuplot-palettes. The C++ code is available in the NEURON ModelDB database (https://

senselab.med.yale.edu/ModelDB/showModel.cshtml?model=190306). The ModelDB accession number for the network model reported in this paper is 190306.

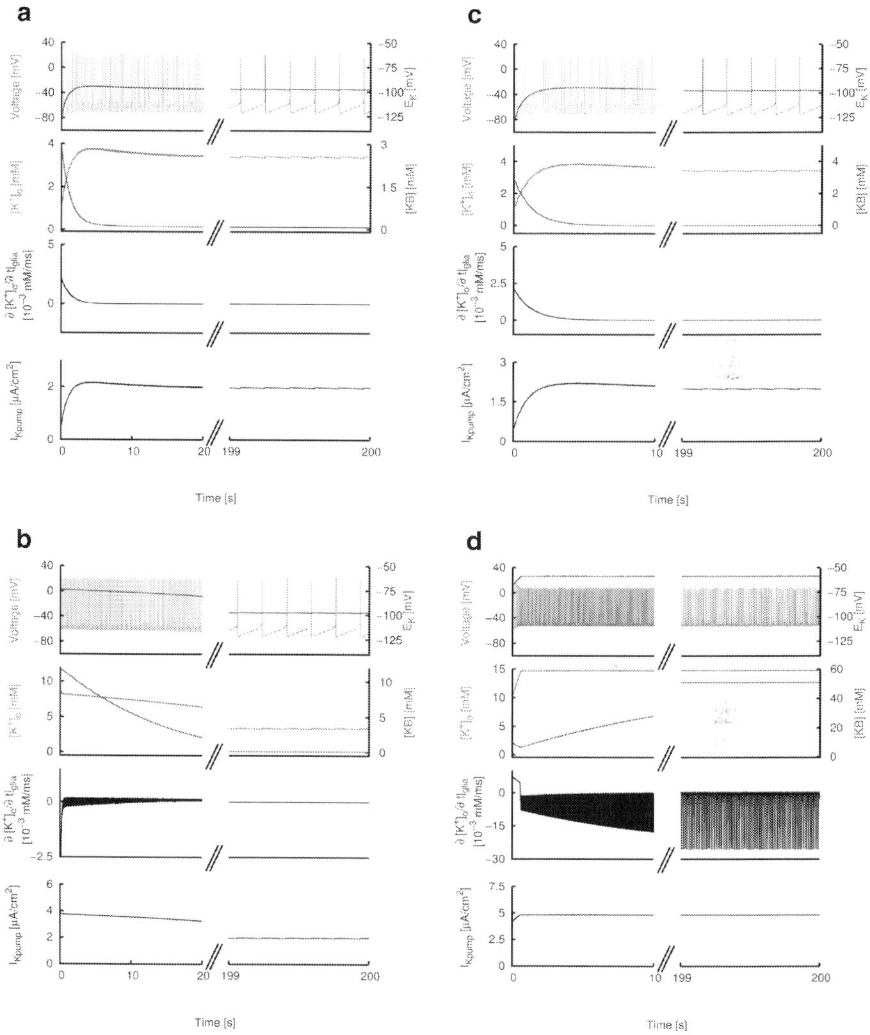

Figure 2. Four single neuron model simulations with physiological and abnormal glial parameters. Bi-stability is observed with abnormal glial parameters. In each of a, b, c and d: Top panel shows the membrane potential (*red*) and potassium reversal potential E_K (*blue*). Second panel shows the extracellular potassium concentration $[K^+]_o$ (*red*) and the glial bound buffer concentration $[KB]$ (*blue*). Third panel shows the rate of change of extracellular potassium concentration due to the glial action $\frac{\partial [K^+]_o}{\partial t}|_{glia}$. Bottom panel shows the potas-

sium pump current I_{Kpump}. (a)"Physiological" parameters. Initial conditions $[K^+]_o(t=0)$=1mM, $[KB](t=0)$=3mM. (b)"Physiological" parameters. Initial conditions $[K^+]_o(t=0)$=10mM, $[KB](t=0)$=10mM. (c)"Pathological" parameters. Initial conditions $[K^+]_o(t=0)$=1mM, $[KB](t=0)$=3mM. (d)"Pathological" parameters. Initial conditions $[K^+]_o(t=0)$=10mM, $[KB](t=0)$=10mM

RESULTS

We build model neocortical networks (Fig. 1) using conductance-based model neurons endowed with glial dynamics for potassium uptake. Based on the observations of human focal seizure data (LFPs and spiking activity in ensembles of single neurons), our aim is to use these model networks to reproduce the neural dynamics during the transition into ictal states of propagated focal seizures. Our presentation is structured as follows. First, based on both single neuron and network simulations (Section 2), we demonstrate that pathological glial uptake of potassium can lead to bi-stable dynamics in the model cortical network. The bi-stable dynamics consist of a "high-activity" state characterized by an elevated $[K^+]_o$ level, and a state of "low-activity" with a low $[K^+]_o$ level. We relate the "high-activity" state to "ictal" events and the "low-activity" state to non-epileptic, normal activity. Although network bistability has been implicated in computational studies as an avenue for seizure transition (Fröhlich et al. 2010), here we demonstrate that abnormal glial buffering can increase the propensity of single neurons (and thus the cortical network) to exhibit bi-stability (Fig. 2). The underlying mechanism for the bistabilty is the imbalance in the reversible reaction of the potassium buffer between the bound and unbound forms (10).

Second, with the seizure transition mechanism in place for the model cortical network, we then examine how certain types of synaptic changes involving mostly inhibitory interneurons and their interaction with potassium extracellular concentration may lead to various types of seizure transitions as observed in the human data. Our strategy is to explore inhibitory synaptic and intrinsic parameters relevant to the occurrence of depolarization block in model inhibitory interneurons. We also note that, in our model, pyramidal neurons are not as prone to depolarization block (in comparison to inhibitory interneurons) because of their Ca^{2+}-mediated potassium conductances (g_{KCa}) and the corresponding after-spike hyperpolarization effects (7).

Variations of parameters related to synaptic inhibition and glial potassium buffering allow us to replicate in detail three types of seizure patterns observed in the human data (gamma seizures and two types of spike-and-wave seizures; Truccolo et al. 2014). Thus, all three types of seizures can each be linked, via the same computational network model, to specific temporal patterns of synaptic activities and glial buffering dynamics during the evolution of propagated ictal states (Figs. 3, 4 and 5). Table 2 summarizes the results of our network simulations in terms of the paths leading to the three types of seizure transitions. All of the primary variables and parameters used in the simulations and analyses below are introduced and defined in Tables 1, 3, 4 and 5.

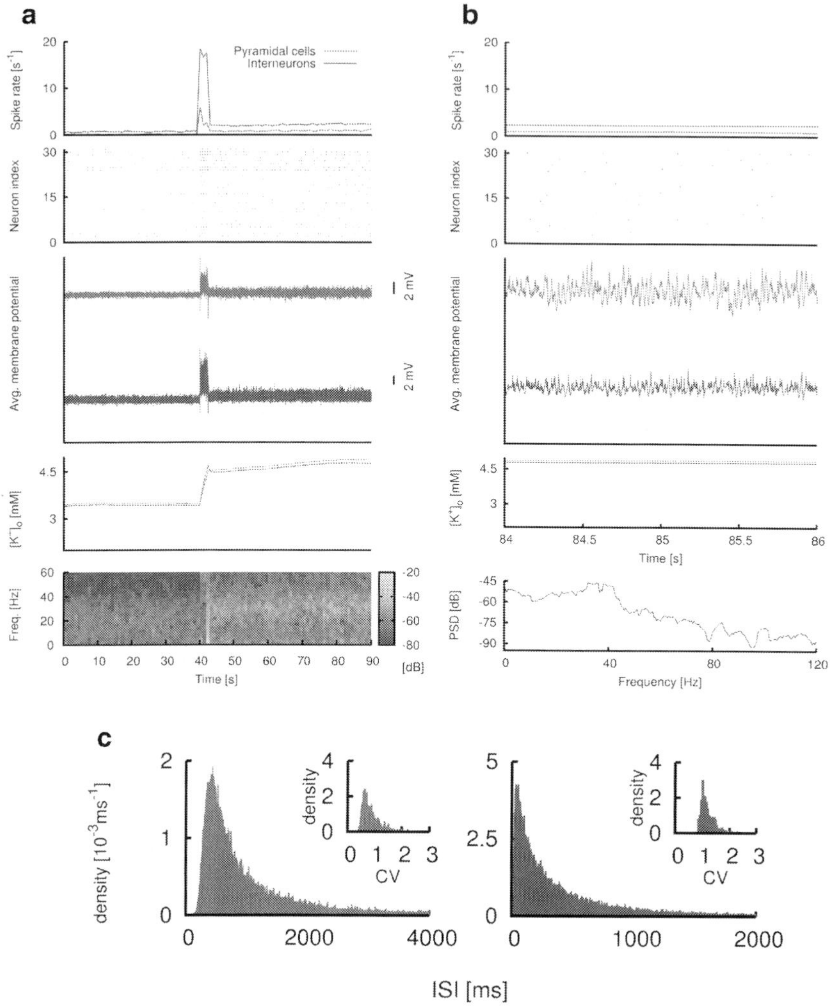

Figutre 3. An example gamma seizure simulation with pathological glial and pump parameters. **(a)**Depiction of neural activity and time course of various biophysical parameters during a 90-second simulation. The network starts with a "low-activity" non-seizure state (first 40 seconds of simulation). A DC stimulation is applied to every neuron in the network between 40 and 42.5 seconds, after which the system is "kick-started" into a high activity state with an elevated level of extracellular potassium (after 42.5 seconds). Topmost panel: average spike rate of pyramidal cells (*red*) and interneurons (blue) during the simulation. Second panel: raster plot of action potentials of neurons in 2 of the 81 simulated minicolumns. These 2 minicolumns are located in the centre of the network. Third panel: average membrane potential values of interneurons (*blue*) and pyramidal cells (*red*) during simulation. Fourth panel: average extracellular

potassium level values for interneurons (blue) and pyramidal cells (*red*) during simulation. Fifth panel:Time-frequency plot of the average membrane potential values of the pyramidal cells (third panel-red). Clear sustained gamma band oscillations (~40 Hz) at the population level emerge with the DC stimulation and persist even after the stimulation is terminated. (b) A 2-second segment from the 90-second simulation in **a**. Same conventions as in **a**, except that the bottom panel is a power spectral density (PSD) plot of the 2-second average membrane potential values of the pyramidal cells. The maximum of the PSD is located in the gamma range. **(c)** Histograms of interspike intervals (ISIs) of all the pyramidal cells (red) and interneurons (blue) during gamma population activity (t≳45t≳45 seconds). Insets: Histograms of the ISI coefficients of variation (CV) for spike trains of individual pyramidal cells (red) and interneurons (blue), also during gamma population activity (t≳45t≳45 seconds)

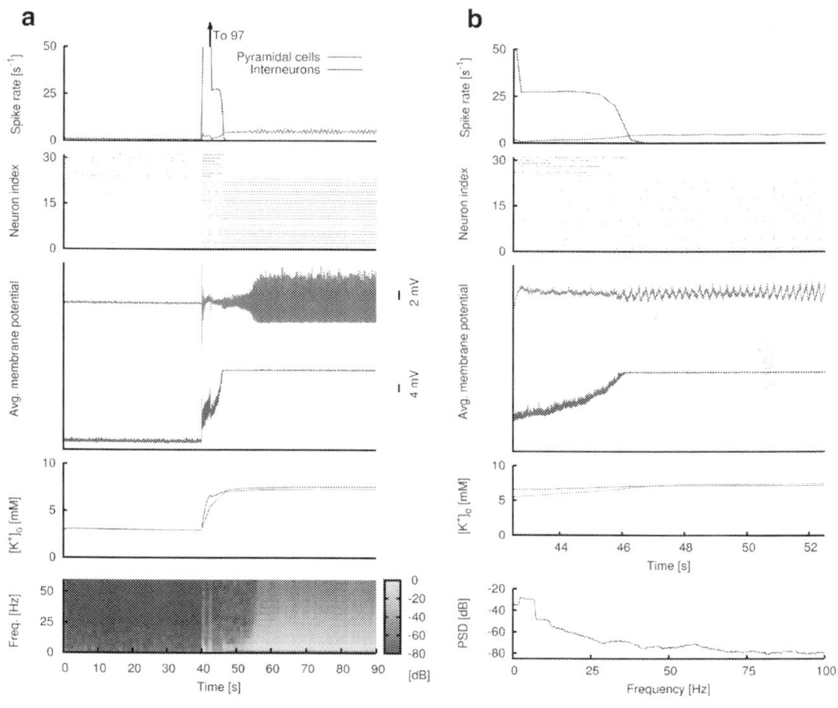

Figure 4. An example simulation showing a transition into a spike-wave-complex (SWC) seizure via low inhibitory synaptic strength. **(a)** Depiction of a 90-second simulation showing several biophysical variables. As in Fig. 3, the network starts with a “low-activity” non-seizure state (first 40 seconds of simulation). DC stimulation is applied to every neuron between 40 and 42.5 seconds. Topmost panel: average spike rate of pyramidal cells (*red*) and interneurons (*blue*) during the simulation. As rhythmic SWC discharges emerge,

the firing rate of pyramidal neurons approaches ~5 Hz. Second panel from top: raster plot of action potentials for neurons in 2 of the simulated 81 minicolumns. The location of these 2 minicolumns is the same as in Fig. 3. Third panel: average membrane potential values of interneurons (blue) and pyramidal cells (*red*) during simulation. Fourth panel: average extracellular potassium level values for interneurons (*blue*) and pyramidal cells (red). Fifth panel: Time-frequency plot of the average membrane potential values for the pyramidal neurons (third panel-red). The power peak at ~5 Hz (after the DC stimulation) corresponds to the emergence and maintenance of rhythmic SWC discharges, even after the DC stimulation is terminated. (b) A 10-second selection from the 90-second simulation in (a) showing the transition (after the DC stimulation) into SWCs. Same convention as in (a), except the bottom panel is a power spectral density (PSD) plot of the 10-second average membrane potential values of the pyramidal cells (third panel-red). The cessation of spiking of interneurons at around *t*=46 seconds (first and second panels) as they enter depolarization block (third panel-blue) is concomitant with the emergence of SWCs of increasing amplitude (third panel-red).

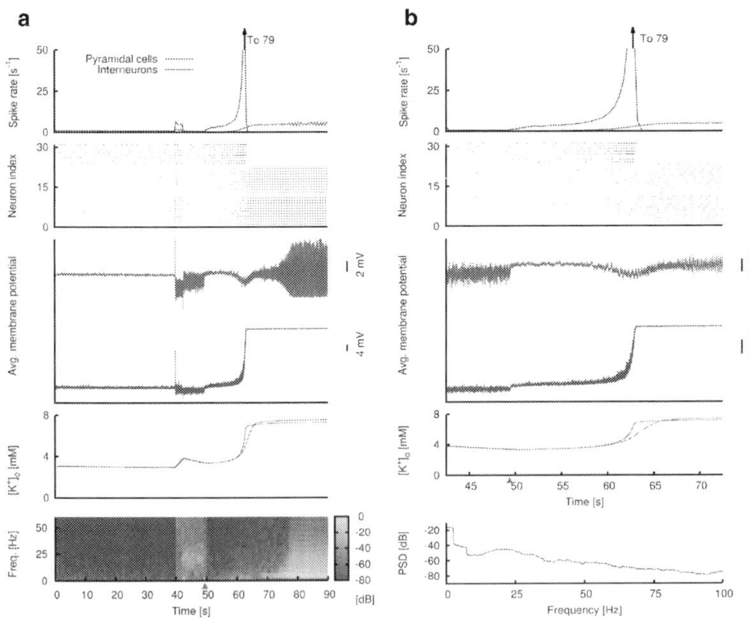

Figure 5. An example simulation showing transition into low voltage fast (~20 Hz) oscillation followed by SWC discharges, obtained via initial high inhibitory synaptic strength and subsequent depletion of inhibition. **(a)** Depiction of a 90-second simulation showing several biophysical variables. The network spends its first 40 seconds in a low-activity "non-epileptic" state. A DC

stimulation is applied to every neuron in the network between 40 and 42.5 seconds. During this DC stimulation period, the effective inhibition of the network is also drastically increased to mimic the build up of inhibitory strength prior to seizure (unlike in Figs. 3 and 4 where the effective inhibition is kept constant throughout the entire simulation). This increased effective inhibition is held constant up to t=49.5 seconds, after which inhibition wears off to zero (starting point of this depletion phase is marked by the red triangle). Topmost panel: average spike rate of pyramidal cells (*red*) and interneurons (*blue*) during the simulation. Second panel:raster plot of action potentials of neurons in 2 of the 81 minicolumns simulated. The location of these 2 minicolumns is the same as in Figs. 3 and 4. Third panel: average membrane potential values of interneurons (*blue*) and pyramidal cells (*red*). Fourth panel: average extracellular potassium level values for interneurons (*blue*) and pyramidal cells (*red*). Fifth panel: time-frequency plot of the average membrane potential values of the pyramidal cells (third panel). (b) A 30-second selection from the 90-second simulation in a showing both the transient low voltage fast (~20 Hz) oscillations and the subsequent emergence of rhythmic SWC discharges. Same convention as in a, except the bottom panel is a power spectral density (PSD) plot of the 30-second average membrane potential values of the pyramidal cells. Transient (~ 20 Hz) oscillations are observed after the DC stimulation but before inhibition wears off starting at t=49.5 seconds (red triangle). As inhibition begins to wear off, the firing rate of interneurons shows an initial transient increase, after which it decreases to zero when the interneurons enter depolarization block (third panel-blue, after around t=50 seconds). The reduced inhibition (due to both active wearing off of inhibition and the cessation of interneuron spiking) and increased $[K^+]_o$ (fourth panel) also lead to the increased firing of pyramidal cells and the emergence of rhythmic SWC discharges (third panel-red, towards the end).

Table 2. Pathways to three main seizure types in the focal seizure model

Seizure type	Synaptic inhibition	Interneuron depolarization block	Representative simulation
Gamma	Strong synaptic inhibition	No	Fig. 3
Spike-wave complex (SWC)	Weak synaptic inhibition	Yes	Fig. 4

Low-voltage fast oscillations followed by SWCs	Strong synaptic inhibition at seizure onset, followed by a weakening of inhibition (e.g. GABA depletion).	Yes	Fig. 5

Table 3. Intrinsic, glial and ionic parameters used for network simulations

Parameter	Description	Units	Values	
			Pyramidal	**Interneurons**
C	Specific neuron membrane capacitance	μF/cm^2	1	1
g_{Na}	Sodium conductance	mS/cm^2	35	35
g_K	Potassium conductance	mS/cm^2	9	S
g_L	Leak conductance	mS/cm^2	0.135	0.1
g_{KCa}	Ca $^{2+}$-activated potassium conductance	mS/cm^2	S	0
g_{Ca}	Calcium conductance	mS/cm^2	0.1	0
E_{Na}	Sodium reversal potential	mV	54	54
E_{Ca}	Calcium reversal potential	mV	120	120
I_{ext}	External applied current to model neurons	μA/cm^2	S	S
$[K^+]_i$	Intracellular potassium concentration	mM	133	133
$[Na^+]_o$	Extracellular sodium concentration	mM	130	130
$[Na^+]_i$	Intracellular sodium concentration	mM	17	17
$[Cl^-]_o$	Extracellular chloride concentration	mM	130	130
$[Cl^-]_i$	Intracellular chloride concentration	mM	8	8
D	Diffusion constant of K^+ in extracellular space (eqn. (12))	cm ^{2}/ms	2.5×10^{-9}	2.5×10^{-9}
I_{Kmax}	Maximal potassium pump current	μA/cm^2	S	S
$[K^+]_{o(eq)}$	Equilibrium extracellular potassium concentration of I_{Kpump}	mM	S	S
$[B]_{max}$	Maximal glial free buffer concentration (eqn. set (11))	mM	500	500
k_b	Backward glial unbinding rate (eqn. set (11))	ms $^{-1}$	0.0008	0.0008
$[K^+]_{o(th)}$	Threshold extracellular potassium concentration of glial buffer (eqn. set (11))	mM	S	S
θ	see eqn. set (11)	mM	S	S

An "S" on the value column denotes that the parameter is dependent on specific simulations. Please refer to the main text for specific values

Table 4: Variables for network simulations

Variable	**Description (one variable for each modeled neuron)**	**Units**
$[K^+]_o$	Extracellular potassium concentration (eqn. (13))	mM
E_K	Potassium reversal potential (eqn. (15))	mV
E_L	Leak reversal potential (eqn. (3))	mV
$[Ca^{2+}]_i$	Intracellular calcium concentration (eqn. (8))	mM
$[B]$	Glial free buffer concentration (eqn. set (11))	mM
$[KB]$	Glial bound buffer concentration ($[KB]\equiv[B]_{max}-[B]$)	mM
k_f	Forward glial binding rate (eqn. sets (10) and (11))	$ms^{-1}\cdot mM^{-1}$
s	Synaptic variable of each neuron (eqn. set (18))	-
g_e	Background excitatory conductance (eqn. set (19))	mS/cm^2
g_i	Background inhibitory conductance (eqn. set (19))	mS/cm^2
V	Membrane potential	mV
h	Sodium current inactivation variable	-
n	Potassium current activation variable	-
m∞m∞	Sodium current fast activation variable	-
t	Time (independent variable)	ms
x,y,z	Spatial coordinates (independent variables eqn. set (12))	cm

Table 5: Synaptic parameters used for network simulations

Parameter	**Description**	**Units**	**Value**
E_i	Inhibitory reversal potential	mV	-72
E_e	Excitatory reversal potential	mV	0
τ_i	Inhibitory synaptic decay time constant	ms	5
τ_e	Excitatory synaptic decay time constant	ms	3
τ_{delay}	Axonal conduction delay	ms	0.5

s_{max}	Maximal value of gating variable per spike	-	1, except for Fig. 5where it varies in time for interneurons
stermmaxsmaxterm	Terminal value of maximal gating variable per spike (Figure 5, interneurons only–equation (21))	-	70
τ_{rise}	see Eq. (21). Figure 5; interneurons only.	ms	62.5
g_{e0}	Mean background excitatory drive to model neurons	mS/ cm ^{2}	S
σ_e	Background excitatory fluctuation level	mS/ cm^2	S
g_{i0}	Mean background inhibitory drive to model neurons	mS/ cm ^{2}	S
σ_i	Background inhibitory fluctuation level	mS/ cm^2	S
gi→isyngsyni→i	Unitary conductance from an inhibitory interneuron to another interneuron	mS/ cm^2	S
ge→esyngsyne→e	Unitary conductance from a pyramidal cell to another pyramidal cell	mS/ cm^2	S
ge→isyngsyne→i	Unitary conductance from a pyramidal cell to another interneuron	mS/ cm^2	S
gi→esyngsyni→e	Unitary conductance from an interneuron to another pyramidal cell	mS/ cm^2	S
P(i→i)P(i→i)	Connection probability from one interneuron to another interneuron	-	0.25
P(e→e)P(e→e)	Connection probability from one pyramidal cell to another pyramidal cell	-	0.15
P(e→i)P(e→i)	Connection probability from one pyramidal cell to another interneuron	-	0.15
P(i→e)P(i→e)	Connection probability from one interneuron to another pyramidal cell	-	0.25
Γm→nΓm→n	Connectivity matrix element (with values of either 1 or 0) between neuron *m* and *n*	-	assigned according to the four probability values above

An "S" on the value column denotes that the parameter is dependent on a specific simulation. Please refer to the main text for specific values

Firing Properties of the Single-Neuron Models under Physiological and Abnormal Glia Uptake Conditions

We used four single pyramidal cell model simulations to show the effects of the potassium pump and glial dynamics on neuronal excitability (Fig. 2). In each of these four simulations, I_{ext}=0.36, $[K^+]_{o(eq)}$=3, $[K^+]_{o(th)}$=15, I_{Kmax}=7, g_{KCa}=5. In this section, we define the "physiological" value θ=−1.15, and the "pathological" value as θ=−0.05 (equation set (11)). Table 3 specifies the units of these parameters. Our aim is to show that the neural system tends to settle into a single steady state for a wide range of initial conditions when the increase in the forward binding rate k_f is smooth with increasing $[K^+]_o$ (as in the physiological case, effected by a larger absolute value of θ – see equation set (11)).

However, when there is a sharper increase in k_f with increasing $[K^+]_o$ (as in the pathological case, effected by a decreased absolute value of θ), some of the initial conditions can lead the neural system into a state where virtually all of the glial buffers are free (i.e. $[KB]\rightarrow 0$). In this case, the system's ability to return potassium to the extracellular space and maintain a target value of extracellular potassium level is impaired. The impairment occurs because Eq. (10) is no longer reversible when $[KB]\rightarrow 0$.

In Fig. 2a and b, we show two single pyramidal neuron model simulations with the same set of "physiological" glial and potassium pump parameters but with different $[K^+]_o$ and $[B]$ initial conditions. In Fig. 2a, we set $[K^+]_o(t=0)$=1 mM and [KB] $(t=0)$=3 mM, and in Fig. 2b, $[K^+]_o(t=0)$=10 mM and [KB] $(t=0)$=10 mM. It is clear that, despite the difference in initial conditions between the two simulations, the model neuron and the glial buffering system in both simulations converge to the same steady state. The steady state is characterized by a dynamic equilibrium established between the glial buffer's binding (k_f) and unbinding (k_b) of the extracellular potassium. This dynamic equilibrium is evidenced by the small but non-vanishing $[KB]$ in the steady state. Moreover, all the steady state variables (including firing rates, $[K^+]_o$ and $[KB]$) are identical between the two simulations, indicating that the steady states of the two simulations are the same. The glial buffering system enforces an upper ceiling for the extracellular potassium level in both simulations.

In Fig. 2c and d, we show the same two simulations with "pathological" glial parameters. In contrast to Fig. 2a and b, we see that the same set of different initial conditions results in distinct steady states for the two simulations. In Fig. 2c, the model neuron and the glial buffer approach a state where there is a lower $[K^+]_o$ and a vanishing value of $[KB]$, while in Fig. 2d, the final $[K^+]_o$ and $[KB]$ values are higher and the neuron is excitable with a higher firing rate. Thus, bi- or multi-stable dynamics at the single neuron level are more easily obtained with a smaller absolute value of θ, resulting into multiple fixed points (steady states) having either a lower $[K^+]_o$ or higher $[K^+]_o$ value.

The main factor behind this bi-stable dynamics is the parameter θ, which determines the level of bias for the forward binding rate k_f against the backward rate k_b (see equation sets (10) and (11)). With a small absolute value of θ, at a critical $[K^+]_o$ level lower than $[K^+]_{o(th)}$, k_b can be overwhelmingly larger than k_f. The large k_b value rapidly lowers $[KB]$. Although this backward unbinding action releases K^+ back to the extracellular space, the I_{Kpump} (9) counteracts this effect and therefore $[K^+]_o$ remains low. Moreover, the small k_f value (due to the bias factor θ) is not sufficient to replenish $[KB]$ via the forward binding action. As a result, the system ends up in a state where virtually all the buffer is in the free state B with a low level of $[K^+]_o$ (as in Fig. 2c).

For the neuron and glial system to exit this state, as we have shown in Fig. 2d, one can either artificially increase the values of $[K^+]_o$ or $[KB]$ (for example, by setting up the initial conditions). Increasing $[K^+]_o$ to a value closer to $[K^+]_{o(th)}$ forces k_b to be less biased against k_f (equation set (11)), so that $[KB]$ can be more efficiently replenished and thus allows a dynamic equilibrium between $[B]$ and $[KB]$. Increasing $[KB]$ artificially serves similar purposes by increasing the backward unbinding release of K^+ to the extracellular space (thus increasing $[K^+]_o$). In essence, decreasing the absolute value of θ reduces the range of $[K^+]_o$ over which the glial buffering system can efficiently regulate, and thus makes the single neuron more susceptible to multi-stable behaviour. We show in the next section that this apparent multistability also carries over to the network simulations with abnormal glial parameters. As mentioned above, we will associate the state in which the $[K^+]_o$ is low with the normal physiological state, and the elevated $[K^+]_o$ level state with the "seizure" state. We also note that, although the above single-neuron analysis was based on pyramidal neurons, the same low and elevated $[K^+]_o$ states occur in the case of FS inhibitory neurons. The effect of elevated $[K^+]_o$ on pyramidal and FS interneuron spiking will, however, differ in the network simulations presented below.

Sustained Gamma Epileptiform Activity Results From an Abnormal but Balanced Glial Dynamics and Inhibitory Synaptic Strength

In this section we show how most features of gamma seizures observed in the human focal epilepsy data (Truccolo et al. 2011, 2014) can be replicated by our network model. These gamma seizures in the human data are characterized by sustained narrow band gamma LFP oscillations (~30−60 Hz). However, few fine temporal synchrony transients exist in the spiking activities of the measured neurons. Neuronal spiking tends to be asynchronous and heterogeneous (Truccolo et al. 2011, 2014). Individual neurons also fire at a lower rate than the gamma frequency of observed LFP oscillations.

Similar neural dynamics is observed in our model simulations (Fig. 3). Critical ingredients for the appearance of sustained gamma band LFP oscillations with asynchronous neuronal firing at a lower rate include high inhibitory conductance values (for both $g_{syn}^{i\to i}$ and $g_{syn}^{i\to e}$) and an axonal conduction time delay (Brunel 2000; Brunel and Wang 2003). We also require a value of $[K^+]_{o(th)}$ that is not too high (see equation set (11)). A moderate $[K^+]_{o(th)}$ value ensures that the model FS inhibitory interneurons have a high excitability in the "epileptic" state, while preventing the interneurons from being overly excited to enter depolarization block. We used $g_{syn}^{e\to e} = 0.0007$, $g_{syn}^{e\to i} = 0.0007$, $g_{syn}^{i\to i} = 0.025$, $g_{syn}^{i\to e} = 0.025$, $g_{e0} = 0.01026$, $\sigma_e = 0.0025$, $g_{i0} = 0.084$, $\sigma_i = 0.02$ for this gamma seizure simulation (Table 5 introduces and defines these synaptic parameters). As for intrinsic and glial properties of model neurons, we chose $\theta=-0.15$, $[K^+]_{o(eq)}=3.6$, $[K^+]_{o(th)}=7.5$, $I_{Kmax}=1.45$, $g_{KCa}=5$ for pyramidal cells and $\theta=-0.15$, $[K^+]_{o(eq)}=3$, $[K^+]_{o(th)}=7.5$, $I_{Kmax} = 1.9$, $g_K = 6.8$ for interneurons. The value of g_K for interneurons was chosen so that they only enter depolarization block at a relatively high level of $[K^+]_o$ ($\gtrsim 8 \gtrsim 8$ mM).

Figure 3a shows the first 90 seconds of the temporal progression of several biophysical variables in the model and the time-frequency spectrum of the average membrane potentials across all the pyramidal neurons in the network (Fig. 3a, lowest panel). We use this average value as a proxy for the LFP activity.

The model cortical network during the first 40 seconds of the simulation is in the "low activity" state (in which $[K^+]_o$ converges to ~3mM). This "low-activity" state at the network level is similar to the single neuron level in Fig. 2c where the replenishment of $[KB]$ is insufficient for the glial buffer to maintain a targeted level of $[K^+]_o$. At time t=40s, a DC stimulation (2.5 μA/cm ^{2} added to I_{ext}) was applied to every neuron in the model network for a short period of 2.5 seconds. This stimulation kick-started the extracellular potassium accumulation process and forced the cortical network to enter into a "high-activity" or "epileptic" state in which the glial buffer attempts to lock into a higher $[K^+]_o$ level (Fig. 3a, 4 th panel), as determined by the parameter $[K^+]_{o(th)}$. Clear gamma band (~40 Hz) oscillations are readily observed in the LFP power spectrum after the DC stimulation (Fig. 3a, lowest panel). Importantly, the gamma activity is sustained, even after DC stimulation is terminated, persisting until the end of the simulation.

Furthermore, at the individual neuron level, both the fast-spiking interneurons and pyramidal cells spike at lower rates of ~ 3 and 1 spike per second, respectively, than the frequency of the gamma LFP activity (see Fig. 3a, topmost panel). The second panels from the top of Fig. 3a and b show raster plots of 2 of the 81 minicolumns in the simulation. The neuron indices in the raster plots are grouped by neuron types (pyramidal or interneuron).

It is also clear that while the LFP proxy (average membrane potential of pyramidal cells) shows gamma oscillations (Fig. 3a and b, 3 rd panels from top), the firing pattern of each individual neuron is not indicative of the global oscillation as observed in the LFP proxy. The density plots of the interspike intervals (ISIs) after the DC stimulation across each of the pyramidal and interneuronal populations (Fig. 3c, main plots) reveal a large variance of ISI values. Spiking in FS interneurons showed high irregularity, with the coefficient of variation (CV) of ISIs concentrating around 1 and higher values, while the CV for pyramidal neurons tended to be lower and more broadly distributed (Fig. 3c, inset plots). Irregularity of pyramidal cell spiking can be increased, nevertheless, by moderately increasing the unitary inhibition from the interneurons to the pyramidal cells. However, we observed a competition between the glial parameters (i.e. θ and $[K^+]_{o(th)}$) and the unitary inhibitory conductance values in terms of sustaining high enough $[K^+]_o$ values to support gamma oscillations. Increasing the unitary inhibitory conductance values affects the sustainability of gamma oscillations. Too strong $g_{syn}^{i \to e}$ or $g_{syn}^{i \to i}$ values make the gamma oscillations less sustainable, with the system immediately settling back into the original "low-activity" state. A rough mean-field estimate of the synaptic currents entering the pyramidal cells after the DC stimulation suggests that the magnitude of the inhibitory current is a few times higher than that of the excitatory current. This average inhibitory synaptic current entering a pyramidal cell can, for example, be estimated by the formula:

$$\left\langle I_{syn}^{i \to e} \right\rangle = P(i \to e) \times g_{syn}^{i \to e} \times N_i \times \langle (V - E_i) \rangle \times \tau_i \times \langle \nu_i \rangle, \tag{20}$$

where v_i is the spike rate of the inhibitory population and N_i is the number of inhibitory interneurons in the network.

Transition Into Spike-And-Wave Complex (SWC) Seizures: Role Of Synaptic Inhibition And Depolarization Block

In this section we show how variations in synaptic inhibition and depolarization block in the examined neuronal network model can reproduce the dynamics of the observed SWC seizures. In this type of seizures, neural activity transitions into high amplitude rhythmic 2–3 Hz LFP discharges. Each SWC event in the LFP lasts for about 300 ms and is characterized by an initial fast "spike" followed by a slow wave potential. Neuronal spiking tends to occur during the LFP "spike" phase and is highly suppressed during the LFP wave phase (e.g. Truccolo et al. 2014). Preliminary examination of neuronal spiking characteristics has indicated that although both putative principal and FS inhibitory interneurons tend to increase their firing rates in the initial stages of spike-wave seizures (Ahmed et al. 2014), putative FS interneurons tend to shut down later, just before the emergence of full spike-wave discharges. Ahmed et al. (2014) observed that this termination of activity is preceded by a progressive decreasing of action potential amplitudes in

FS interneurons – an indication that FS interneurons enter depolarization block, resulting eventually in the cessation of their activity.

Figure 4 shows simulation results for the case where the overall inhibitory synaptic strength is kept at a low level during the entire simulation, leading to SWC seizures. Because the aim is to have FS interneurons to eventually enter depolarization block, we set a higher $[K^+]_{o(th)}$ and a lower g_K value for the interneurons than in the gamma seizure simulation (Fig. 3). Higher $[K^+]_{o(th)}$ values increase FS interneuron excitability, while lower g_K values lead FS interneurons to enter depolarization block at a lower $[K^+]_o$ level. Parameters for model neurons and glial properties of model neurons corresponded to: θ=−0.15, $[K^+]_{o(eq)}$=3.6, $[K^+]_{o(th)}$=8.5, I_{Kmax}=1.8, g_{KCa}=12 for pyramidal cells and θ=−0.15, $[K^+]_{o(eq)}$=3, $[K^+]_{o(th)}$=7.75, I_{Kmax}=1.9, g_K=3.5 for interneurons. Unitary inhibitory conductance values $g_{syn}^{i\rightarrow e}$ or $g_{syn}^{i\rightarrow i}$ were set here to only one-tenth of that in the gamma seizure simulation (Fig. 3). Synaptic parameters corresponded to: $g_{syn}^{e\rightarrow e} = 0.0007$, $g_{syn}^{e\rightarrow i} = 0.0007$, $g_{syn}^{i\rightarrow i} = 0.0025$, $g_{syn}^{i\rightarrow e} = 0.0025$, , g_{e0}=0.01016, $_e$=0.0025, g_{i0}=0.084, $_i$=0.02.

As before, the simulation starts at a "low-activity" state (t=0−40 seconds) and a DC stimulation (1.6 μA/cm ^{2}) is delivered between t=40 and 42.5 seconds, "kicking" the system into a high-activity "epileptic" state. Since the overall inhibitory synaptic strength is low, but $[K^+]_{o(th)}$ is higher than that in the gamma seizure case, FS inhibitory interneurons fire at a higher frequency after the DC stimulation than in Fig. 3, creating a positive feedback which further increases $[K^+]_o$. When $[K^+]_o$ reaches a critical value, inhibitory interneurons enter depolarization block, which eventually leads to cessation of inhibition (at around t=46 seconds in Fig. 4). We note that, in contrast to the gamma seizures, the attainment of the critical value for $[K^+]_o$ is possible here because $[K^+]_{o(th)}$ is set to a higher level, such that the glial potassium ceiling is higher.

After cessation of synaptic inhibition, the network dynamics is dominated by the interaction amongst pyramidal cells and evolve into rhythmic SWC discharges. Throughout, abnormal glial buffering maintains the high $[K^+]_o$ required for keeping the interneurons in the depolarization block regime, thus creating a long-lasting "seizure" state. In the model simulations, I_{KCa} adaptation currents in the pyramidal cells and sufficiently strong excitatory-excitatory $(g_{syn}^{e\rightarrow e})$ couplings (Van

Vreeswijk and Hansel 2001; Dur-e-Ahmad et al. 2012; Nicola and Campbell 2013b; Ferguson et al. 2015) are critical for both the "spike" and "wave" firing suppression phases of the the SWC discharges.

Transition into Seizures Consisting Of Initial Low-Voltage Fast LFP Oscillations Followed By Spike-And-Wave Discharges

Next, we demonstrate the generation of the second type of SWC seizures. In this case, the emergence of rhythmic SWC discharges is preceded by low voltage higher frequency (~10−20 Hz) LFP oscillations. Neuronal spiking activity during these fast oscillations tends to be asynchronous as in the previously examined gamma seizures. In other words, while there is a clear fast frequency oscillation at the LFP level, individual neurons spike at a lower rate without obvious synchrony with the global (i.e. LFP) oscillation patterns. The similarity of the firing dynamics of individual neurons between these transient oscillations and gamma seizures leads us to speculate that both phenomena might share a common synaptic and glial mechanism. As shown in Fig. 5, the seizure begins with low voltage fast (~20 Hz) LFP oscillations which eventually evolve into low-frequency rhythmic SWC discharges. Similarly to the gamma seizures (Fig. 3), the initial fast oscillations in the model require a high overall inhibitory synaptic strength, while the evolution and maintenance of spike-wave discharges require the opposite.

To reproduce this type of seizure dynamics, the following sequence of events was implemented in the model: a substantial increase in inhibitory synaptic strength at seizure transition (leading to low voltage fast oscillations) is followed by a rapid decrease, which is hypothesized here to occur via GABA depletion (Zhang et al. 2012). Once synaptic inhibition is below normal levels, the same path as in the first type of SWC seizures follows: inhibitory interneurons enter depolarization block and the network evolves into rhythmic SWC discharges supported mostly by the population of pyramidal neurons. Specifically, the synaptic parameters $g_{syn}^{i\rightarrow i}$ and $g_{syn}^{i\rightarrow e}$ were set to constant values during the entire simulation.

To increase the effective inhibitory synaptic strength during seizure transition (i.e. during the DC stimulation phase, t=40−42.5s), we let the maximal value of the gating variable per spike, s_{max}, for interneurons (equation set (18)), vary temporally instead of being a constant as in Figs. 3 and 4. The value of s_{max} for interneurons was set to 1 at the beginning of the simulation (Table 5) and kept at this value during the "low activity" phase (t=0–40s). After that, s_{max} for interneurons asymptotically approached a terminal value stermmaxsmaxterm during the DC stimulation phase according to the following formula:

$$s_{max}(t+\delta t) = s_{max}(t) \times \left(1 + \left(\frac{s_{max}^{term}}{s_{max}(t)} - 1\right) \times \frac{\delta t}{\tau_{rise}}\right), \tag{21}$$

where τ_{rise}=62.5ms is the time constant of the increase in s_{max} and $s_{max}^{term} = 70$ is the asymptotic terminal value of s_{max} (i.e. 70 times the value of original s_{max} before the DC stimulation phase).

The increased s_{max} value for interneurons was then held constant during the fast oscillations phase (from 42.5 to 49.5 seconds). After that it was decreased to simulate the loss of inhibition. We set $s_{max} \rightarrow 0.1 \times s_{max}$ for individual interneuron after each spike of the same interneuron during this "depletion" phase. (The start of this depletion phase is marked by a red triangle in Fig. 5.) Eventually the inhibition approaches zero, the "depleted" state.

In summary, Fig. 5 shows the emergence of clear low voltage fast (~ 20 Hz) LFP oscillations right after the onset of the DC stimulation. These fast oscillations disappear shortly after the start of the inhibition "depletion" phase (marked by a red triangle in Fig. 5). The strong build up of inhibitory synaptic strength (up to 70 times the initial strength) during the DC stimulation is necessary because it slows the rise of $[K^+]_o$ during DC stimulation and keeps the interneurons from entering depolarization block. Moreover, this very high level of synaptic inhibition keeps the population frequency in the ~10–20 Hz range. A lower, but still large inhibitory strength would have led to the emergence of gamma oscillations, as in Fig. 3.

The initial decrease in $[K^+]_o$ during the transient fast oscillations (Fig. 5a, fourth plot from top) is due to the strong inhibition which limits the firing of inhibitory interneurons. As inhibition begins to wear off at t=49.5 seconds, the trajectory of $[K^+]_o$ reverses its course because both pyramidal cells and interneurons fire more as a result of decreased inhibitory synaptic strength. Finally, when $[K^+]_o$ reaches a sufficiently high level (~7 mM for this simulation), the interneurons enter depolarization block and inhibition ceases, as clearly shown by the complete termination of interneuron firing (Fig. 5b, raster plot at around t=63 seconds). After cessation of inhibition, SWC discharges emerge supported only by the activity of interacting pyramidal neurons. With the exception of the time varying nature of s_{max} for interneurons and a DC stimulation here set to 2.5 instead of 1.6 μA/cm^2, the numerical values of the synaptic (including $g_{syn}^{i\rightarrow i}$ and $g_{syn}^{i\rightarrow e}$), intrinsic and glial parameters were the same as those for the first type of SWC seizures (Fig. 4).

DISCUSSION

Our study is motivated by recent microelectrode array recordings of ensembles of single neurons and high-density LFP activity in human neocortex during propagated focal seizures (Truccolo et al. 2011; Schevon et al. 2012; Ahmed et al. 2014; Truccolo et al. 2014; Wagner et al. 2015). These recordings suggest different ways via which seizures starting in a focal site can recruit more distal neocortical areas. We used a biophysical cortical network model consisting of conductance-based neurons, coupled with glial buffer for $[K^+]_o$, to examine potential mechanisms underlying transitions into ictal states in recruited neocortical areas during seizure spread. Through the analyses and simulations of single neuron and network models, we have demonstrated that neocortical networks can be made susceptible to seizures as a result of abnormal glial potassium buffering and changes in synaptic inhibition.

First, at the single neuron level, the imbalance in the conversion between potassium buffer in the bound and unbound forms leads to bi-stability such that one state has a lower $[K^+]_o$ level (thus lower excitability) than the other. This bi-stability also carries over to the network level. Perturbation either by noise or by DC stimulation can "kick-start" the high $[K^+]_o$ hyper-excitable state from the low $[K^+]_o$ state. Second, we then showed that the three major types of seizure transitions as observed in the human microelectrode array recordings can be reproduced when these pathological glial potassium dynamics interact with various synaptic and intrinsic parameter settings for inhibitory interneurons.

In our simulations, transitions into gamma (~30−60 Hz) ictal activity were obtained with a high inhibitory conductance between cortical neurons. Such high inhibitory synaptic strength was required to prevent the interneurons from entering depolarization block during the seizure state. Moreover, since high inhibitory strength tends to decrease $[K^+]_o$, the long-lasting gamma seizure state was the result of a subtle balance between high inhibitory conductance values and the accumulation of $[K^+]_o$ due to abnormal glial potassium buffering dynamics.

For transitions into SWC ictal activity, the cortical network can either experience a generally low value of inhibitory conductance between cortical neurons, or an initial high inhibitory synaptic strength followed by inhibition breakdown. In the first scenario, the transition into SWC was direct and involved interneurons entering depolarization block because of the low inhibitory synaptic strength. As soon as the interneurons entered depolarization block, inhibition ceased. At that point, neuronal spiking activity supported only by interacting pyramidal cells led to recurring SWC discharges. In the second SWC scenario, there was a buildup of inhibitory synaptic strength prior to the transition into spike-wave discharges. The result of this inhibition buildup is the occurrence of transient low-amplitude fast oscillations around 10–20 Hz. This buildup of inhibitory synaptic strength was followed by a wearing off of inhibition (e.g. through GABA depletion), thus providing a low inhibition environment for interneurons to enter depolarization block and the population of pyramidal cells to eventually sustain recurring SWC discharges.

Abnormal Glial Buffering and Its Relationship to Seizure

In this work we have established a computational model of seizure through the occurrence of bi-stability at the single neuron level. This bi-stability is the result of a hypothesized pathological glial potassium dynamics in which there is an imbalance between potassium buffers in their bound and unbound forms. Bi- (or multi)stability, either at the single neuron or at the network level, has been implicated in numerous physiological and pathological conditions in the brain (Van Ooyen et al. 1992; Sasaki et al. 2007; Freyer et al. 2009; Fröhlich et al. 2010; Anderson et al. 2012; Ho et al. 2012; He 2014; Ho et al. 2014; Hübel et al. 2014). In the particular case of epileptic seizures, Fröhlich et al. (2010) have demonstrated in computational models that it is possible for network bistability to exist even when the glial potassium dynamics are within a physiological range (through a higher absolute value of θ, the parameter determining the bias favouring forward versus backward binding rate). In this scenario, Fröhlich et al. (2010) associate higher $[K^+]_o$ states with epileptic seizures. In the models used here, we did not observe bi-stability either at the single neuron or network level when θ was set to physiological values. We only observed bi-stability with pathological (i.e. smaller in absolute value) values of θ. We have singled out (via single neuron simulations) the differential in transition rates between $[B]$ and $[KB]$ as the important factor underlying such bi-stability. The fact that Fröhlich et al. (2010) used two-compartment neuron models, while we used single-compartment, might explain this difference between the two studies. Moreover, different synaptic connection parameters were used by Fröhlich et al. (2010). Different synaptic connectivity might affect the length and strength of perturbation required to elicit the bi-stable behaviour, if at all possible. A more thorough examination of how glial potassium buffering may affect neuronal dynamics should involve detailed bifurcation analyses of the combined neuron-glial system (Touboul 2008; Nicola and Campbell 2013a; Kim and Nykamp 2014; Nicola and Campbell 2014). Such analyses could determine the intrinsic and glial parameters that allow multistability. From the existence and stability of these states, one would expect to gain some insight into how resilient the network is to seizure initiation and propagation. Although we have focused on their role on seizure propagation, potassium and glial buffering may also play an important role in seizure termination (Kramer et al. 2012; González-Ramírez et al. 2015).

Changes in Inhibitory and Excitatory Conductances

Our focus on changes in inhibition was motivated by several previous (*in vivo/vitro*) animal models (Trevelyan et al. 2006; Zhang et al. 2012; Grasse et al. 2013; žiburkus et al. 2013; Uva et al. 2015) and *in vitro* studies of human epileptic cortical tissue (e.g. D'Antuono et al. 2004), where inhibitory interneuron activity appears to change significantly during preictal and initial periods of the seizure, before major changes in principal cells are detected. Recent work by žiburkus et al. (2013) has also examined in detail how changes in both inhibitory and excitatory conductances may lead to imbalances in excitation/inhibition preceding and during seizure like events in hippocampal slices under the potassium channel blocker 4-aminopyridine and reduced extracellular magnesium.

In that respect, we emphasize that the changes in inhibition as implemented in our simulations did not only affect inhibitory conductances. Changes in inhibition in our model were implemented in part by systematically varying the levels of unitary synaptic inhibition, which in turn affected several other network properties such as the number of neurons generating spikes at any given time, and thus firing rates in both populations of pyramidal neurons and FS interneurons. In this way, by varying unitary synaptic inhibition, the E-I balance in the simulated networks also varied dynamically, with both excitatory and inhibitory conductances being altered. In addition, after transitions into spike-wave discharges in the simulated SWC seizures (Figs. 4 and 5), these discharges were supported only by the synaptic interactions among pyramidal neurons.

Furthermore, we note that there are a couple of important differences between the animal model in žiburkus et al. (2006, 2013) and our model simulations. First, apart from temporary depolarization block periods, the activity of inhibitory neurons remained throughout the seizures (i.e. until seizure termination) in žiburkus et al.'s model. In our computational model, it is only during gamma seizures that the spiking activity of FS inhibitory interneurons is preserved throughout the seizure and depolarization block is absent. By contrast, FS interneurons enter depolarization block and remain in that state during the SWC seizures. Second, while in žiburkus et al.'s study transient depolarization block events in pyramidal neurons are also present and play an important role, in our computational model pyramidal neurons tend to be robust to depolarization block both in SWC and gamma seizures. This robustness resulted primarily from the slow calcium-mediated potassium afterhyperpolarization (AHP) currents (I_{KCa}, Eq. (7)) in the model pyramidal neurons, which prevented them from evolving into regimes of high enough firing rates required for depolarization block. These features of the biophysical model examined here seem consistent with the dynamics of propagated seizures (in neocortical patches distal from putative onset areas) observed in our human data. First, both inhibitory and excitatory spiking activity appears to be preserved throughout the gamma seizures (Truccolo et al. 2011). Second, in SWC seizures, spiking in putative FS interneurons appears to terminate before the transition into large amplitude spike wave discharges. This cessation of FS interneuron spiking activity is accompanied by signatures of depolarization block, while no similar signatures were found in putative principal cells (Ahmed et al. 2014).

Network Inhibition and its Relationship to Seizure Propagation

Using biophysical neuronal network models of neocortical patches, we have demonstrated how FS inhibitory interneurons may play an important role in three types of observed human seizures. In particular, we have quantified the role of FS interneurons by varying the inhibitory synaptic strength and controlling how easily interneurons enter depolarization block. For the two types of SWC seizures modeled here (Figs. 4 and 5), ultimately it is the failure of inhibition (when the FS inhibitory interneurons enter depolarization block) that leads the network to transition into SWC activity. We examined two main scenarios that led to the failure of inhibition, resulting in SWC seizures. In the first scenario, FS inhibitory interneurons enter depolarization block when there is an overall low inhibitory synaptic strength. When interneurons are excited during seizure initiation, the lower inhibitory synaptic strength allows interneurons to fire at a higher frequency, thus hastening the $[K^+]_o$ accumulation process and leading eventually to depolarization block (Fig. 4). In the second scenario, an initially large inhibitory synaptic strength is followed by GABA depletion (Zhang et al. 2012). Initially, interneurons are mutually inhibited by the large inhibitory synaptic strength and their firing frequencies are too low to trigger $[K^+]_o$ accumulation. If such level of inhibition was not perturbed, seizure propagation would fail. The decreased effective inhibitory strength resulting from subsequent GABA depletion allows for higher firing rates in interneurons (and principal cells) to develop, again leading eventually to depolarization block in inhibitory interneurons (Fig. 5).

Some previous studies have emphasized the role of feedforward inhibition in seizure propagation in neocortex (Trevelyan et al. 2006; Schevon et al. 2012; for a review see Paz and Huguenard 2015). The main idea is that the propagation of initially localized ictal activity to more distal cortical sites involves predominantly feedforward inhibition. This feedforward inhibitory drive could be a mechanism for the initial strengthening of inhibition during seizure propagation assumed in our model in two of the three seizure propagation scenarios (gamma and SWC with preceding fast oscillations). Furthermore, Trevelyan and colleagues (Trevelyan et al. 2006; Schevon et al. 2012) have emphasized the role of an "inhibitory restraint" or "veto" mechanism during neocortical propagation of focal seizures. Inhibitory veto of seizure propagation has been shown in low Mg^{2+} mouse models of epilepsy (Trevelyan et al. 2006) and has been argued to be present also in human focal seizures (Schevon et al. 2012). This inhibitory veto mechanism would lead to "penumbra areas" ahead of the ictal wavefront. The failure of such inhibitory veto would allow the successful propagation of the ictal wavefront into new recruited ictal areas. Otherwise, the inhibition would be strong enough to contain the seizure spread. In this way, one could also argue that the gamma seizures examined here would represent a failure in seizure propagation. This possibility remains an open and important experimental question. Given that these sustained gamma oscillations reflect abnormal dynamics during secondarily generalized focal seizures, we currently consider this activity as propagated seizures. The gamma seizures in our model (Fig. 3) represent a situation in which there is a balance between enhanced inhibition and $[K^+]_o$ accumulation. In other words, the seizure propagates while inhibitory interneuronal activity is preserved. In a related experimental study, our group (Lu et al. 2015) has shown that constant, strong enough depolarization driven by optogenetic stimulation can induce sustained (~50 Hz) gamma oscillations in healthy primate motor cortex.

In view of the strong inhibition in simulations of gamma seizures (Fig. 3), we do not preclude the possibility that some of the "high-activity" states in the simulated gamma seizures may in fact be long-lasting transients as opposed to asymptotic stable states. In other words, the elevated $[K^+]_o$ in the "high-activity" states may nevertheless settle back to a much lower value after a long enough time. However, this issue has little effect on our main claims in this paper, given the long time scale (at least in the order of minutes) over which $[K^+]_o$ remains elevated and increasing.

Model Limitations and Future Work

Our analyses and simulations show how different types of seizure transitions observed in intracortical recordings of propagated human focal seizures can be accounted for by variations in synaptic inhibition and extracellular potassium concentration. Several limitations of the model, however, warrant future improvement. First, although the interaction of $[K^{+}]_{o}$ and inhibitory synaptic dynamics can account for the phenomena observed in our human epilepsy data (Ahmed et al. 2014), several other ionic or synaptic mechanisms are possible and have been studied in different animal models of epilepsy (e.g. Alfonsa et al. 2015).

Second, our model exploration was restricted to the transition into ictal states. In particular, we did not examine how seizures terminate, which is also a complex topic and may involve multiple ionic and synaptic components (such as glutamate depletion–Lado and Moshé 2008; GABA upregulation–Wen et al. 2015; fluctuations in ionic concentrations–Krishnan and Bazhenov 2011; Kramer et al. 2012). A more detailed model, including more complex ionic dynamics, will be required to address this issue. Furthermore, gap junction effects were not included in the model. Although electrical synapses have been suggested as an underlying mechanism for interneuronal bursting (Skinner et al. 1999) and synchronous population gamma oscillations (Traub et al. 2001), neither of these two features appeared to be prominent in our human data where neuronal spiking during gamma seizures is largely asynchronous. Thus, it remains an open question whether gap junction effects are critical to account for transitions to and maintenance of gamma seizures. Nevertheless, gap junctions may still play an important role during SWC discharges as previously hypothesized by Traub and colleagues (e.g. Traub et al. 2005). We hope to address these open issues in the future.

Third, in the hypothesized contributing factors to seizure transitions examined here, we have remained agnostic about what ultimately leads to transient dysfunctions in inhibitory activity and in potassium glial-buffering processes. We note that several other factors involving changes in oxygen, Na+/K+ ATP pumps and cell volume (e.g. Ingram et al. 2014; Wei et al. 2014a), for example, can also contribute to elevations in extracellular potassium. We also emphasize that the goal in this study is to replicate the dynamics observed in propagated seizures, in other words, ictal dynamics in neocortical patches distal to the putative seizure onset zones. As described in Truccolo et al. (2011, 2014) and Wagner et al. (2015), our recorded neocortical patches were close, but distal to the putative seizure onset zones. The dynamics involving seizure initiation in seizure onset/focus areas might involve different mechanisms.

As stated above, we have focused on fast time-scale changes in synaptic inhibition which have been reported in several previous animal studies and *in vitro* studies of human epileptic cortical tissue. Although these studies have provided some initial evidence for transient changes in inhibitory interneuron activity preceding seizures, it remains a major open question how these dysfunctions in interneuron network activity (and potassium glial buffering) would arise in the longer time scale of epileptogenesis.

NOTES

Acknowledgments

The authors would like to thank the editors, reviewer, Drs. F. Skinner, Liang Zhang, Felipe Gerhard and Timothée Proix for constructive comments on the manuscript. We also thank Drs. Sydney Cash, Omar Ahmed and Mark Kramer for valuable discussions. This research was conducted using computational resources and services at the Center for Computation and Visualization, Brown University.

REFERENCES

1. Ahmed, O. J., Kramer, M. A., Truccolo, W., Naftulin, J. S., Donoghue, J. A., Eskandar, E. N., Cosgrove, G. R., Blum, A. S., Potter, N. S., Hochberg, L. R., & Cash, S. S. (2014). Inhibitory single neuron control of seizures and epileptic traveling waves in humans. *BMC Neuroscience*, *15*(Suppl 1), F3.
2. Alfonsa, H., Merricks, E. M., Codadu, N. K., Cunningham, M. O., Deisseroth, K., Racca, C., & Trevelyan, A. J. (2015). The contribution of raised intraneuronal chloride to epileptic network activity. *The Journal of Neuroscience*, *35*(20), 7715–7726.
3. Anderson, W. S., Kudela, P., Cho, J., Bergey, G. K., & Franaszczuk, P. J. (2007). Studies of stimulus parameters for seizure disruption using neural network simulations. *Biological Cybernetics*, *97*, 173–194.
4. Anderson, W., Azhar, F., Kudela, P., Bergey, G., & Franaszczuk, P. (2012). Epileptic seizures from abnormal networks: why some

seizures defy predictability. *Epilepsy Research*, *99*, 202–213.

5. Bazhenov, M., Timofeev, I., Steriade, M., & Sejnowski, T. J. (2004). Potassium model for slow (2-3 Hz) *in vivo* neocortical paroxysmal oscillations. *Journal of Neurophysiology*, *92*, 1116–1132.
6. Brunel, N. (2000). Dynamics of sparsely connected networks of excitatory and inhibitory spiking neurons. *Journal of Computational Neuroscience*, *8*, 183–208.
7. Brunel, N., & Wang, X. J. (2003). What determines the frequency of fast network oscillations with irregular neural discharges? I. Synaptic dynamics and excitation-inhibition balance. *Journal of Neurophysiology*, *90*, 415–430.
8. Buxhoeveden, D. P., & Casanova, M. F. (2002). The minicolumn hypothesis in neuroscience. *Brain: A Journal of Neurology*, *125*, 935–951.
9. Cammarota, M., Losi, G., Chiavegato, A., Zonta, M., & Carmignoto, G. (2013). Fast spiking interneuron control of seizure propagation in a cortical slice model of focal epilepsy. *The Journal of Physiology*, *591*(4), 807–822.
10. Cressman, J.R. Jr, Ullah, G., Ziburkus, J., Schiff, S., & Barreto, E. (2009). The influence of sodium and potassium dynamics on excitability, seizures, and the stability of persistent states: I. Single neuron dyamics. *Journal of Computational Neuroscience*, *26*, 159–170.
11. D'Antuono, M., Louvel, J., Köhling, R., Mattia, D., Bernasconi, A., Olivier, A., Turak, B., Devaux, A., Pumain, R., & Avoli, M. (2004). Gabaa receptor-dependent synchronization leads to ictogenesis in the human dysplastic cortex. *Brain: A Journal of Neurology*, *127*, 1626–1640.
12. Dur-e-Ahmad, M., Nicola, W., Campbell, S., & Skinner, F. (2012). Network bursting using experimentally constrained single compartment CA3 hippocampal neuron models with adaptation. *Journal of Computational Neuroscience*, *33*(1), 21–40.
13. Ermentrout, B. (2002). Simulating, analyzing, and animating dynamical systems: A guide to XPPAUT for researchers and students (SIAM).
14. Ferguson, K., Njap, F., Nicola, W., Skinner, F., & Campbell, S. (2015).

Examining the limits of cellular adaptation bursting mechanisms in biologically-based excitatory networks of the hippocampus. *Journal of Computational Neuroscience, 39*, 289–309.

15. Freyer, F., Aquino, K., Robinson, P. A., Ritter, P., & Breakspear, M. (2009). Bistability and non-Gaussian fluctuations in spontaneous cortical activity. *The Journal of Neuroscience, 29*(26), 8512–8524.
16. Fröhlich, F., Bazhenov, M., Iragui-Madoz, V., & Sejnowski, T. J. (2008). Potassium dynamics in the epileptic cortex: New insights on an old topic. *The Neuroscientist, 14*, 422.
17. Fröhlich, F., Sejnowski, T. J., & Bazhenov, M. (2010). Network bistability mediates spontaneous transitions between normal and pathological brain states. *The Journal of Neuroscience, 30*(32), 10734–10743.
18. González-Ramírez, L.R., Ahmed, O. J., Cash, S. S., Wayne, C. E., & Kramer, M. A. (2015). A biologically constrained, mathematical model of cortical wave propagation preceding seizure termination. *PLoS Computational Biology, 11*(2), e1004065.
19. Grasse, D., Karunakaran, S., & Moxon, K. (2013). Neuronal synchrony and the transition to spontaneous seizures. *Experimental Neurology, 248*, 72–84.
20. He, B. J. (2014). Scale-free brain activity: past, present, and future. *Trends in Cognitive Sciences, 18*, 480–487.
21. Ho, E. C. Y., Strüber, M., Bartos, M., Zhang, L., & Skinner, F. K. (2012). Inhibitory networks of fast-spiking interneurons generate slow population activities due to excitatory fluctuations and network multistability. *The Journal of Neuroscience, 32*(29), 9931–9946.
22. Ho, E. C. Y., Eubanks, J. H., Zhang, L., & Skinner, F. K. (2014). Network models predict that reduced excitatory fluctuations can give rise to hippocampal network hyper-excitability in meCP2-null mice. *PloS One, 9*, e91148.
23. Hübel, N., Scholl, E., & Dahlem, M. A. (2014). Bistable dynamics underlying excitability of ion homeostasis in neuron models. *PLoS Computational Biology, 10*, e1003551.
24. Ingram, J., Zhang, C., Cressman, J., Hazra, A., Wei, Y., Koo, Y., žIburkus, J., Kopelman, R., Xu, J., & Schiff, S. (2014). Oxygen and seizure dynamics: I. experiments. *Journal of Neurophysiology, 112*,

205–212.

25. Jirsa, V. K., Stacey, W. C., Quilichini, P. P., Ivanov, A. I., & Bernard, C. (2014). On the nature of seizure dynamics. *Brain: A Journal of Neurology*, *133*, 2210–2230.
26. Kager, H., Wadman, W. J., & Somjen, G. G. (2000). Simulated seizures and spreading depression in a neuron model incorporating interstitial space and ion concentrations. *Journal of Neurophysiology*, *84*, 495–512.
27. Kim, C., & Nykamp, D. (2014). Dynamics of a network of excitatory and inhibitory neurons induced by depolarization block. *BMC Neuroscience*, *15*(Suppl 1), P76.Pub
28. Kramer, M. A., Truccolo, W., Eden, U. T., Lepage, K. Q., Hochberg, L. R., Eskandar, E. N., Madsen, J. R., Lee, J. W., Maheshwari, A., Halgren, E., Chu, C. J., & Cash, S. S. (2012). Human seizures self-terminate across spatial scales via a critical transition. *Proc Natl Acad Sci U S A*, *109*, 21116–21121.
29. Krishnan, G. P., & Bazhenov, M. (2011). Ionic dynamics mediate spontaneous termination of seizures and postictal depression state. *The Journal of Neuroscience*, *31*(24), 8870–8882.
30. Krishnan, G., Filatov, G., & Bazhenov, M. (2013). Dynamics of high-frequency synchronization during seizures. *Journal of Neurophysiology*, *109*, 2423–2437.
31. Lado, F. A., & Moshé, S.L. (2008). How do seizures stop? *Epilepsia*, *49*, 1651–1664.
32. Lu, Y., Truccolo, W., Wagner, F., Vargas-Irwin, C., Ozden, I., Zimmermann, J., May, T., Agha, N., Wang, J., & Nurmikko, A. (2015). Optogenetically induced spatiotemporal gamma oscillations and neuronal spiking activity in primate motor cortex. *Journal of Neurophysiology*, *113*, 3574–3587.
33. Naze, S., Bernard, C., & Jirsa, V. (2015). Computational modeling of seizure dynamics using coupled neuronal networks: Factors shaping epileptiform activity. *PLoS Computational Biology*, *11*, e1004209.
34. Nicola, W., & Campbell, S. (2013a). Bifurcations of large networks of two-dimensional integrate and fire neurons. *Journal of Computational Neuroscience*, *35*, 87–108.

35. Nicola, W., & Campbell, S. A. (2013b). Mean-field models for heterogeneous networks of two-dimensional integrate and fire neurons. *Frontiers in Computational Neuroscience*, *7*(184), 1–25.
36. Nicola, W., & Campbell, S. (2014). Non-smooth bifurcations of mean field systems of two-dimensional integrate and fire neurons. arXiv:1408.4767.
37. Park, E. H., & Durand, D. M. (2006). Role of potassium lateral diffusion in non-synaptic epilepsy: a computational study. *Journal of Theoretical Biology*, *238*, 666–682.
38. Paz, J., & Huguenard, J. R. (2015). Microcircuits and their interactions in epilepsy: is the focus out of focus? *Nature Neuroscience*, *18*, 351–359.
39. Perucca, P., Dubeau, F., & Gotman, J. (2014). Intracranial electroencephalographic seizure-onset patterns: effect of underlying pathology. *Brain: A Journal of Neurology*, *137*, 183–196.
40. Richardson, K., Fanselow, E., & Connors, B. (2008). Neocortical anatomy and physiology. In Engel, J., & Pedley, T. (Eds.) *Epilepsy: A Comprehensive Textbook* (pp. 323–336): Lippincott-Williams & Wilkins.
41. Rudolph, M., Piwkowska, Z., Badoual, M., Bal, T., & Destexhe, A. (2004). A method to estimate synaptic conductances from membrane potential fluctuations. *Journal of Neurophysiology*, *91*(6), 2884–2896.
42. Sasaki, T., Matsuki, N., & Ikegaya, Y. (2007). Metastability of active CA3 networks. *The Journal of Neuroscience*, *27*(3), 517–528.
43. Schevon, C. A., Weiss, S. A., McKhann, G. Jr, Goodman, R. R., Yuste, R., Emerson, R. G., & Trevelyan, A. J. (2012). Evidence Of an inhibitory restraint of seizure activity in humans. *Nature Communications*, *3*(1060), 1–11.
44. Skinner, F. K. (2006). Conductance-based models. Scholarpedia 1:1408 revision #125663.
45. Skinner, F. K., Zhang, L., Perez-Velazquez, J. L., & Carlen, P. L. (1999). Bursting in inhibitory interneuronal networks: a role for gap-junctional coupling. *Journal of Neurophysiology*, *81*, 1274–1283.

46. Sritharan, D., & Sarma, S. V. (2014). Fragility in dynamic networks: application to neural networks in the epileptic cortex. *Neural Computation*, *26*(10), 2294–2327.
47. Thurman, D. J., Beghi, E., Begley, C. E., Berg, A. T., Buchhalter, J. R., Ding, D., Hesdorffer, D. C., Hauser, W. A., Kazis, L., Kobau, R., Kroner, B., Labiner, D., Liow, K., Logroscino, G., Medina, M. T., Newton, C. R., Parko, K., Paschal, A., Preux, P. M., Sander, J. W., Selassie, A., Theodore, W., Tomson, T., & Wiebe, S. (2011). Standards for epidemiologic studies and surveillance of epilepsy. *Epilepsia*, *52*, 2–26.
48. Touboul, J. (2008). Bifurcation analysis of a general class of nonlinear integrate-and-fire neurons. *SIAM Journal of Applied Mathematics*, *68*, 1045–1079.
49. Traub, R. D., Kopell, N., Bibbig, A., Buhl, E. H., LeBeau, F. E. N., & Whittington, M. A. (2001). Gap junctions between interneuron dendrites can enhance synchrony of gamma oscillations in distributed networks. *The Journal of Neuroscience*, *21*(23), 9478–9486.
50. Traub, R. D., Contreras, D., Cunningham, M. O., Murray, H., LeBeau, F. E. N., Roopun, A., Bibbig, A., Wilent, W. B., Higley, M. J., & Whittington, M. A. (2005). Single-column thalamocortical network model exhibiting gamma oscillations, sleep spindles, and epileptogenic bursts. *Journal of Neurophysiology*, *93*(4), 2194–2232.
51. Trevelyan, A. J., Sussillo, D., Watson, B. O., & Yuste, R. (2006). Modular propagation of epileptiform activity: evidence for an inhibitory veto in neocortex. *The Journal of Neuroscience*, *26*(48), 12447–12455.
52. Truccolo, W., Donoghue, J. A., Hochberg, L. R., Eskandar, E. N., Madsen, J. R., Anderson, W. S., Brown, E. N., Halgren, E., & Cash, S. S. (2011). Single-neuron dynamics in human focal epilepsy. *Nature Neuroscience*, *14*(5), 635–643.
53. Truccolo, W., Ahmed, O. J., Harrison, M. T., Eskandar, E. N., Cosgrove, G. R., Madsen, J. R., Blum, A. S., Potter, N. S., Hochberg, L. R., & Cash, S. S. (2014). Neuronal ensemble synchrony during human focal seizures. *The Journal of Neuroscience*, *34*(30), 9927–9944.

54. Ullah, G., Cressman, J.R. Jr, Barreto, E., & Schiff, S. J. (2009). The influence of sodium and potassium dynamics on excitability, seizures, and the stability of persistent states: II. Network and glial dynamics. *Journal of Computational Neuroscience*, *29*, 171–183.
55. Uva, L., Breschi, G. L., Gnatkovsky, V., Taverna, S., & de Curtis, M. (2015). Synchronous inhibitory potentials precede seizure-like events in acute models of focal limbic seizures. *The Journal of Neuroscience*, *35*, 3048–3055.
56. Van Ooyen, A., Van Pelt, J., Corner, M. A., & Lopes da Silva, F. H. (1992). The emergence of long-lasting transients of activity in simple neural networks. *Biological Cybernetics*, *67*, 269–277.
57. Van Vreeswijk, C., & Hansel, D. (2001). Patterns of synchrony in neural networks with spike adaptation. *Neural Computation*, *13*, 959–992.
58. Wagner, F., Eskandar, E., Cosgrove, G., Madsen, J., Blum, A., Potter, N., Hochberg, L., Cash, S., & Truccolo, W. (2015). Microscale spatiotemporal dynamics during neocortical propagation of human focal seizures. *Neuroimage*, *122*, 114–130.
59. Wang, Y., Goodfellow, M., Taylor, P. N., & Baier, G. (2014). Dynamic mechanisms of neocortical focal seizure onset. *PLoS Computational Biology*, *10*, e1003787.
60. Wang, X. J., & Buzsáki, G. (1996). Gamma oscillation by synaptic inhibition in a hippocampal interneuronal network model. *The Journal of Neuroscience*, *16*(20), 6402–6413.
61. Wei, Y., Ullah, G., & Schiff, S. (2014a). Oxygen and seizure dynamics: II. computational modeling. *Journal of Neurophysiology*, *112*, 213–223.
62. Wei, Y., Ullah, G., & Schiff, S. (2014b). Unification of neuronal spikes, seizures, and spreading depression. *The Journal of Neuroscience*, *34*, 11733–11743.
63. Wen, B., Qian, H., Feng, J., Ge, R. J., Xu, X., Cui, Z. Q., Zhu, R. Y., Pan, L. S., Lin, Z. P., & Wang, J. H. (2015). A portion of inhibitory neurons in human temporal lobe epilepsy are functionally upregulated: an endogenous mechanism for seizure termination. *CNS Neuroscience & Therapeutics*, *21*, 204–214.

64. Zhang, Z. J., Koifman, J., Shin, D. S., Ye, H., Florez, C. M., Zhang, L., Valiante, T. A., & Carlen, P. L. (2012). Transition to seizure: Ictal discharge is preceded by exhausted presynaptic GABA release in the hippocampal CA3 region. *The Journal of Neuroscience*, *32*(7), 2499–2512.

65. žiburkus, J., Cressman, J., Barreto, E., & Schiff, S. (2006). Interneuron and pyramidal cell interplay during *in vitro* seizure-like events. *Journal of Neurophysiology*, *95*, 3948–3954.

66. žiburkus, J., Cressman, J. R., & Schiff, S. J. (2013). Seizures as imbalanced up states: excitatory and inhibitory conductances during seizure-like events. *Journal of Neurophysiology*, *109*, 1296–1306.

SECTION V:
DYNAMICS OF COGNITION

9 THE DYNAMIC BRAIN: FROM SPIKING NEURONS TO NEURAL MASSES AND CORTICAL FIELDS

Gustavo Deco[1], Viktor K. Jirsa[2,3], Peter A. Robinson[4,5,6], Michael Breakspear[7,8], Karl Friston[9]

[1] Institució Catalana de Recerca i Estudis Avançats (ICREA), Universitat Pompeu Fabra, Department of Technology, Computational Neuroscience, Barcelona, Spain
[2] Theoretical Neuroscience Group, Institut Sciences de Mouvement, Marseille, France
[3] Center for Complex Systems and Brain Sciences, Department of Physics, Florida Atlantic University, Boca, Florida, United States of America
[4] School of Physics, University of Sydney, Sydney, New South Wales, Australia
[5] Brain Dynamics Center, Westmead Millennium Institute, Westmead Hospital and University of Sydney, Westmead, New South Wales, Australia
[6] Faculty of Medicine, University of Sydney, Sydney, New South Wales, Australia
[7] School of Psychiatry, University of New South Wales, Sydney, and The Black Dog Institute, Randwick, New South Wales, Australia
[8] School of Physics, University of Sydney, Sydney, New South Wales, Australia
[9] Wellcome Trust Centre for Neuroimaging, University College London, London, United Kingdom

ABSTRACT

The cortex is a complex system, characterized by its dynamics and architecture, which underlie many functions such as action, perception, learning, language, and cognition. Its structural architecture has been studied for more than a hundred years; however, its dynamics have been addressed much less thoroughly. In this paper, we review and integrate, in a unifying framework, a variety of computational approaches that have been used to characterize the dynamics of the cortex, as evidenced at different levels of measurement. Computational models at different space–time scales help us understand the fundamental mechanisms that underpin neural processes and relate these processes to neuroscience data. Modeling at the single neuron level is necessary because this is the level at which information is exchanged between the computing elements of the brain; the neurons. Mesoscopic models tell us how neural elements interact to yield emergent behavior at the level of microcolumns and cortical columns. Macroscopic models can inform us about whole brain dynamics and interactions between large-scale neural systems such as cortical regions, the thalamus, and brain stem. Each level of description relates uniquely to neuroscience data, from single-unit recordings, through local field potentials to functional magnetic resonance imaging (fMRI), electroencephalogram (EEG), and magnetoencephalogram (MEG). Models of the cortex can establish which types of large-scale neuronal networks can perform computations and characterize their emergent properties. Mean-field and related formulations of dynamics also play an essential and complementary role as forward models that can be inverted given empirical data. This makes dynamic models critical in integrating theory and experiments. We argue that elaborating principled and informed models is a prerequisite for grounding empirical neuroscience in a cogent theoretical framework, commensurate with the achievements in the physical sciences.

INTRODUCTION

The brain appears to adhere to two fundamental principles of functional organization, functional integration and functional specialization, where the integration within and among specialized areas is mediated by connections among them. The distinction relates to that between localisationism and connectionism that dominated thinking about cortical function in the nineteenth century. Since the early anatomic theories of Gall, the identification of a particular brain region with a specific function has become a central theme in neuroscience. In this paper, we address how distributed and specialized neuronal responses are realized in terms of microscopic brain dynamics; we do this by showing how neuronal systems, with many degrees of freedom, can be reduced to lower dimensional systems that exhibit adaptive behaviors.

It is commonly accepted that the information processing underlying brain functions, like sensory, motor, and cognitive functions, is carried out by large groups of interconnected neurons [1]–[4]. Neurons are the cells responsible for encoding, transmitting, and integrating signals originating inside or outside the nervous system. The transmission of information within and between neurons involves changes in the so-called resting membrane potential, the electrical potential of the neurons at rest, when compared to the extracellular space. The inputs one neuron receives at the synapses from other neurons cause transient changes in its resting membrane potential, called postsynaptic potentials. These changes in potential are mediated by the flux of ions between the intracellular and extracellular space. The flux of ions is made possible through ion channels present in the membrane. The ion channels open or close depending on the membrane potential and on substances released by the neurons, namely neurotransmitters, which bind to receptors on the cell's membrane and hyperpolarize or depolarize the cell. When the postsynaptic potential reaches a threshold, the neuron produces an impulse. The impulses or spikes, called action potentials, are characterized by a certain amplitude and duration and are the units of information transmission at the interneuronal level. Information is thought to be encoded in terms of the frequency of the action potentials, called spiking or firing rate (i.e., rate coding), as well as in the timing of action potentials (i.e., temporal coding).

One way to investigate the biological basis of information processing in the brain is to study the response of neurons to stimulation. This can be done in experimental animals using implanted electrodes to record the rates and timing of action potentials. However, this invasive approach is generally not possible in humans. To study brain function in humans, techniques allowing the indirect study of neuronal activity have been developed. An example is functional magnetic resonance imaging (fMRI), measuring regional changes in metabolism and blood flow associated with changes in brain activity. This approach to measuring regional differences in brain activity is possible because at a macroscopic level the cortex is organized into spatially segregated regions known to have functionally specialized roles. A technique such as fMRI allows the mapping of brain regions associated with a particular task or task component. Understanding the fundamental principles underlying higher brain functions requires the integration of different levels of experimental investigation in cognitive neuroscience (from single neurons, neuroanatomy, neurophysiology, and neuroimaging, to neuropsychology and behavior) via a unifying theoretical framework that captures the neural dynamics inherent in the elaboration of cognitive processes. In this paper, we review and integrate a variety of computational approaches that have been used to characterize the dynamics of the cortex, as evidenced at different levels of measurement.

The paper is structured as follows. The central theme of this review is that the activity in populations of neurons can be understood by reducing the degrees of freedom from many to few, hence resolving an otherwise intractable computational problem. The most striking achievement in this regard is the reduction of a large population of spiking neurons to a distribution function describing their probabilistic evolution—that is, a function that captures the likely distribution of neuronal states at a given time. In turn, this can be further reduced to a single variable describing the mean firing rate. This reduction is covered first, in the next section. In the section entitled Neural Modes and Masses, we return to the full probability distribution function and show how it can be represented by a set of scalars that parameterize it parsimoniously. These parameters are equivalent to the moments of the distribution. In many instances, a few—possibly even one (equivalent to the center of mass)—are sufficient to summarize activity. These are known as Neural Mass Models. These models capture the dynamics of a neuronal population. Naturally, it is useful to understand how neuronal activity unfolds on the spatially continuous cortical sheet. This can be addressed with neural field models; involving differential operators with both temporal and spatial terms. That is, neuronal activity depends on its current state as well as spatial gradients, which allow its spread horizontally across the cortical surface. These models are covered in the Neural Field Modelssection. In Numerical Simulations: Ensemble Activity from Neuronal to Whole Brain Scale, we provide numerical simulations of neuronal ensemble dynamics across a hierarchy of spatial and temporal scales. At the microscopic scale, we simulate an entire array of spiking neurons in response to a sensory-evoked synaptic current. By comparing the response to that of a mesoscopic neural mass model, we show what is gained and what is lost by abstracting to a more tractable set of evolution equations. The spread of activity across the cortical surface, in a neural field model, is also illustrated. Finally, in the section entitled Cognitive and Clinical Applications, we illustrate applications of neural ensemble modeling in health and disease; namely, decision-making, auditory scene analysis, and absence seizures.

A summary of the notation for all the main dynamical variables and physiological parameters is given in Table 1.

Table 1. List of notation and symbols

Quantity	Symbol	SI Unit
Neural membrane potential	V	V
Neural membrane capacitive current	I	A
Time elapsed since most recent action potential	T	s
Neural membrane capacitance	C	F
Neural membrane resistance	R	ω
Neural membrane time constant	$\tau = RC$	s
Neural leak or resting potential	V_L	V
Neural firing threshold	θ	V
Neural action potential spike	δ	
Neural refractory period	T_{ref}	s
Neural reset (post-firing) membrane potential	V_{reset}	V
Synaptic efficacy of cell *j* onto cell *i*	J_{ij}	A s^{-1}
Neural membrane phase space state vector	ν	varies
Neural ensemble probability density	$p(\nu,t)$	varies
Neural ensemble dynamics (flow)	$f(\nu,t)$	varies
Neural ensemble random fluctuations (dispersion)	$D(\nu,t)$	varies
Neural ensemble jacobian	Q	varies
Neural ensemble average synaptic efficacy	$\langle J \rangle_J$	A s^{-1}
Neural ensemble mean firing rate	$Q(t)$	s^{-1}
Neural ensemble infinitesimal depolarization	$\varepsilon = dV(t)$	V
Neural ensemble mean membrane potential (drift)	$\mu = \dot{\mu}_\nu$	V
Neural ensemble mean capacitive current	$\mu_a = \dot{\mu}_\nu$	A
Neural ensemble membrane potential variance (diffusion)	σ^2	$(V)^2$
Neural ensemble probability density flux	F	varies
Neural population transfer function	$\phi(\nu,\sigma)$	s^{-1}
Neural population probability basis functions (modes)	$\eta(\nu)$	varies
Neural mass synaptic gain time constant	κ	s^{-1}
Neural mass synaptic response coefficient	γ	dimensionless
Neural field local membrane potential in population *a*	$V_a(r,t)$	V
Neural field local firing rate in population *a*	$Q_a(r,t)$	s^{-1}
Mean number of synapses on neuron *a* from neurons *b*	N_{ab}	
Mean time-integrated strength of the response of V_a per incoming spike from neurons *b*	s_{ab}	μ V s
Average rate of incoming spikes (pulse density) between populations *a* and *b*	ϕ_{ab}	s^{-1}
Discrete time delay between populations *a* and *b*	τ_{ab}	s
Coupling strength between neural populations *a* and *b*	$\nu_{ab} = N_{ab}s_{ab}$	V s
Mean decay rate of the soma response to a delta-function synaptic input	α_{ab}	s^{-1}
Mean rise rate of the soma response to a delta-function synaptic input	β_{ab}	s^{-1}
Firing threshold for channels of type *a*	θ_a	V
Characteristic range of axons, including dendritic arborization	r, r_{ab}	m
Characteristic action potential propagation velocity	c, c_{ab}	m s^{-1}
Temporal damping coefficient in the absence of pulse regeneration	$\gamma_a = c_a/r_a$	s^{-1}
Steady state sigmoid slope in population *a*	ρ_a	$(V\,s)^{-1}$
Macroscopic observable	$\psi(k,\omega) = \mathbf{M}Q$	
Spatiotemporal measurement matrix	$\mathbf{M}$	
Autonomous (uncoupled) growth/decay rate of neural mass	ζ	s^{-1}

The first column gives a brief description of the parameter, with its symbol listed in the second. The unit of each quantity is given in the third column.
doi:10.1371/journal.pcbi.1000092.t001

MEAN-FIELD MODELS

This section provides an overview of mean-field models of neuronal dynamics and their derivation from models of spiking neurons. These models have a long history spanning a half-century (e.g., [5]) and are formulated using concepts from statistical physics. In this section, we try to clarify some key concepts and show how they relate to each other. Models are essential for neuroscience, in the sense that the most interesting questions pertain to neuronal mechanisms and processes that are not directly observable. This means that questions about neuronal function are generally addressed by inference on models or their parameters, where the model links neuronal processes that are hidden from our direct observation. Broadly speaking, models are used to generate data, to study emergent behaviors, or they can be used as forward or observation models, which are inverted given empirical data. This inversion allows one to select the best model (given some data) and make probabilistic comments about the parameters of that model. Mean-field models are suited to data which reflect the behavior of a population of neurons, such as the electroencephalogram (EEG), magnetoencephalogram (MEG), and fMRI. The most prevalent models of neuronal populations or ensembles are based upon something called the mean-field approximation. The mean-field approximation is used extensively in statistical physics and is essentially a technique that finesses an otherwise computationally or analytically intractable problem. An exemplary approach, owing to Boltzmann and Maxwell, is the approximation of the motion of molecules in a gas by mean-field terms such as temperature and pressure.

Ensemble Density Models

Ensemble models attempt to model the dynamics of large (theoretically infinite) populations of neurons. Any single neuron could have a number of attributes; for example, post-synaptic membrane depolarization, V, capacitive current, I, or the time since the last action potential, T. Each attribute induces a dimension in the phase space of a neuron; in our example the phase space would be three dimensional and the state of each neuron would correspond to a point $\nu=\{V,I,T\} \in \Re^3$ or particle in phase space. Imagine a very large number of neurons that populate phase space with a density $p(\nu,t)$. As the state of each neuron evolves, the points will flow through phase space, and the ensemble density $p(\nu,t)$ will evolve until it reaches some steady state or equilibrium. p(v,t) is a scalar function returning the probability density at each point in phase space. It is the evolution of the density per se that is characterized in ensemble density methods. These models are particularly attractive because the density dynamics conform to a simple equation: the Fokker-Planck equation

$$\dot{p} = -\nabla\cdot(f - D\nabla)p \equiv \frac{\partial p}{\partial t} = tr\left(-\frac{\partial(fp)}{\partial v} + \frac{\partial}{\partial v}\left(D\frac{\partial p}{\partial v}\right)\right). \qquad (1)$$

This equation comprises a flow and a dispersion term; these terms embed the assumptions about the dynamics (phase flow, f(v,t)) and random fluctuations (dispersion, D(v,t)) that constitute our model at the neuronal level. This level of description is usually framed as a (stochastic) differential equation (Langevin equation) that describes how the states evolve as functions of each other and some random fluctuations with

$$dv = f(v)dt + \sigma d\omega, \qquad (2)$$

where, $D=\frac{1}{2}\sigma^2$ and ω is a standard Wiener process; i.e., $w(t)-w(t+\Delta t)\sim N(0, \Delta t)$. Even if the dynamics of each neuron are complicated, or indeed chaotic, the density dynamics remain simple, linear, and deterministic. In fact, we can write the density dynamics in terms of a linear operator or Jacobian Q

$$\dot{p} = Qp$$
$$Q = \nabla\cdot(D\nabla - f). \qquad (3)$$

In summary, for any model of neuronal dynamics, specified as a stochastic differential equation, there is a deterministic linear equation that can be integrated to generate ensemble dynamics. In what follows, we will explain in detail the arguments that take us from the spiking behavior of individual neurons to the mean-field dynamics described by the Fokker-Planck equation. We will consider the relationship between density dynamics and neural mass models and how these can be extended to cover spatiotemporal dynamics in the brain.

From Spiking Neurons to Mean-Field Models

The functional specialization of the brain emerges from the collective network dynamics of cortical circuits. The computational units of these circuits are spiking neurons, which transform a large set of inputs, received from different neurons, into an output spike train that constitutes the output signal of the neuron. This means that the spatiotemporal spike patterns produced by neural circuits convey information among neurons; this is the microscopic level on which the brain's representations and computations rest [6]. We assume that the nonstationary temporal evolution of the spiking dynamics can be captured by one-compartment, point-like models of neurons, such as the leaky integrate-and-fire (LIF) model [7] used below. Other models relevant for systems neuroscience can be found in [4],[8],[9]. In the LIF model, each neuron i can be fully described in terms of a single internal variable, namely the depolarization $V_i(t)$ of the neural membrane. The basic circuit of a LIF model consists of a capacitor, C, in parallel with a resistor, R, driven by a synaptic current (excitatory or inhibitory postsynaptic potential, EPSP or IPSP, respectively). When the voltage across the capacitor reaches a threshold θ, the circuit is shunted (reset) and a δ pulse (spike) is generated and transmitted to other neurons. The subthreshold membrane potential of each neuron evolves according to a simple RC circuit, with a time constant τ=RC given by the following equation:

$$\tau \frac{dV_i(t)}{dt} = -[V_i(t) - V_L] + RI_i(t), \tag{4}$$

where $I_i(t)$ is the total synaptic current flow into the cell i and V_L is the leak or resting potential of the cell in the absence of external afferent inputs. In order to simplify the analysis, we neglect the dynamics of the afferent neurons (see [10] for extensions considering detailed synaptic dynamics such as AMPA, NMDA, and GABA). The total synaptic current coming into the cell i is therefore given by the sum of the contributions of δ-spikes produced at presynaptic neurons. Let us assume that N neurons synapse onto cell i and that J_{ij} is the efficacy of synapse j, then the total synaptic afferent current is given by

$$RI_i(t) = \tau \sum_{j=1}^{N} J_{ij} \sum_k \delta\left(t - t_j^{(k)}\right), \tag{5}$$

where $t_j^{(k)}$ is the emission time of the kth spike from the jth presynaptic neuron. The subthreshold dynamical Equation 4, given the input current (from Equation 5), can be integrated, and yields

$$V_i(t) = V_L + \sum_{j=1}^{N} J_{ij} \int_0^t e^{-s/\tau} \sum_k \delta\left(t - s - t_j^{(k)}\right) ds,$$

$$= V_L + \sum_{j=1}^{N} J_{ij} e^{-\left(t - t_j^{(k)}\right)/\tau} \sum_k H\left(t - t_j^{(k)}\right). \qquad (6, 7)$$

if the neuron i is initially (t=0) at the resting potential ($V_i(0)=V_L$). In Equation 7, H(t) is the Heaviside function (H(t)=1 if t>0, and H(t)=0 if t<0). Thus, the incoming presynaptic δ-pulse from other neurons is basically low-pass filtered to produce an EPSP or IPSP in the post-synaptic cell. Nevertheless, the integrate-and-fire (IF) model is not only defined by the subthreshold dynamics but includes a reset after each spike generation, which makes the whole dynamics highly nonlinear. In what follows, we present a theoretical framework which is capable of dealing with this.

The Population Density Approach

Realistic neuronal networks comprise a large number of neurons (e.g., a cortical column has $O(10^4)$–$O(10^8)$ neurons) which are massively interconnected (on average, a neuron makes contact with $O(10^4)$ other neurons). The underlying dynamics of such networks can be described explicitly by the set of coupled differential equations (Equation 4) above. Direct simulations of these equations yield a complex spatiotemporal pattern, covering the individual trajectory of the internal state of each neuron in the network. This type of direct simulation is computationally expensive, making it very difficult to analyze how the underlying connectivity relates to various dynamics. In fact, most key features of brain operation seem to emerge from the interplay of the components; rather than being generated by each component individually. One way to overcome these difficulties is by adopting the population density approach, using the Fokker-Planck formalism (e.g., [11]). As noted above, the Fokker-Planck equation summarizes the flow and dispersion of states over phase space in a way that is a natural summary of population dynamics in genetics (e.g., [12]) and neurobiology (e.g., [13],[14]).

In what follows, we derive the Fokker-Planck equation for neuronal dynamics that are specified in terms of spiking neurons. This derivation is a little dense but illustrates the approximating assumptions and level of detail that can be captured by density dynamics. The approach we focus on was introduced by [15] (see also [16],[17]). In this approach, individual IF neurons are grouped together into populations of statistically similar neurons. A statistical description of each population is given by a probability density function that expresses the distribution of neuronal states (i.e., membrane potential) over the population. In general, neurons with the same state V(t) at a given time t have a different history because of random fluctuations in the input current I(t). The main source of randomness is from fluctuations in recurrent currents (resulting from "quenched" randomness in the connectivity and transmission delays) and fluctuations in the external currents. The key assumption in the population density approach is that the afferent input currents impinging on neurons in one population are uncorrelated. Thus, neurons sharing the same state V(t) in a population are indistinguishable. Consequently, the dynamics are described by the evolution of the probability density function:

$$p(\upsilon,t)d\upsilon = Prob\{V(t) \in [\upsilon,\upsilon+d\upsilon]\}, \tag{8}$$

which expresses the population density, which is the fraction of neurons at time t that have a membrane potential V(t) in the interval [ν,ν+dν]. The evolution of the population density is given by the Chapman-Kolmogorov equation

$$p(\upsilon,t+dt) = \int_{-\infty}^{+\infty} p(\upsilon-\varepsilon,t)\rho(\varepsilon|\upsilon-\varepsilon)d\varepsilon, \tag{9}$$

where $\rho(\varepsilon|\nu)=\mathrm{Prob}\{V(t+dt)=\nu+\varepsilon|V(t)=\nu\}$ is the conditional probability that generates an infinitesimal change $\varepsilon=V(t+dt)-V(t)$ in the infinitesimal interval dt. The Chapman-Kolmogorov equation can be written in a differential form by performing a Taylor expansion in $p(\nu',t)$ $\rho(\varepsilon|\nu')$ around $\nu'=\nu$; i.e.,

$$p(\upsilon',t)\rho(\varepsilon|\upsilon') = \sum_{k=0}^{\infty} \frac{(-\varepsilon)^k}{k!} \frac{\partial^k}{\partial \upsilon'^k} [p(\upsilon',t)\rho(\varepsilon|\upsilon')]|_{\upsilon'=\upsilon}. \tag{10}$$

In the derivation of the last equation, we have assumed that $p(\nu',t)$ and $\rho(\varepsilon|\nu')$ are infinitely many times differentiable in ν. Inserting this expansion in Equation 9, and replacing the time derivative in ν′ by the equivalent time derivative in ν, we obtain

$$p(\upsilon,t+dt) = p(\upsilon,t)\int_{-\infty}^{+\infty} \rho(\varepsilon|\upsilon)d\varepsilon - \frac{\partial}{\partial \upsilon}\left[p(\upsilon,t)\left(\int_{-\infty}^{+\infty} \varepsilon\rho(\varepsilon|\upsilon)d\varepsilon\right)\right]$$

$$+\frac{1}{2}\frac{\partial^2}{\partial v^2}\left[p(v,t)\left(\int_{-\infty}^{+\infty}\varepsilon^2\rho(\varepsilon|v)d\varepsilon\right)\right]+\ldots,$$
$$=\sum_{k=0}^{\infty}\frac{(-1)^k}{k!}\frac{\partial^k}{\partial v^k}\left[p(v,t)\langle\varepsilon^k\rangle_v\right], \tag{11,12}$$

where $\langle\ldots\rangle_v$ denotes the average with respect to $\rho(\varepsilon|\ v)$ at a given v. Finally, taking the limit for $dt \rightarrow 0$, we obtain:

$$\frac{\partial p(v,t)}{\partial t}=\sum_{k=1}^{\infty}\frac{(-1)^k}{k!}\frac{\partial^k}{\partial v^k}\left[p(v,t)\lim_{dt\to 0}\frac{1}{dt}\langle\varepsilon^k\rangle_v\right]. \tag{13}$$

Equation 13 is known as the Kramers-Moyal expansion of the original integral Chapman-Kolmogorov equation (Equation 9). It expresses the time evolution of the population density in differential form.

The Diffusion Approximation

The temporal evolution of the population density as given by Equation 13 requires the moments $\langle\varepsilon^k\rangle_v$ due to the afferent current during the interval dt. These moments can be calculated by the mean-field approximation. In this approximation, the currents impinging on each neuron in a population have the same statistics, because as we mentioned above, the history of these currents is uncorrelated. The mean-field approximation entails replacing the time-averaged discharge rate of individual cells with a common time-dependent population activity (ensemble average). This assumes ergodicity for all neurons in the population. The mean-field technique allows us to discard the index denoting the identity of any single neuron and express the infinitesimal change, dV(t), in the membrane potential of all neurons as:

$$dV(t)=\langle J\rangle_J NQ(t)dt-\frac{V(t)-V_L}{\tau}dt, \tag{14}$$

where N is the number of neurons, and Q(t) is the mean population firing rate. This is determined by the proportion of active neurons by counting the number of spikes $n_{spikes}(t,t+dt)$ in a small time interval dt and dividing by N and by dt [18]; i.e.,

$$Q(t)=\lim_{dt\to 0}\frac{n_{spikes}(t,t+dt)}{Ndt}. \tag{15}$$

In Equation 14, $\langle J \rangle_J$ denotes the average of the synaptic weights in the population. The moments of the infinitesimal depolarization, $\varepsilon = dV(t)$, can now be calculated easily from Equation 14. The first two moments in the Kramers-Moyal expansion are called drift and diffusion coefficients, respectively, and they are given by:

$$M^{(1)} = \lim_{dt \to 0} \frac{1}{dt} \langle \varepsilon \rangle_\upsilon = \langle J \rangle_J NQ(t) - \frac{\upsilon - V_L}{\tau} = \frac{\mu(t)}{\tau} - \frac{\upsilon - V_L}{\tau}, \quad (16)$$

$$M^{(2)} = \lim_{dt \to 0} \frac{1}{dt} \langle \varepsilon^2 \rangle_\upsilon = \langle J^2 \rangle_J NQ(t) = \frac{\sigma(t)^2}{\tau}. \quad (17)$$

In general, keeping only the leading term linear in dt, it is easy to prove that for k>1,

$$\langle \varepsilon^k \rangle_\upsilon = \langle J^k \rangle_J NQ(t)dt + O\left(dt^2\right), \quad (18)$$

and hence,

$$M^{(k)} = \lim_{dt \to 0} \frac{1}{dt} \langle \varepsilon^k \rangle_\upsilon = \langle J^k \rangle_J NQ(t). \quad (19)$$

The diffusion approximation arises when we neglect high-order (k>2) terms. The diffusion approximation is exact in the limit of infinitely large networks, i.e., $N \to \infty$, if the synaptic efficacies scale appropriately with network size, such that $J \to 0$ but $NJ^2 \to$ const. In other words, the diffusion approximation is appropriate, if the minimal kick step, J, is very small but the overall firing rate is very large. In this case, all moments higher than two become negligible, in relation to the drift (μ) and diffusion (σ^2) coefficients.

The diffusion approximation allows us to omit all higher orders k>2 in the Kramers-Moyal expansion. The resulting differential equation describing the temporal evolution of the population density is called the Fokker-Planck equation, and reads

$$\frac{\partial p(\upsilon,t)}{\partial t} = \frac{1}{2\tau}\sigma^2(t)\frac{\partial^2 p(\upsilon,t)}{\partial \upsilon^2} + \frac{\partial}{\partial \upsilon}\left[\left(\frac{\upsilon - V_L - \mu(t)}{\tau}\right)p(\upsilon,t)\right]. \quad (20)$$

In the particular case that the drift is linear and the diffusion coefficient, $\sigma^2(t)$, is given by a constant, the Fokker-Planck equation describes a well-known stochastic process called the Ornstein-Uhlenbeck process [19]. Thus, under the diffusion approximation, the Fokker-Planck equation (Equation 20) expresses an Ornstein-Uhlenbeck process. The Ornstein-Uhlenbeck process describes the temporal evolution of the membrane potential V(t) when the input afferent currents are given by

$$RI(t) = \mu(t) + \sigma\sqrt{\tau}\omega(t), \tag{21}$$

where ω(t) is a white noise process. Under the diffusion approximation, Equation 21 can also be interpreted (by means of the Central Limit Theorem), as the case in which the sum of many Poisson processes (Equation 5) becomes a normal random variable with mean μ(t) and variance σ^2.

The Mean-Field Model

The simulation of a network of IF neurons allows one to study the dynamical behavior of the neuronal spiking rates. Alternatively, the integration of the non-stationary solutions of the Fokker-Planck equation (Equation 20) also describes the dynamical behavior of the network, and this would allow the explicit simulation of neuronal and cortical activity (single cells, EEG, fMRI) and behavior (e.g., performance and reaction time). However, these simulations are computationally expensive and their results probabilistic, which makes them unsuitable for systematic explorations of parameter space. However, the stationary solutions of the Fokker-Planck equation (Equation 20) represent the stationary solutions of the original IF neuronal system. This allows one to construct bifurcation diagrams to understand the nonlinear mechanisms underlying equilibrium dynamics. This is an essential role of the mean-field approximation: to simplify analyses through the stationary solutions of the Fokker-Planck equation for a population density under the diffusion approximation (Ornstein-Uhlenbeck process) in a self-consistent form. In what follows, we consider stationary solutions for ensemble dynamics.

The Fokker-Planck equation describing the Ornstein-Uhlenbeck process, with $\mu=\langle J\rangle_J NQ(t)$ and $\sigma^2=\langle J^2\rangle_J NQ(t)$, can be rewritten as a continuity equation:

$$\frac{\partial p(\upsilon,t)}{\partial t} = -\frac{\partial F(\upsilon,t)}{\partial \upsilon}, \tag{22}$$

where F is the flux of probability defined as follows:

$$F(\upsilon,t) = -\frac{\upsilon - V_L - \mu}{\tau} p(\upsilon,t) - \frac{\sigma^2}{2\tau}\frac{\partial p(\upsilon,t)}{\partial \upsilon}. \tag{23}$$

The stationary solution should satisfy the following boundary condition:

$$p(\theta,t)=0, \tag{24}$$

And

$$\frac{\partial p(\theta,t)}{\partial \upsilon}=-\frac{2Q\tau}{\sigma^2}, \tag{25}$$

which expresses the fact that the probability current at threshold gives, by a self-consistent arguments, the average firing rate, Q, of the population. Furthermore, at ν→−4 the probability density vanishes fast enough to be integrable; i.e.,

$$\lim_{\upsilon\to-\infty} p(\upsilon,t)=0, \tag{26}$$

And

$$\lim_{\upsilon\to-\infty} \upsilon p(\upsilon,t)=0. \tag{27}$$

In addition, the probability mass leaving the threshold at time t has to be re-injected at the reset potential at time $t+t_{ref}$ (where t_{ref} is the refractory period of the neurons), which can be accommodated by rewriting Equation 22 as follows:

$$\frac{\partial p(\upsilon,t)}{\partial t}=-\frac{\partial}{\partial \upsilon}\left[F(\upsilon,t)+Q(t-t_{ref})H(\upsilon-V_{reset})\right], \tag{28}$$

where H(.) is the Heaviside function. The solution of Equation 28 satisfying the boundary conditions (Equations 24–27) is:

$$p_s(\upsilon)=\frac{2Q\tau}{\sigma}\exp\left(-\frac{\upsilon-V_L-\mu}{\sigma^2}\right)\int_{\frac{\upsilon-V_L-\mu}{\sigma}}^{\frac{\theta-V_L-\mu}{\sigma}} H\left(x-\frac{V_{reset}-V_L-\mu}{\sigma}\right)e^{x^2}dx. \tag{29}$$

Taking into account the fraction of neurons, Qt_{ref}, in the refractory period and the normalization of the mass probability,

$$\int_{-\infty}^{\theta} p_s(\upsilon)d\upsilon+Qt_{ref}=1. \tag{30}$$

Finally, substituting Equation 29 into Equation 30, and solving for Q, we obtain the population transfer function, φ, of Ricciardi [13]:

$$Q=\left[t_{ref}+\tau\sqrt{\pi}\int_{\frac{V_{reset}-V_L-\mu}{\sigma}}^{\frac{\theta-V_L-\mu}{\sigma}} e^{x^2}\{1+\mathrm{erf}(x)\}dx\right]^{-1}=\phi(\mu,\sigma), \tag{31}$$

Where $\mathrm{erf}(x)=2/\sqrt{\pi}\int_0^x e^{-y^2}dy$.

The stationary dynamics of each population can be described by the population transfer function, which provides the average population rate as a function of the average input current. This can be generalized easily for more than one population. The network is partitioned into populations of neurons whose input currents share the same statistical properties and fire spikes independently at the same rate. The set of stationary, self-reproducing rates, Q_i, for different populations, i, in the network can be found by solving a set of coupled self-consistency equations, given by:

$$Q_i=\phi(\mu_i,\sigma_i) \tag{32}$$

To solve the equations defined by Equation 32 for all i, we integrate the differential equation below, describing the approximate dynamics of the system, which has fixed-point solutions corresponding to Equation 32:

$$\tau_x\frac{dQ_x}{dt}=-Q_x+\phi(\mu_x,\sigma_x). \tag{33}$$

This enables a posteriori selection of parameters, which induce the emergent behavior that we are looking for. One can then perform full nonstationary simulations using these parameters in the full IF scheme to generate true dynamics. The mean-field approach ensures that these dynamics will converge to a stationary attractor that is consistent with the steady-state dynamics we require [10],[20]. In our case, the derived transfer function, φ, corresponds consistently to the assumptions of the simple LIF model described in the From Spiking-Neurons to Mean-Field Models section. Further extension for more complex and realistic models are possible. For example, an extended mean-field framework, which is consistent with the IF and realistic synaptic equations that considers both the fast and slow glutamatergic excitatory synaptic dynamics (AMPA and NMDA) and the dynamics of GABA inhibitory synapses, can be found in [10]. Before turning to neural mass models, we consider some applications of mean-field modeling that will be reprised in the last section.

Competition and Cooperation

How are different cortical representations integrated to form a coherent stream of perception, cognition, and action? The brain is characterized by a massive recurrent connectivity between cortical areas, which suggests that integration of partial representations might be mediated by cross talk via interareal connections. Based on this view [21], and neurophysiological evidence [22], it has been hypothesized that each cortical area represents a set of alternative hypotheses, encoded in the activities of cell assemblies. Representations of conflicting hypotheses compete with each other; however, each area represents only a part of the environment or internal state. In order to arrive at a coherent global representation, different cortical areas bias each others' internal representations by communicating their current states to other areas, thereby favoring certain sets of local hypotheses over others. By recurrently biasing each others' competitive internal dynamics, the neocortical system arrives at a global representation in which each area's state is maximally consistent with those of the other areas. This view has been referred to as the biased-competition hypothesis. In addition to this competition-centered view, a cooperation-centered picture of brain dynamics, where global representations find their neural correlate in assemblies of coactivated neurons, has been formulated [21],[23]. Coactivation is achieved by increased connectivity among the members of each assembly. Reverberatory communication between the members of the assembly then leads to persistent activation to engender temporally extended representations.

The mean-field approach has been applied to biased-competition and cooperation networks and has been used to model single neuronal responses, fMRI activation patterns, psychophysical measurements, effects of pharmacological agents, and effects of local cortical lesions [6], [24]–[33]. In the section entitled Cognitive and Clinical Applications, we present one of these examples, in the context of decision-making.

NEURAL MODES AND MASSES

The Fokker-Planck equation, (Equation 1), is a rather beautiful and simple expression that prescribes the evolution of ensemble dynamics, given any initial conditions and equations of motion that embed our neuronal model. However, it does not specify how to encode or parameterize the density itself. There are several approaches to this. These include binning the phase space and using a discrete approximation to a continuous density. However, this can lead to a vast number of differential equations, especially if there are multiple states for each population. One solution to this is to reduce the number of states (i.e., dimension of the phase space) to render the integration of the Fokker-Planck more tractable. One elegant example of this reduction can be found in [34]. Here, population dynamics are described by a set of one-dimensional partial differential equations in terms of the distributions of the refractory density (where the refractory state is defined by the time elapsed since the last action potential). This furnishes realistic simulations of the population activity of hippocampal pyramidal neurons, based on something known as the refractory density equation and a single-neuron threshold model. The threshold model is a conductance-based model with adaptation-providing currents.

An alternative approach to dimension reduction is to approximate the ensemble densities with a linear superposition of probabilistic modes or basis functions $\eta(\nu)$ that cover phase space. In this section, we overview this modal approach to ensemble dynamics, initially in the general setting and then in the specific case, where the dynamics can be captured by the activity of a single node.

Moments and Modes of Density Dynamics

Instead of characterising the density dynamics explicitly, one can summarize it in terms of coefficients parameterising the expression of modes:

$$p = \sum_i \mu_i \eta(\nu) = \eta\mu. \quad \textbf{(34)}$$

where $\mu=\eta^{-}p$, η^{-} being the generalized inverse of the matrix encoding the basis set of modes.

A useful choice for the basis functions are the eigenfunctions (i.e., eigen vectors) of the Fokker-Planck operator, Q [17], where $Q\eta=\eta\lambda \Rightarrow \eta^{-} Q\eta=\lambda$ and λ is a leading-diagonal matrix of eigenvalues. Because the Fokker-Planck operator conserves probability mass, all its real eigenvalues are zero or negative. In the absence of mean-field effects, the biorthogonality of the eigenfunctions effectively uncouples the dynamics of the modes they represent

$$\dot{p} = Qp \Rightarrow,$$

$$\eta^{-}\dot{p} = \eta^{-}Q\eta\mu, \Rightarrow$$

$$\dot{\mu}_i = \lambda_i \mu_i. \tag{35}$$

The last expression means that, following perturbation, each mode decays exponentially, to disclose the equilibrium mode, η_0, that has a zero eigenvalue. Because the eigenvalues are complex (due to the fact that the Jacobian is not symmetric), the decay is oscillatory in nature, with a frequency that is proportional to the imaginary part of the eigenvalue and a rate constant proportional to the real part. The key thing about this parameterisation is that most modes will decay or dissipate very quickly. This means we only have to consider a small number of modes, whose temporal evaluation can be evaluated simply with

$$\dot{\mu} = \lambda\mu, \Rightarrow$$

$$\dot{\mu}(t) = \exp(t\lambda)\mu(0), \Rightarrow$$

$$p(v,t) \approx \eta \exp(t\eta^{-}Q\eta)\eta^{-}\rho(0). \tag{36}$$

See [35] for an example of this approach, in which the ensuing nonlinear differential equations were used in a forward model of observed data. In summary, we can formulate the ensemble dynamics of any neuronal system, given its equations of motion, using the equation above. This specifies how the coefficients of probability modes would evolve from any initial state or following a perturbation to the neuronal states. It furnishes a set of coupled differential equations that can be integrated to form predictions of real data or to generate emergent behaviors. We have introduced parameterisation in terms of probability modes because it provides a graceful link to neural mass models.

Neural Mass Models

Neural mass models can be regarded as a special case of ensemble density models, where we summarize our description of the ensemble density with a single number. Early examples can be found in the work of [5],[36],[37]. The term mass action model was coined by [38] as an alternative to density dynamics. These models can be motivated as a description in terms of the expected values of neuronal states, μ, under the assumption that the equilibrium density has a point mass (i.e., a delta function). This is one perspective on why these simple mean-field models are called neural mass models. In short, we replace the full ensemble density with a mass at a particular point and then summarize the density dynamics by the location of that mass. What we are left with is a set of nonlinear differential equations describing the evolution of this mode. But what have we thrown away? In the full nonlinear Fokker-Planck formulation, different phase functions or probability density moments could couple to each other; both within and between populations or ensembles. For example, this means that the average depolarisation in one ensemble could be affected by the dispersion or variance of depolarisation in another. In neural mass models, we ignore this possibility because we can only couple the expectations or first moments. There are several devices that are used to compensate for this simplification. Perhaps the most ubiquitous is the use of a sigmoid function, $\varsigma(\mu_v)$, relating expected depolarisation to expected firing rate [38]. This implicitly encodes variability in the postsynaptic depolarisation, relative to the potential at which the neuron would fire. A common form for neural mass equations of motion posits a second order differential equation for expected voltage, or, equivalently, two coupled first order equations, $\mu=\{\mu_v,\mu_i\}$ where

$$
\begin{aligned}
&\left(\frac{1}{\gamma^2}\frac{\partial^2}{\partial t^2}+\frac{2}{\gamma}\frac{\partial}{\partial t}+1\right)\mu_v=\varsigma(\mu_v),\\
&\dot{\mu}_v=\mu_a,\\
&\dot{\mu}_a=\kappa^2\varsigma(\mu_v)-2\gamma\mu_a-\gamma^2\mu_v,\\
&\varsigma(\mu_v)=\frac{2\kappa}{1+\exp(-r\mu_v)}-\kappa.
\end{aligned}
\tag{37}
$$

Here μ_a can be regarded as capacitive current. The constant γ controls the rise time of voltage, in response to inputs (see also the Neural Field Models section). These differential equations can be expressed as a convolution of inputs, $\varsigma(\mu_v)$, to give the expected depolarization, μ_v; i.e., the convolution of the input signal with an impulse response kernel W(t)

$$\mu_v(t) = \int W(t-t')\varsigma[\mu_v(t)]dt',$$
$$W(t) = \gamma^2 t\exp(-\gamma t). \tag{38}$$

The input is commonly construed to be a firing rate (or pulse density) and is a sigmoid function, ς, of mean voltage of the same or another ensemble. The coupling constant, κ, scales the amplitude of this mean-field effect. This form of neural mass model has been used extensively to model electrophysiological recordings (e.g., [39]–[41]) and has been used recently as the basis of a generative model for event-related potentials that can be inverted using real data [42].

In summary, neural mass models are special cases of ensemble density models that are furnished by ignoring all but the expectation or mean of the ensemble density. This affords a considerable simplification of the dynamics and allows one to focus on the behavior of a large number of ensembles, without having to worry about an explosion in the number of dimensions or differential equations one has to integrate. The final sort of model we will consider is the generalisation of neural mass models that allow for states that are functionals of position on the cortical sheet. These are referred to as neural field models and are discussed in the following sections.

NEURAL FIELD MODELS

The density dynamics and neural mass models above covered state the attributes of point processes, such as EEG sources, neurons, or neuronal compartments. An important extension of these models speaks to the fact that neuronal dynamics play out on a spatially extended cortical sheet. In other words, states like the depolarisation of an excitatory ensemble in the granular layer of cortex can be regarded as a continuum or field, which is a function of space, x, and time, $\mu(t) \rightarrow \mu(x,t)$. This allows one to formulate the dynamics of the expected field in terms of partial differential equations in space and time. These are essentially wave equations that accommodate lateral interactions. Although we consider neural field models last, they were among the first mean-field models of neuronal dynamics [43],[44]. Key forms for neural field equations were proposed and analysed by [45]–[47]. These models were generalized by [48],[49] who, critically, considered delays in the propagation of spikes over space. The introduction of propagation delays leads to dynamics that are very reminiscent of those observed empirically.

Typically, neural field models can be construed as a spatiotemporal convolution (c.f., Equation 38) that can be written in terms of a Green›s function; e.g.,

$$\mu_v(x,t) = \int W(x-x',t-t')\varsigma[\mu_v(x',t')]dx'dt',$$
$$W(x-x',t-t') = \delta\left(t-t'-\frac{|x-x'|}{c}\right)\frac{\exp\left(\frac{-|x-x'|}{\gamma}\right)}{2\gamma}, \quad (39)$$

where $|x-x'|$ is the distance between the spatial locations x and x′, c is the characteristic speed of spike propagation, and γ reflects the spatial decay of lateral interactions. The corresponding second order equations of motion are a neural wave equation (see [48],[49] and below)

$$\left[\frac{1}{\gamma^2}\frac{\partial}{\partial^2} + \frac{2}{\gamma}\frac{\partial}{\partial t} + 1 - r^2\nabla^2\right]\mu_v = \left(1 + \frac{1}{\gamma}\frac{\partial}{\partial t}\right)\varsigma(\mu_v), \quad (40)$$

where $\gamma = c/r$ and ∇^2 is the Laplacian. The formal similarity with the neural mass model in (37) is self-evident. These sorts of models have been extremely useful in modeling spatiotemporally extended dynamics (e.g., [50]–[53]). The generic form of neural field dynamics can be written as (see also [53]):

$$\dot{\mu} = f(\mu) + \alpha\int_{\Gamma} W(|x-x'|)\varsigma[\mu(x',T_c)]dx' + h, \quad (41)$$

where $\mu = \mu(x,t)$ is the neural field, capturing the neural mass activity at time t and position x. $f(\mu)$ captures the local dynamics of the neural field, and $T_c = t - |x-x'|/c$ is the time delay due to signal propagation. h is a constant threshold value and Γ is the spatial domain of the neural field, where $x \in \Gamma = [0,L]$. The kernel $W(|x-x'|)$ denotes the connectivity function, which is translationally invariant in space, i.e., the probability that two neural masses are connected depends only on the distance between them. If we neglect the local dynamics $f(\mu)$, $\dot{\mu} = 0$, and use an exponential kernel as in Equation 39, we recover Equations 39 and 40. This approximation is valid when the axonal delays contribute mostly to the dynamics, for instance in large-scale networks, when the local dynamics are much faster than the network dynamics. It is easy to show that most realistic connectivity kernels provide a neural wave equation like Equation 40; this is due to the fact that the connectivity must remain integrable. As above, the parameter c is the propagation velocity of action potentials traveling down an axon. If $f(\mu)=-\zeta\,\mu$, then the constant $\zeta>0\in\Re$ represents the growth rate of the neural mass. $\alpha>0$ is a scaling constant. Under instantaneous interactions, $c\rightarrow\infty$, single population models with locally excitatory and laterally inhibitory connectivity can support global periodic stationary patterns in one dimension as well as single or multiple localized solutions (bumps and multi-bumps) [47]. This class of models are also sometimes referred to as continuous attractor neural networks (CANN). When the firing rate, ς, is a Heaviside step function, [45] was able to construct an explicit one-bump solution of the form

$$\mu(x)=\int_0^a W(|x-x'|)dx', \quad \mu(0)=h=\mu(a), \tag{42}$$

where the value a corresponds to the width of the bump. Amari also identified criteria to determine if only one bump, multiple bumps, or periodic solutions exist and if they are stable. This simple mathematical model can be extended naturally to accommodate multiple populations and cortical sheets, spike frequency adaptation, neuromodulation, slow ionic currents, and more sophisticated forms of synaptic and dendritic processing as described in the review articles [4],[54],[55]. Spatially localized bump solutions are equivalent to persistent activity and have been linked to working memory in prefrontal cortex [56],[57]. During behavioral tasks, this persistent elevated neuronal firing can last for tens of seconds after the stimulus is no longer present. Such persistent activity appears to maintain a representation of the stimulus until the response task is completed. Local recurrent circuitry has received the most attention, but other theoretical mechanisms for the maintenance of persistent activity, including local recurrent synaptic feedback and intrinsic cellular bistability [58],[59], have been put forward. The latter will be captured by specific choices of the local dynamics, $f(\mu)$, in Equation 41; for instance, [60] choose a cubic-shaped function of the firing rate, which, under appropriate parameters, allows for intrinsic bistability. Single bump solutions have been used for neural modeling of the head-direction system [61]–[64], place cells [65]–[68], movement initiation [69], and feature selectivity in visual cortex, where bump formation is related to the tuning of a particular neuron›s response [70]. Here the neural fields maintain the firing of its neurons to represent any location along a continuous physical dimension such as head direction, spatial location, or spatial view. The mathematical analysis of the neural field models is typically performed with linear stability theory, weakly nonlinear perturbation analysis, and numerical simulations. With more than one population, nonstationary (traveling) patterns are also possible. In two dimensions, many other interesting patterns can occur, such as spiral waves [71], target waves, and doubly periodic patterns. These latter patterns take the form of stripes and checkerboard-like patterns, and have been linked to drug-induced visual hallucinations [72]. For smooth sigmoidal firing rates, no closed-form spatially localized solutions are known, though much insight into the form of multibump solutions has been obtained using techniques first developed for the study of fourth-order pattern forming systems [73]. Moreover, in systems with mixed (excitatory and inhibitory) connectivity or excitatory systems with adaptive currents, solitary traveling pulses are also possible. The bifurcation structure of traveling waves in neural fields can be analysed using a so-called Evans function and has recently been explored in great detail [74].

Much experimental evidence, supporting the existence of neural fields, has been accumulated (see [53] for a summary). Most of these results are furnished by slice studies of pharmacologically treated tissue, taken from the cortex [75]–[77], hippocampus [78], and thalamus [79]. In brain slices, these waves can take the form of synchronous discharges, as seen during epileptic seizures [80], and spreading excitation associated with sensory processing [81]. For traveling waves, the propagation speed depends on the threshold, h, which has been established indirectly in real neural tissue (rat cortical slices bathed in the GABA-A blocker picrotoxin) by [82]. These experiments exploit the fact that (i) cortical neurons have long apical dendrites and are easily polarized by an electric field, and (ii) that epileptiform bursts can be initiated by stimulation. A positive (negative) electric field applied across the slice increased (decreased) the speed of wave propagation, consistent with the theoretical predictions of neural field theory, assuming that a positive (negative) electric field reduces (increases) the threshold, h, in Equation 42.

Recent Developments in Neural Field Models

More and more physiological constraints have been incorporated into neural field models of the type discussed here (see Equations 39 and 40). These include features such as separate excitatory and inhibitory neural populations (pyramidal cells and interneurons), nonlinear neural responses, synaptic, dendritic, cell-body, and axonal dynamics, and corticothalamic feedback [38], [43], [44], [48], [50], [83]–[87]. A key feature of recent models is that they use parameters that are of functional significance for EEG generation and other aspects of brain function; for example, synaptic time constants, amount of neurotransmitter release or reuptake, and the speed of signal propagation along dendrites. Inferences can also be made about the parameters of the nonlinear IF response at the cell body, and about speeds, ranges, and time delays of subsequent axonal propagation, both within the cortex and on extracortical paths (e.g., via the thalamus). It is also possible to estimate quantities that parametrize volume conduction in tissues overlying the cortex, which affect EEG measurements [88], or hemodynamic responses that determine the blood oxygen level–dependent (BOLD) signals [89]. Each of these parameters is constrained by physiological and anatomical measurements, or, in a few cases, by other types of modeling. A key aim in modeling is to strike a balance between having too few parameters to be realistic, and too many for the data to be able to constrain them effectively.

Recent work in this area has resulted in numerous quantitatively verified predictions about brain electrical activity, including EEG time series [86],[87],[90], spectra [50],[86],[87],[90],[91], coherence and correlations, evoked response potentials (ERPs) [87], and seizure dynamics [86],[90],[92]. Inversion of these models has also furnished estimates of underlying physiological parameters and their variations across the brain, in different states of arousal and pathophysiology [86],[93],[94].

There are several interesting aspects to these modeling initiatives, which generalize the variants discussed in earlier sections: (i) synaptic and dendritic dynamics and summation of synaptic inputs to determine potentials at the cell body (soma), (ii) generation of pulses at the axonal hillock, and (iii) propagation of pulses within and between neural populations. We now look more closely at these key issues.

Synaptodendritic Dynamics and the Soma Potential

Assume that the brain contains multiple populations of neurons, indexed by the subscript a, which labels simultaneously the structure in which a given population lies (e.g., a particular nucleus) and the type of neuron (e.g., interneuron, pyramidal cell). Then the spatially continuous soma potential, V_a, is the sum of contributions, V_{ab}, arriving as a result of activity at each type of (mainly) dendritic synapse b, where b indexes both the incoming neural population and the neurotransmitter type of the receptor. (Note that V_a is linearly related to the current reaching the soma, and to μ in earlier sections.) Thus we write

$$V_a(\mathbf{r},t) = \sum_b V_{ab}(\mathbf{r},t), \tag{43}$$

where **r**=(x,y) denotes the spatial coordinates, t the time. The summation is assumed to be linear, and all potentials are measured relative to the resting potential [95]. For moderate perturbations relative to a steady state, the value of the resting potential can be subsumed into the values of other parameters [95]. As above, the cortex is approximated as a 2-D sheet and **r**is assumed to be the actual position in the case of the cortex; other structures, such as the thalamus, are linked to the cortex via a primary topographic map. This map links points in a one-to-one manner between structures; i.e., we assign the same value of r to such points. Hence, in structures other than the cortex, this dimensional map coordinate, **r**, denotes a rescaled physical dimension (i.e. the physical coordinate multiplied by the ratio of the cortical scale to the structure's scale), a point that should be remembered when interpreting values of spatial parameters in these structures.

The subpotentials, V_{ab}, respond in different ways to incoming spikes, depending on their synaptic dynamics (ion-channel kinetics, diffusion in the synaptic cleft, etc.), and on subsequent signal dispersion in the dendrites. The resulting soma response to a delta-function input at the synapse can be approximated via the differential equation [50].

$$D_{ab}(r,t)V_{ab}(r,t) = \nu_{ab}\phi_{ab}(r,t-\tau_{ab}),$$

$$D_{ab}(r,t) = \frac{1}{\alpha_{ab}(r,t)\beta_{ab}(r,t)}\frac{d^2}{dt^2} + \left[\frac{1}{\alpha_{ab}(r,t)} + \frac{1}{\beta_{ab}(r,t)}\right]\frac{d}{dt} + 1,$$

$$\nu_{ab} = N_{ab}s_{ab}, \qquad (44,45,46)$$

where ν_{ab} is a coupling strength, N_{ab} is the mean number of synapses on neuron a from neurons b, s_{ab} is the mean time-integrated strength of the response of V per incoming spike, and θ_{ab} is the average rate of incoming spikes (allowing for the possibility of a discrete time delay, τ_{ab}, between populations b and a in addition to any delays due to spreading within populations). The parameter α_{ab} is the mean decay rate of the soma response to a delta-function synaptic input, while β_{ab} is the mean rise rate: this biexponential form has been found to be a good approximation [8],[50],[96],[97]. If the α_{ab} and β_{ab} are independent of b (which is not generally the case), then the subscript b on D_{ab} can be omitted and V_a itself satisfies Equation 44 with the right side of Equation 44 replaced by the sum of P_{ab} over b. This approximation is also valid if α and β are interpreted as effective values, averaged over subpopulations.

Pulse Generation

In cells with voltage-gated ion channels, action potentials are produced at the axonal hillock when the soma potential exceeds some threshold θ_a. When averaged over a population of neurons, with normal response characteristics, a reasonable approximation for the firing rate, Q, is

$$Q_a(\mathbf{r},t) = Q_{a\max}S_a[V_a(\mathbf{r},t)], \qquad (47)$$

where Q_{amax} is the maximum firing rate and S_a is a monotonic increasing sigmoidal function that approaches zero as $V_a \to -\infty$ and unity as $V_a \to \infty$. A commonly used approximation is

$$S_a[V_a(\mathbf{r},t)] = \frac{1}{1+\exp[-\{V_a(\mathbf{r},t)-\theta_a(\mathbf{r},t)\}/\tilde{\sigma}_a(\mathbf{r},t)]}, \qquad (48)$$

where θ_a is the firing threshold for channels of type a and $\sigma_a = \tilde{\sigma}_a\pi/\sqrt{3}$ is the standard deviation of the threshold over the population.

Axonal Propagation

Spatiotemporal propagation of pulses within and between populations determines the values of φ_{ab}. If we indicate the firing rate Q_a for the cell type a by a subscript, then φ_{ab} can be expressed in terms of the firing rate at other locations and earlier times. If we assume linear propagation, signals propagate as described by the neural field equation (Equation 40).

$$\left[\frac{1}{\gamma_{ab}^2}\frac{\partial^2}{\partial t^2}+\frac{2}{\gamma_{ab}}\frac{\partial}{\partial t}+1-r_{ab}^2\nabla^2\right]\phi_{ab}(\mathbf{r},t)=Q_b(\mathbf{r},t), \tag{49}$$

where, as per Equation 40, r_{ab} is the characteristic range of axons, including dendritic arborization, c_{ab} is the characteristic velocity of signals in these axons, and $\gamma_{ab}=c_{ab} / r_{ab}$ is the resulting temporal damping coefficient in the absence of pulse regeneration. Note that, in comparision to the use of the terms μ_v and $\varsigma(\mu_v)$ in Equation 40, the present wave-equation is formalized in relationship to population-specific pulse densities, φ_{ab}, and firing rates, $Q_{a,b}$. By employing population-specific fields and parameters, it allows each population to generate a family of outgoing fields that propagate to different populations in different ways. Equation 49 is also satisfied if φ_{ab} is replaced by the free propagator $\Gamma_{ab}^{(0)}(\mathbf{r}-\mathbf{r}',t-t')$ and the right side is replaced by a source of the form $\delta(\mathbf{r}-\mathbf{r}')\delta(t-t')$. In Fourier space, this gives

$$\Gamma_{ab}^{(0)}(\mathbf{k},\omega)=\frac{1}{\left(k^2+q_{0ab}^2\right)r_{ab}^2},$$

$$q_{0ab}^2 r_{ab}^2=(1-i\omega/\gamma_{ab})^2, \tag{50, 51}$$

where $\boldsymbol{k}=(k_x,k_y)$ is the wave vector and ω is the angular frequency. Critically, the neural field Equation 49 enables very diffuse (i.e., not topographically specific) connections between populations to be handled straightforwardly, simply by increasing r_{ab} while reducing γ_{ab}, thereby allowing influences to propagate long distances with little damping.

Parameters and Modulations

The above equations contain a number of parameters encoding physiology and anatomy (e.g., coupling strengths, firing thresholds, time delays, velocities, etc.). In general, these can vary in space, due to differences among brain regions, and in time, due to effects like habituation, facilitation, and adaptation. In brief, time-dependent effects can be included in neural field models by adding dynamical equations for the evolution of the parameters. Typically, these take a form in which parameter changes are driven by firing rates or voltages, with appropriate time constants. The simplest such formulation is [95]

$$x(\mathbf{r},t) = x^{(0)} + x^{(1)} H(t) \otimes \left[y(\mathbf{r},t) - y^{(0)} \right], \tag{52}$$

where x is the evolving parameter, y is the quantity that drives the evolution, $x^{(0)}$ and $y^{(0)}$ are steady state values, and $x^{(1)}$ is a constant that describes the strength of feedback. The symbol ⊗ indicates a convolution of the driver with the temporal response function H(t), which incorporates the time constants of the dynamics. If we use the normalized form

$$H(t) = \eta^n t^{n-1} \exp(-\eta t), \tag{53}$$

then we find the differential equivalent of Equation 53:

$$\left(\frac{1}{\eta} \frac{d}{dt} + 1 \right)^n \left[x(\mathbf{r},t) - x^{(0)} \right] = x^{(1)} \left[y(\mathbf{r},t) - y^{(0)} \right]. \tag{54}$$

Steady states and Dynamics

Here, we first discuss how to find the steady states of neural field models. Important phenomena have been studied by linearizing these models around their steady state solutions. Hence, we discuss linear properties of such models, including how to make predictions of observable quantities from them; including transfer functions, spectra, and correlation and coherence functions. In doing this, we assume for simplicity that all the model parameters are constant in time and space, although it is possible to relax this assumption at some cost in complexity. Linear predictions from neural field models have accounted successfully for a range of experimental phenomena, as mentioned above. Nonlinear dynamics of such models have also been discussed in the literature, resulting in successful predictions of epileptic dynamics, for example [86],[92], but are not considered here (but see the Cognitive and Clinical Applications section).

Steady states and global dynamics. Previous work has shown that many properties of neuronal dynamics can be obtained by regarding activity changes as perturbations of a steady state [86]. Spatially uniform steady states can be obtained by solving the preceding equations with all time and space derivatives set to zero, assuming that the parameters are spatially constant. The spatially uniform steady states are thus the solutions of the set of equations

$$Q_a = S_a\left(\sum_b \nu_{ab} Q_b\right), \tag{55}$$

which are generally transcendental in form.

Linear equations for activity. Of the relevant equations above, all but Equation 48 are linear in Q. Equation 48 can be linearized by replacing the sigmoid, S_a, by its slope, ρ_a, at the steady state value of V_a; we also approximate this quantity as constant. If we Fourier transform the resulting set of linear equations, we find for the fluctuating parts

$$Q_a(\mathbf{k},\omega) = \rho_a V_a(\mathbf{k},\omega),$$

$$V_a(\mathbf{k},\omega) = \sum_b V_{ab}(\mathbf{k},\omega),$$

$$V_{ab}(\mathbf{k},\omega) = L_{ab}(\omega)\nu_{ab}e^{i\omega\tau_{ab}}\phi_{ab}(\mathbf{k},\omega),$$

$$L_{ab}(\omega) = (1 - i\omega/\alpha_{ab})^{-1}(1 - i\omega/\beta_{ab})^{-1},$$

$$\phi_{ab}(\mathbf{k},\omega) = \Gamma_{ab}^{(0)}(\mathbf{k},\omega)Q_b(\mathbf{k},\omega), \tag{56,57,58,59,60}$$

where $\Gamma_{ab}^{(0)}$ is given by Equation 50 and we have assumed that all the parameters of the equations (but not the fields of activity) are constant on the timescales of interest. Note that we have assumed the system to be unbounded in order to employ a continuous Fourier transform here. The case of bounded systems with discrete spatial eigenmodes can be treated analogously.

For any given spatial wavenumber, k, and temporal frequency, ω, Equations 56–60 can be rearranged to obtain

$$Q_a(\mathbf{k},\omega) = \sum_b J_{ab}(\omega)\Gamma_{ab}^{(0)}(\mathbf{k},\omega)Q_b(\mathbf{k},\omega),$$

$$J_{ab}(\omega) = L_{ab}(\omega)G_{ab}e^{i\omega\tau_{ab}}, \tag{61,62}$$

where the gains are defined by $G_{ab}=\rho_a\nu_{ab}$.

If there are N′ neural populations and J′ stimulus sources, and we assume that there is no feedback of stimuli on themselves, or of the brain on stimuli, then we can write Equation 61 as

$$\sum_b A_{ab}(\mathbf{k},\omega)Q_b(\mathbf{k},\omega) = \sum_j B_{aj}(\mathbf{k},\omega)N_j(\mathbf{k},\omega),$$

$$A_{ab}(\mathbf{k},\omega) = \delta_{ab} - B_{ab}(\mathbf{k},\omega),$$

$$B_{ab}(\mathbf{k},\omega) = J_{ab}(\omega)\Gamma^{(0)}_{ab}(\mathbf{k},\omega), \quad (63,64,65)$$

where the sum on the left of Equation 63 extends only over populations in the brain, while the sum on the right covers only stimulus sources, denoted by j. Q_j is written as N_j to make the distinction between population firing rates and incoming stimulus rates absolutely clear. We can now write Equation 63 in matrix form as

$$AQ = BN, \quad (66)$$

where **A** is an N′×N′ matrix, ***Q*** is an N′-element column vector, **B** is an N′×J′ matrix, and **N** is a J′-element column vector. We can simply invert Equation 66 to find ***Q*** in terms of the stimuli **N**:

$$\mathbf{Q} = \mathbf{TN}, \quad (67)$$

where $\mathbf{T}=\mathbf{A}^{-1}\mathbf{B}$ is the N′×J′ transfer matrix of the system. The element T_{aj} is the response of Q_a to a change in N_j at the same frequency and wave vector.

Observables. A measurable scalar quantity ψ, such as an EEG scalp voltage or voltage difference, can generally be approximated by a linear function of the firing rates, Q_a. For example, a scalp potential may involve contributions from several populations, with various weights (that may include filtering by volume conduction effects). In this case, at given ω,

$$\psi(k,\omega) = \boldsymbol{MQ} = \boldsymbol{MTN}, \quad (68)$$

where **M** is an N′-element row vector of complex-valued measurement coefficients that encode spatiotemporal filtering characteristics, phase shifts, etc. For example, the coefficients of the matrix **M** can be chosen such that $\psi(\mathbf{k},\omega)=\varphi_{ab}(\mathbf{k},\omega)$. Further classes of measurement functions are those relating the neural activity to, for example, local field potentials, multiunit activity, the blood oxygen level–dependent (BOLD) response that forms the basis of functional magnetic resonance imaging (fMRI), the metabolic responses underlying positron emission tomography

(PET), or single-photon emission computed tomography (SPECT). In what follows, we will implicitly absorb **M** into **T** for simplicity.

Dispersion and stability. The dispersion relation of linear waves in the system is given by

$$\det A(\mathbf{k},\omega)=0, \qquad (69)$$

and the system is stable at a particular real **k** if all the frequency roots of this equation have negative imaginary parts. If the steady state is stable for all **k**, spectra and other properties of the linear perturbations can be self-consistently defined; otherwise a fully nonlinear analysis is needed.

Spectra. The power spectral density of ψ at **k** and ω is

$$P(\mathbf{k},\omega)=|\psi(\mathbf{k},\omega)|^2=|\mathbf{TN}|^2. \qquad (70)$$

The frequency and wavenumber spectra are then

$$P(\omega)=\int\frac{d^2\mathbf{k}}{(2\pi)^2}P(\mathbf{k},\omega).$$

$$P(\mathbf{k})=\int\frac{d\omega}{2\pi}P(\mathbf{k},\omega). \qquad (71,72)$$

A position-dependent frequency cross-spectrum can be calculated from Equation 70:

$$P(\mathbf{r},\mathbf{r}',\omega)=\langle\psi(\mathbf{r},\omega)\psi^*(\mathbf{r}',\omega)\rangle, \qquad (73)$$

where the angle brackets denote an average over multiple trials and/or over the phase of the exogenous stimuli that drive the system. The spectrum at a particular point, **r**, is P(**r**,**r**′,ω).

Correlation and coherence functions. In steady state, the two-point correlation function can be obtained from Equation 73 via the Wiener-Khinchtine theorem, giving

$$C(\mathbf{r},\mathbf{r}',T)=\int\frac{d\omega}{2\pi}e^{-i\omega T}P(\mathbf{r},\mathbf{r}',\omega). \qquad (74)$$

In the case where the system is statistically uniform, Equation 74 depends only on the separation **R**=**r**′−**r**, giving

$$C(\mathbf{R},T)=\int\frac{d\omega}{2\pi}\int\frac{d^2\mathbf{k}}{(2\pi)^2}e^{i\mathbf{k}\cdot\mathbf{R}-i\omega T}|\psi(\mathbf{k},\omega)|^2, \qquad (75)$$

Where

$$\psi(\mathbf{r},t) = \int \frac{d\omega}{2\pi} \int \frac{d^2\mathbf{k}}{(2\pi)^2} e^{i\mathbf{k}\cdot\mathbf{r} - i\omega t}\mathbf{T}(\mathbf{k},\omega)\mathbf{N}(\mathbf{k},\omega) \quad (76)$$

has been used and the arguments of **T** and **N** have been shown for emphasis. At **R**=0, Equation 74 becomes the Fourier transform of the local power spectrum. In terms of the above expressions, the normalized correlation function and the coherence function, which are both used widely in the literature, are

$$\rho(\mathbf{r},\mathbf{r}',T) = \frac{C(\mathbf{r},\mathbf{r}',T)}{[C(\mathbf{r},\mathbf{r}',T)C(\mathbf{r},\mathbf{r}',T)]^{1/2}},$$

$$\gamma(\mathbf{r},\mathbf{r}',\omega) = \frac{P(\mathbf{r},\mathbf{r}',\omega)}{[P(b,f,r,\mathbf{r}',\omega)P(\mathbf{r},\mathbf{r}',\omega)]^{1/2}}, \quad (77,78)$$

respectively.

Time series and evoked potentials. The time series of ψ at a given point can be obtained via the transfer function, by first calculating the Fourier form of the stimuli that generate it; these can be background noise sources, discrete impulses, or sinusoidal drives, for example. In the case of an impulsive stimulus, the resulting ERP is obtained by setting

$$\mathbf{N}(\mathbf{k},\omega) = 1. \quad (79)$$

Similarly, for a spatially uniform sinusoidal drive, the resulting steady state evoked potential (SSEP) is obtained by using

$$\mathbf{N}(\mathbf{k},\omega) = \pi\left[\delta(\omega-\omega_0)e^{i\phi} + \delta(\omega+\omega_0)e^{-i\phi}\right], \quad (80)$$

where ω_0 is the drive frequency and φ is its phase.

Case of one long-range population. An important case, in many applications, is the situation where spatial spreading of activity is dominated by the axons of one population, typically because they have the longest range, are most numerous, or have the highest axonal velocity. In this case, one can ignore the **k** dependence in the other propagators, and it becomes possible to express the transfer function with elements of the form

$$T_{aj} = \frac{A_{aj}(\omega)}{k^2 + q_{ab}^2(\omega)}, \quad (81)$$

where q_{ab}^2 is typically a complicated expression depending on the various $J_{ab}(\omega)$.

Heterogeneous Connectivity in Neural Fields

The brain's network dynamics depend on the connectivity within individual areas, as well as generic and specific patterns of connectivity among cortical and subcortical areas [4],[9],[98]. Intrinsic or intracortical fibers are confined to cortical gray matter in which the cortical neurons reside; these intrinsic connections define the local connectivity within an area. Intracortical fibers are mostly unmyelinated and extend laterally up to 1 cm (in the human brain) with excitatory and inhibitory connections. Their distribution is mostly invariant under spatial translations (homogeneous) [84],[99], which fits the assumptions on the connectivity function in neural fields so far. On the other hand, the corticocortical (extrinsic) fiber system contains fibers which leave the gray matter and connect distant areas (up to 20 cm [84]). This fiber system is myelinated, which increases the transmission speed by an order of magnitude, and is not invariant under spatial translations (heterogeneous); in fact it is patchy [99]. Due to finite transmission speeds, time delays of interareal communication can reach 50–100 ms [84], which is not negligible. Several studies have focused on spatially continuous neural fields, which describe the temporal change of neural activity on local scales, typically within a brain area (see [4],[54],[55] for reviews), assuming homogeneous connectivity and time delays. As discussed in the previous section, early attempts include neural field theories which approximate the large-scale components of the connectivity matrix as translationally invariant and decaying over space [45],[48],[50]. These approaches have been successful in capturing key phenomena of large-scale brain dynamics, including characteristic EEG power spectra [45],[50], epilepsy [92], and MEG activity during sensorimotor coordination [49]. Here we review extensions of these efforts and address network stability under variation of (i) intracortical (intrinsic) connectivity, (ii) transmission speed, and (iii) length of corticocortical (extrinsic) fibers. All three anatomical attributes undergo characteristic changes during the development of the human brain and its function, as well changing in the aged and diseased brain (see [9] for an overview). As a first step, we can split the connectivity function, W, into two parts, the homogeneous connectivity, $W_{hom}(|x-y|)$, which depends only on the distance, and the heterogeneous connectivity, $W_{het}(x,y)$, which captures the effects of the extrinsic fiber system (for an alternative approach with applications to visual gamma phenomena, see [100]–[102]). We can then rewrite the neural field equation as follows:

$$\dot{\mu} = -\zeta\mu + \alpha \int_{\Gamma} W_{hom}(|x-y|)\varsigma[\mu(y, T_{c^{hom}})]dy + \int_{\Gamma} W_{het}(x,y)\varsigma[\mu(y, T_c)]dy, \tag{82}$$

where $T_c = t - |x-y|/c$ and c is the propagation speed through the heterogeneous corticocortical or extrinsic connections. These fibers are myelinated and hence to be distinguished from the typically unmyelinated (hence slower) intracortical fibers. The latter intrinsic fibers have a transmission speed of c^{hom} and a transmission delay $T_{c^{hom}} = t - |x-y|/c^{hom}$. If W_{het} describes the connectivity of n areas, then it can always be written as a sum of two-point connections via

$$W_{het}(x,y) = \sum_{i,j=1}^{n} v_{ij}\delta(x-x_i)\delta(y-x_j), \quad i \neq j, \tag{83}$$

where $v_{ij} \in \Re$ again represents the coupling strength between areas at x_i and x_j. The fixed point solution is given by $\mu_0(x)$ with $\dot{\mu}_0(x) = 0$. To gain insight into the linear stability of this equilibrium solution $\mu_0(x)$, we perform a mode expansion of $\mu(x,t)$ into a set of spatial basis functions $\{\varphi_k(x)\}$ such that

$$\mu(x,t) = \mu_0(x) + \int_{-\infty}^{\infty} dk\, \xi_k(t)\phi_k(x), \tag{84}$$

where $\xi_k(t)$ is the time-dependent amplitude related to the spatial basis function $\varphi_k(x)$. The adjoint set of spatial biorthogonal basis functions is denoted by $\{\phi_k^{\dagger}(x)\}$. It will be generally true (except in degenerate cases) that only one spatial pattern will become unstable first. For simplicity, we consider the stationary solution $\mu_0(x)=0$ to be the rest state and consider its deviations $\mu(x,t)=\xi_{k'}(t)\varphi_{k'}(x)$+c.c., where c.c. denotes the complex conjugate. Then the linear stability of each temporal mode is given by

$$(\partial_t+\zeta)\xi_k(t)=\tilde{\alpha}\int_\Gamma dy\int_\Gamma dx\, W_{hom}\varphi_k^{\dagger}(x)\varphi_k(y)\xi_k(T_{c^{hom}})$$
$$+\sum_{i,j=1}^{n}\tilde{v}_{ij}\varphi_k^{\dagger}(x_i)\varphi_k(x_j)\xi_k(t-d/c),\qquad (85)$$

where $d=|x_i-x_j|>0$. Also, $\tilde{\alpha}=\alpha(\partial S/\partial\mu)$ and $\tilde{v}_{ij}=v_{ij}(\partial S/\partial\mu)$ (but for simplicity, we drop the tilde in $\tilde{v}_{ij}$ from now on).

Let us pause for a moment and reflect upon the significance of Equation 85. Equation 85 describes the rate of (temporal) change, $\partial_t\xi_k(t)$, of its corresponding spatial neural activation pattern, $\varphi_k(x)$. This pattern will change as a function of its own spatial configuration, $\varphi_k(x)$, the connections (W_{hom} and v_{ij}), and, last but not least, the transmission times of information exchange, $T_{c^{hom}}$ and d/c. If the rate of change, $\partial_t\xi_k(t)$, is positive, then the particular pattern $\varphi_k(x)$ is unstable, otherwise it is stable. In other words, Equation 85 identifies quantitatively how a particular neural activation is impacted by its local and global connectivity in a biologically realistic environment, including signal exchange with finite and varying (intracortical versus corticocortical) transmission speeds. Every treatment of the interplay of anatomical connectivity (local and global connections) and functional connectivity (network dynamics) will have to be represented in the form of Equation 85 or a variation thereof. In this sense, we have here achieved our goal stated in the introduction of this section.

To illustrate the effects of interplay between anatomical and functional connectivity, we discuss a simple example following [103],[104]. We assume that there exists only a single corticocortical fiber with terminals at locations x_1 and x_2, that is n=2. Then we have an architecture as shown in Figure 1. Our objective is to identify the stability boundaries of the rest state activity, here the equilibrium solution $\mu_0(x)=0$.

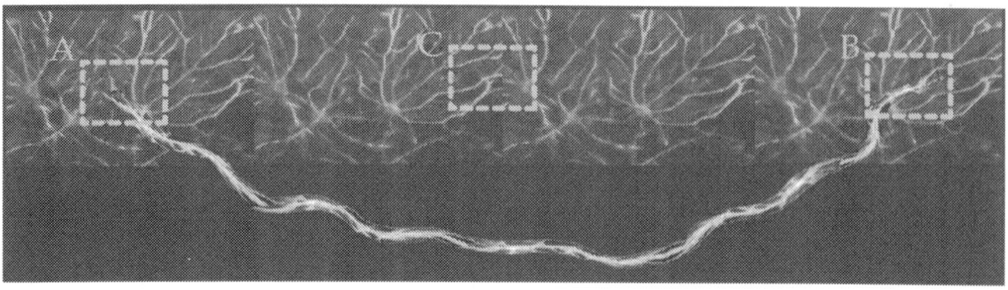

Figure 1. Anatomical connectivity, $W=W_{hom}+W_{het}$, comprising homogeneous and heterogeneous connections. The intracortical connections are illustrated as densely connected fibers in the upper sheet and define the homogeneous connectivity W_{hom}. A single fiber connects the two distant regimes (A) and (B) and contributes to the heterogeneous connectivity, W_{het}, whereas regime (C) has only homogeneous connections.

We will consider eigenfunctions of the form $\varphi_k(x)=e^{ikx}$. Changing the variables such that $z=y-x$ and assuming a solution of the form $\xi_k(t)=e^{\lambda t}$, $\lambda \in C$, the stability condition can then be determined by the following characteristic equation:

$$\lambda+\zeta=\tilde{\alpha}\int_{\Gamma} W_{hom}(|z|)e^{-\lambda|z|/c^{hom}}e^{ikz}dz+$$

$$\frac{1}{2}\left(v_{12}e^{ikd}+v_{21}e^{-ikd}\right)e^{-\lambda d/c}, \tag{86}$$

Linear stability of Equation 86 is obtained if $Re[\lambda]<0$ and is lost, according to [105], at $Re[\lambda]=0$, that is $\lambda=i\omega$. Figure 2 shows various connectivity kernels, W_{hom}, that are often found in the literature.

Qubbaj and Jirsa [104] discussed the properties of the characteristic Equation 86 in detail, considering separately the special cases of symmetric and asymmetric connectivity, W. The characteristic equation defines the critical boundary in the parameter space of v_{12}, v_{21}, c, c^{hom}, at which the resting activity, $\mu_0(x)=0$, becomes unstable. Recall that c and c^{hom} are the conduction velocities along extrinsic and intrinsic axons, respectively.

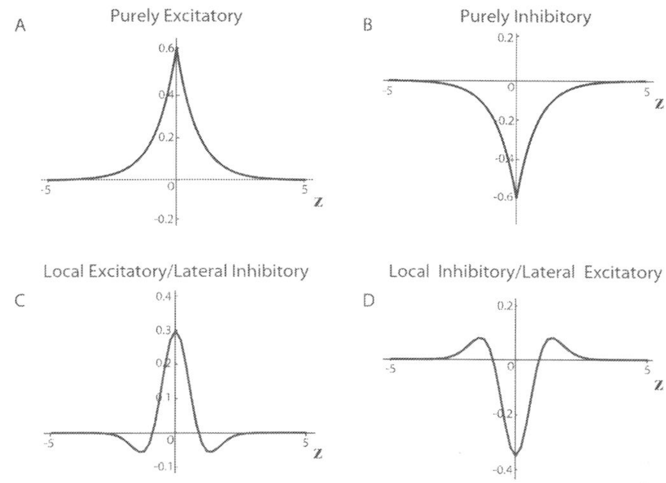

Figure 2. Typical homogeneous connectivity kernels, $W_{hom}(z)$, used for local architectures plotted as a function of spatial distance z. Purely excitatory connectivity is plotted in (A); purely inhibitory in (B); center-on, surround-off in (C); and center-off, surround-on in (D). The connectivity kernel in (C) is the most widely used in computational neuroscience.

The general result of [104] can be represented as a critical surface separating stable from unstable regimes as shown in Figure 3. Here the critical transmission delay, $\tau=d/c$, through the heterogeneous fiber is plotted as a function of the real and imaginary part of the eigenvalue of the connectivity, W. Essentially a heterogeneous fiber with symmetric weights, $\nu_{21}=\nu_{12}=\nu$, has only real eigenvalues, whereas asymmetries result in imaginary components. We find that for positive values of ν greater than a critical value, the system becomes unstable through a non-oscillatory instability for all values of c, c^{hom}, (bold line in Figure 3). Within the cylindrical component of the surface, the equilibrium of the system remains always stable for all values of c, c^{hom}, and hence a time delay shows no effect. In the other regimes of the critical surface, the system typically destabilizes via an oscillatory instability, $\omega\neq0$, and is sensitive to time delays. The surface shown in Figure 3 represents the lower bound of stability with $c/c^{hom}=1$. Increases of the ratio, c/c^{hom}, equivalent to increases of degree of myelination, result in a larger enclosed volume by this surface, i.e., increase of stability. The largest stability region is that for a purely inhibitory kernel followed by that of a local inhibitory and lateral excitatory kernel. The next largest involves a local excitatory and lateral inhibitory kernel. The smallest stability region is obtained for a purely excitatory kernel.

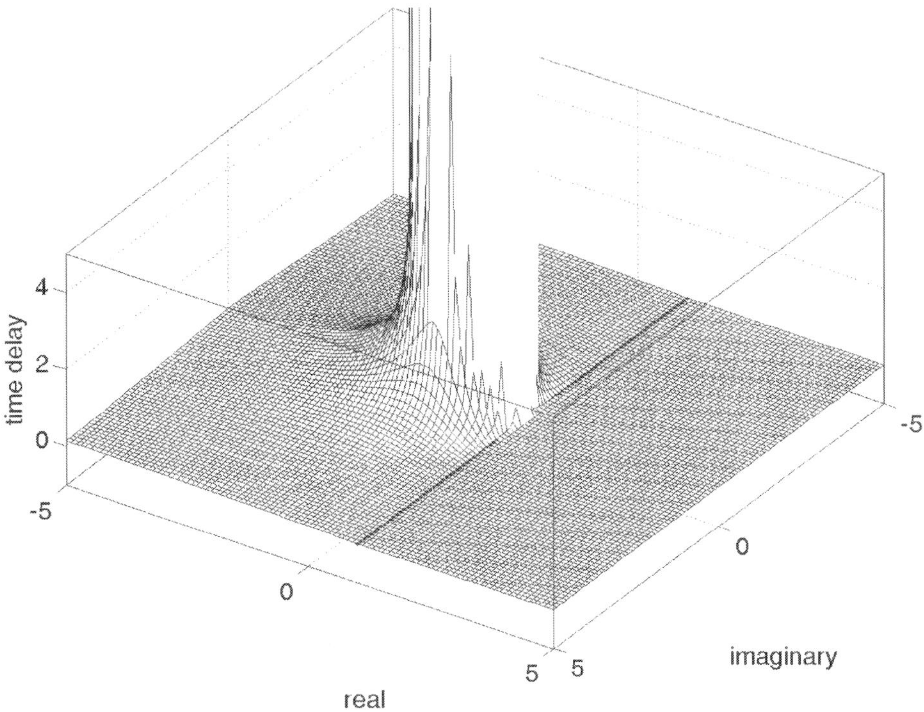

Figure 3. Minimal stable regions for the equilibrium state of a neural field as a function of its connectivity and time delay τ=*d*/*c*. The critical surface, at which the equilibrium state undergoes an instability, is plotted as a function of the real and imaginary part of the eigenvalue of its connectivity, W. Regimes below the surface indicate stability, above instability. The vertical axis shows the time delay via transmission along the heterogeneous fiber.

A surprising result is that all changes of the extrinsic pathways have the same qualitative effect on the stability of the network, independent of the local intrinsic architecture. This is not trivial, since despite the fact that extrinsic pathways are always excitatory the net effect on the network dynamics could have been inhibitory, if the local architecture is dominated by inhibition. Hence qualitatively different results on the total stability could have been expected. Such is not the case, as we have shown here. Obviously the local architecture has quantitative effects on the overall network stability, but not qualitatively differentiated effects. Purely inhibitory local architectures are most stable, purely excitatory architectures are the least stable. The biologically realistic and interesting architectures, with mixed excitatory and inhibitory contributions, play an intermediate role. When the stability of the network's fixed point solution is lost, this loss may occur through an oscillatory instability or a nonoscillatory solution. The loss of stability for the nonoscillatory solution is never affected by the transmission speeds, a direct physical consequence of its zero frequency allowing time for all parts of the system to evolve in unison. The only route to a non-oscillatory instability is through the increase of the heterogeneous connection strength. For oscillatory instabilities, the situation is completely different. An increase of heterogeneous transmission speeds always causes a stabilization of the global network state. These results are summarized in Figure 4.

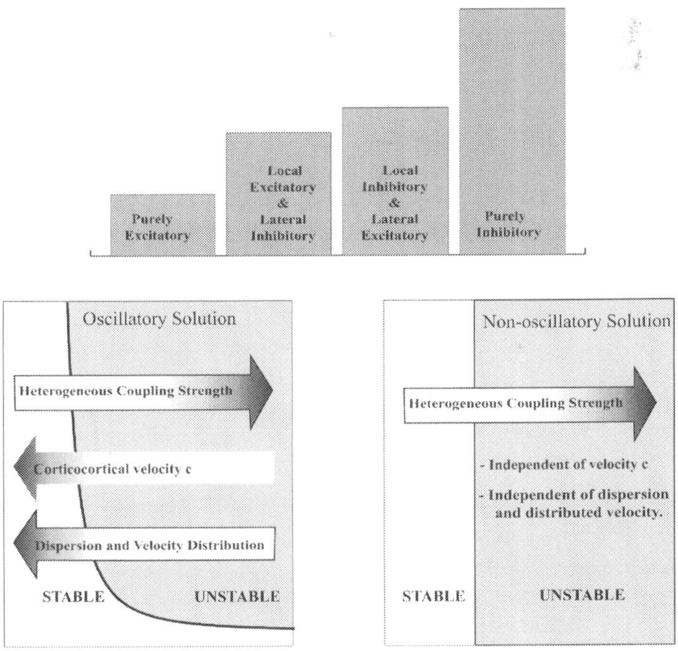

Figure 4. Summary of the stability changes of a neural field with mixed (local/global) connectivity. (Top) The relative size of stability area for different connectivity kernels. (Bottom) Illustration of change of stability as a function of various factors. Gradient within the arrows indicates the increase of the parameter indicated by each arrow. The direction of the arrow refers to the effect of the related factor on the stability change. The bold line separating stable and unstable regions indicates the course of the critical surface as the time delay changes.

NUMERICAL SIMULATIONS: ENSEMBLE ACTIVITY FROM NEURONAL TO WHOLE BRAIN SCALES

This section illustrates neuronal ensemble activity at microscopic, mesoscopic, and macroscopic spatial scales through numeric simulations. Our objective is to highlight some of the key notions of ensemble dynamics and to illustrate relationships between dynamics at different spatial scales.

Ensemble Dynamics at the Microscopic Scale

To illustrate ensemble dynamics from first principles, we directly simulate a network of coupled neurons which obey deterministic evolution rules and receive both stochastic and deterministic inputs. The system is constructed to embody, at a microscopic level, the response of the olfactory bulb to sensory inputs, as originally formulated by Freeman [38], [106]–[108]. Specifically, in the absence of a sensory input, neurons fire sporadically due to background stochastic inputs. The presence of additional synaptic currents due to a sensory input (e.g., inhaled odor) evokes a bifurcation onto a limit cycle or chaotic attractor. Note that in this section we simulate dynamics at the scale of coupled individual neurons. We can derive directly the predicted ensemble mean response by simply summing over all neurons. We compare this with an explicit model of neural mass dynamics at the mesoscopic scale in the subsequent section.

Microscopic Evolution Equations

Each neuron is modeled as a planar reduction of the Hodgkin-Huxley model [109],[110], namely,

$$\frac{dV_i(t)}{dt} = \sum_{ion} f_{ion}[V_i(t)] + I, \tag{87}$$

where f_{ion} introduces conductance-determined transmembrane currents through voltage-dependent channels, ion={Na^+,K^+} and I are synaptic currents. The planar reduction has slow potassium channel kinetics but fast sodium channels, whose states vary directly with transmembrane potential [111]. Synaptic currents are modeled, for the present purposes, to arise from three sources,

$$I = \sum_j H_c\left[V_j\left(t - \tau_j\right)\right] + I_{noise} + I_{sensory}. \tag{88}$$

The first term represents recurrent feedback from neurons within the ensemble due to their own firing. The coupling term, H_c, incorporates both the nature of the (all-to-all) within-ensemble coupling and the EPSP with parametric strength c. For the present purposes, the EPSP consists of a brief steady current whenever the presynaptic neuron is depolarized. The external currents, I_{noise}, introduce stochastic inputs (e.g., from brain stem inputs) and are modeled as a constant flow with a superimposed Poisson train of discrete pulses. The final term, $I_{sensory}$, models sensory input, consisting of a constant synaptic current to a subset of neurons, whenever the sensory stimulus is present. Hence this system permits an exploration of the relative impact of the flow (deterministic) and diffusive (stochastic) effects as embodied at the ensemble level by the Fokker-Planck equation (Equation 20) at the neuronal network level. The Nernst potentials, conductances, and background current are set so that, in the absence of noise and sensory inputs, each neuron rests just below a saddle-node bifurcation to a limit cycle [109]. This implies that neurons are spontaneously at rest (quiescent) but depolarize with a small perturbation. If the perturbation is due to a stochastic train, then the neuron fires randomly at an average rate proportional to the stochastic inputs. However, following a small increase in the constant flow term, due to a sensory input, $I_{sensory}$, the quiescent state becomes unstable and the neuron evolves on a (noise-modulated) limit cycle. Figure 5 shows a stochastically driven neuron (A) compared to a noise-modulated periodic neuron (B). In the former case, the activity is dominated by the stochastic terms. In the latter case, the limit cycle dynamics dominate, although the stochastic inputs modulate the depolarization amplitude.

A

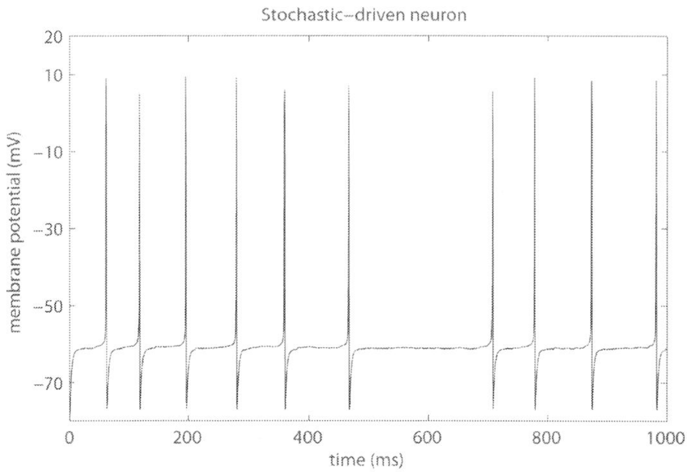

B

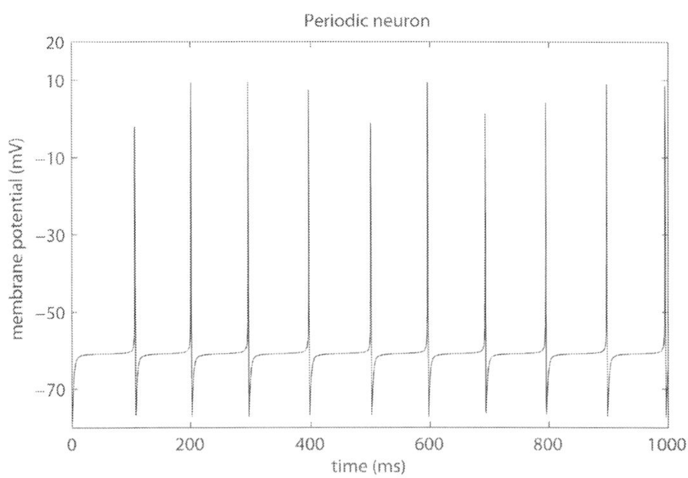

Figure 5. Planar spiking neuron. (A) Stochastically perturbed fixed point. (B) Limit cycle attractor.

Microscopic Dynamics

Figure 6 shows the results of simulating an ensemble of 250 neurons with a sensory input to all neurons between t=1,000 ms to t=3,000 ms. Figure 6A shows a raster plot of the neural spike timing whilst Figure 6B shows the simulated local field potential from the ensemble (=total current flow across all neurons).

As constructed, the effect of the input is to effect a bifurcation in each neuron from stochastic to limit cycle dynamics. The secondary effect of the appearance of limit cycle dynamics is to suppress the impact of the spatially uncorrelated stochastic inputs. Hence the neurons show an evolution towards phase locking, which was not present prior to the stimulus.

As evident in Figure 6B, the increased firing synchrony leads in turn to a marked increase in the simulated local field potentials as individual neurons begin to contribute concurrent ion currents. Once the stimulus ends, there is a brief quiescent phase because all of the neurons have just fired and require a short train of stochastic inputs before they commence firing again. Interestingly, there is evidence of damped mean-field oscillations in the ensemble following stimulus termination, abating after some further 800 ms.

To underscore the observation that the mean synaptic currents evidence an emergent phenomenon, and not merely the super-position of a bursting neuron, the time series of a single neuron is provided in Figure 6C. Clearly no burst is evident at this scale.

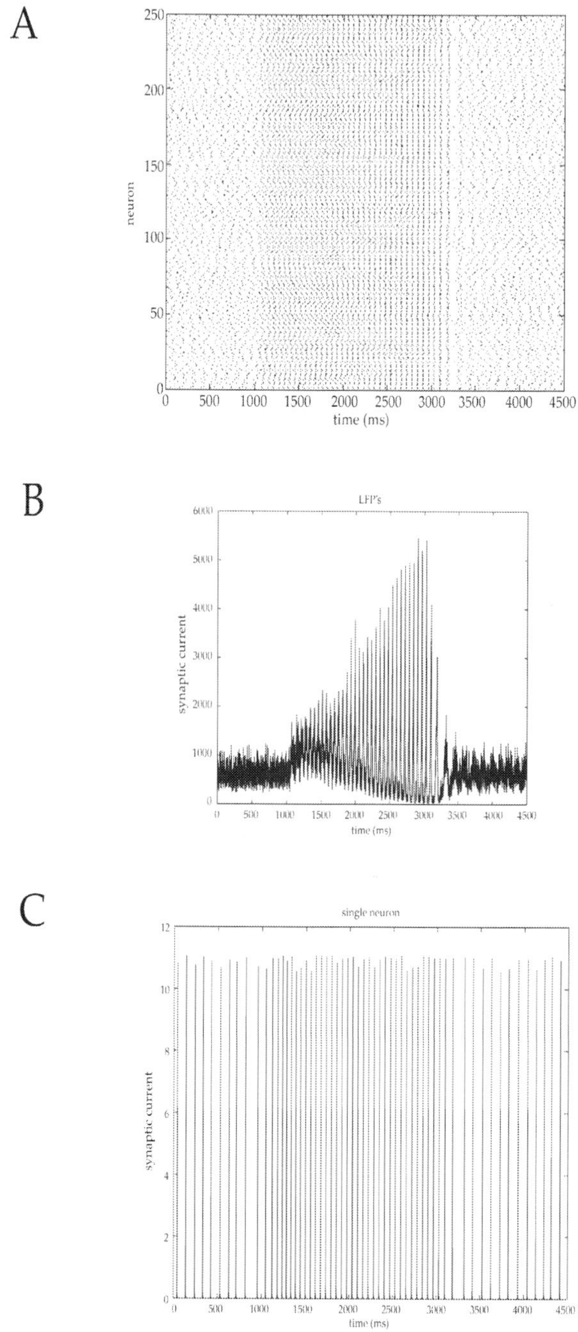

Figure 6. Results of simulating an ensemble of 250 neurons with sensory evoked synaptic currents to all neurons between *t*=1,000 ms and t=3,000

ms. (A) Raster plot. (B) Mean synaptic currents. (C) Time series of a single neuron. The effect of the input is to effect a bifurcation in each neuron from stochastic to limit cycle dynamics (phase locking), suppressing the impact of the spatially uncorrelated stochastic inputs. As evident in (A), the increased firing synchrony leads in turn to a marked increase in the simulated local field potentials. The mean synaptic currents evidence an emergent phenomenon, and not merely the superposition of a bursting neuron, as can be seen in (C): clearly no burst is evident at this scale.

The impact of the stimulus input on the density of the ensemble is shown in Figure 7, which shows the spike-timing difference of all neurons in the ensemble with respect to a randomly chosen seed-neuron. The mean spike-timing difference is 0 ms throughout the simulation.

This is because the system has complete symmetry, so that all neurons fire, on average, symmetrically before or after any other neuron. However, as evident in Figure 7A, the variance in relative spike-timing decreases dramatically during the stimulus interval. Of note is that the ensemble variance does not simply step down with the onset of the stimulus, but rather dynamically diminishes throughout the presence of the stimulus. When this occurs, the mean-field term continues to increase in amplitude. Figure 7B shows the evolution of the kurtosis (normalized so that a Gaussian distribution has a kurtosis of zero).

Prior to the stimulus, and reflecting the weak network coupling, the ensemble has a mesokurtotic (broad) distribution. It increases markedly following the stimulus onset, implying a dynamical evolution towards a leptokurtotic (peaked) distribution. That is, although the parameter values are static, the ensemble mean, variance, and kurtosis evolve dynamically in an inter-related fashion.

Hence this system exhibits time-dependent interdependence between its first, second, and fourth moments. This is the sort of coupling (between moments of the ensemble density) that neural mass models do not capture.

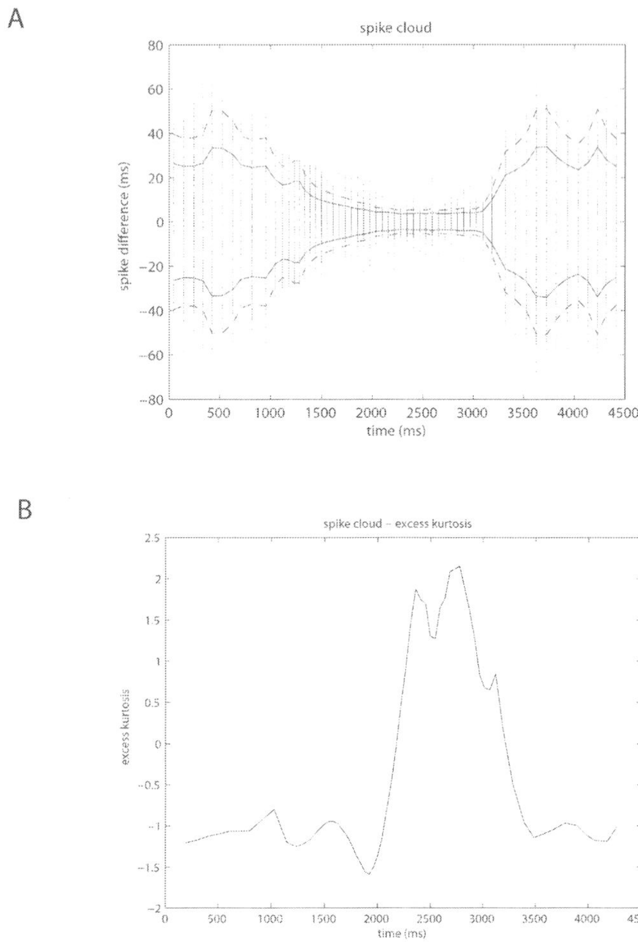

Figure 7. Contraction of spike-timing differences due to synaptic inputs. A seed neuron is chosen at random and the interneuron spike difference for all other neurons is plotted each time it spikes. (A) Solid and dashed lines show ±1 and ±1.5 standard deviations of the ensemble spike timing. (B) The normalized fourth moment (excess kurtosis) derived from a moving frame.

It is important to note that the spatiotemporal structure of the noise remains constant throughout the simulation, as does the intra-ensemble coupling. Hence the appearance of phase locking is an emergent feature of the dynamics and has not been imposed. A dynamic contraction of the ensemble cloud occurs whether the pre-existing noise continues unchanged during the stimulus input—hence increasing the firing rate of each neuron—or diminishes so that, in the absence of coupling, firing rates are kept the same on average. In the latter case (as in Figure 5), there is simply a change from stochastic to periodic firing. The ensemble cloud is visualized directly in Figure 8. The upper row shows the first return map for the ensemble over five consecutive time steps. For each neuron, this is defined as the inter-spike delay at time t=Tplotted against the inter-spike delay for the subsequent spike at t=T+1. Six such first return state-space values are plotted for all neurons. To control for changes in spike rate, these plots are normalized to the average firing rate. Values for the seed neuron used in Figure 7 are plotted in red. The left column shows the ensemble state, prior to the stimulus current. The right column shows the intra-stimulus activity. The contraction of the ensemble is seen clearly. In addition, the first return map confirms that individual neurons have stochastic dynamics prior to the stimulus, which change to periodic (i.e., a fixed point in the first return map) during the stimulus. The lower row of Figure 7 shows corresponding probability distributions of the inter-neuron spike-timing differences. This reiterates that not only does the distribution contract, but as the mean-field dynamics become strongly nonlinear, the ensemble kurtosis increases markedly from sub- to super-Gaussian.

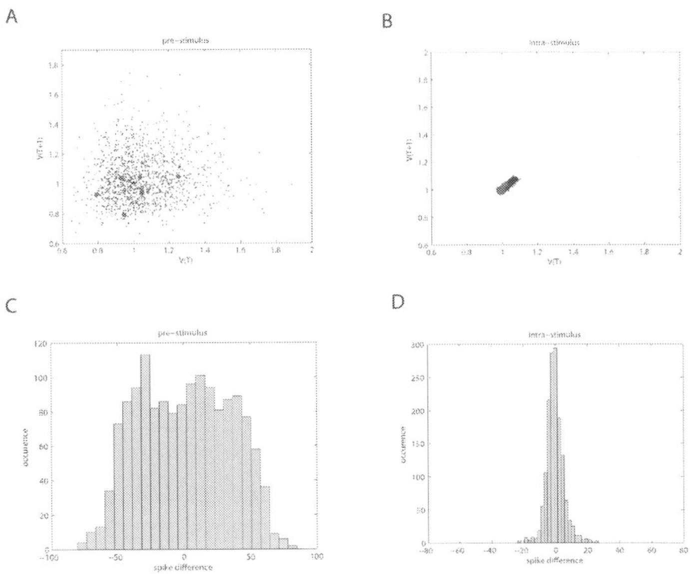

Figure 8. Contraction of spike-timing differences due to synaptic inputs. The left column shows the ensemble state, prior to the stimulus current. The right column shows the intrastimulus activity. Top row: First return map for

the cloud interspike delay over five consecutive time steps, before (A) and following (B) synaptic input. The plots are normalized to the average firing rate to control for changes in spike rate. Values for the seed neuron used in Figure 7 are plotted in red. Lower row (C,D): the corresponding spike timing histograms. The ensemble kurtosis increases markedly from sub- to super-Gaussian.

Neural Mass Dynamics at the Mesoscopic Scale

Whilst such simulations are illustrative, they are computationally intensive; even when limited to just 250 neurons at <5 s of integration time. As discussed in The Mean-Field Model section, it is possible to study a reduced model representing only the mean ensemble dynamics. This is essentially achieved by generalizing parameter values (such as ion channel thresholds) from individual point values to population likelihood values. Freeman [38] additionally introduced synaptic effects through convolving the inputs with a suitable response kernel as presented in Equation 37. For the simple illustration here, we do not introduce synaptic filtering.

Mesoscopic Evolution Equations

For the present purpose, we simulate a single mass with both excitatory and inhibitory neurons [52],[112]. Expected mean states of the ensemble excitatory neurons $\mu_e=\langle V_e\rangle$ are, as above, based upon Morris Lecar planar dynamics, with slow potassium channel dynamics. Inhibitory neurons, μ_i, respond passively to input from excitatory neurons and feedback to induce additional outward (rectifying) currents in excitatory cells. In the microscopic system considered above, interneuron coupling was via a direct pulse during presynaptic depolarization. At the mesoscopic scale, neuronal coupling is via neural firing pulse densities, ς_a, which capture the expected neuronal firing rate, given the mean neuronal transmembrane potential $\varsigma(\mu_a)$ for a=e,i. Assuming a Gaussian distribution of individual neuronal firing thresholds, one obtains a symmetric sigmoid-shaped function for ς_a as per The Mean-Field Model section. The dynamics are thus of the form

$$\frac{d\mu_e(t)}{dt} = \sum_{ion} f_{ion}[\mu_e(t)] + v_{ee}G(\varsigma_e) + v_{ie}G(\varsigma_i)$$

$$+ v_{ne}I_{noise} + v_{se}I_{sensory}, \quad (89)$$

$$\frac{d\mu_i(t)}{dt} = \nu_{ei}G(\varsigma_e) + \nu_{ni}I_{noise}, \tag{90}$$

where the function G represents the coupling between mean firing rates and induced synaptic currents. By targeting either Na^+ or Ca^{++} currents and including (postsynaptic) voltage-dependent effects, this function can incorporate, to a first-order approximation, a variable proportion of AMPA or NMDA-like kinetics [52]. The coefficients $\nu_{\alpha\beta}$ represent the synaptic density between excitatory (e) and inhibitory (i) populations or from the stochastic/noise (n) or sensory (s) inputs. Note that both populations receive stochastic inputs but only the excitatory population receives the sensory input $I_{sensory}$. The functions f_{ion} are the same as for the microscopic system including the slow potassium channel—although they are now parameterized by population-wide estimates.

Mesoscopic Dynamics

Figure 9 shows the response of a single neural mass to sensory evoked synaptic currents with the same temporal timing as for the microscopic system. Prior to the stimulus, the system is in a stable fixed point regimen. The stochastic inputs act as perturbations around this point, giving the time series a noisy appearance, consistent with the prestimulus microscopic ensemble activity. However, the mechanisms are quite distinct: Individual neurons within the microscopic ensemble fired stochastically, but at uncorrelated times. Hence, at the level of the ensemble, such individual events contribute in a piecemeal fashion. That is, although individual neurons exhibit nonlinear dynamics, the ensemble mean dynamics are (linearly) stable to the stochastic inputs until the background current is increased. In the mesoscopic case, the system as a whole is stable to small perturbations prior to the stimulus current. The temporally uncorrelated stochastic inputs are effectively filtered by the response properties of the system around this fixed point to yield the simulated activity.

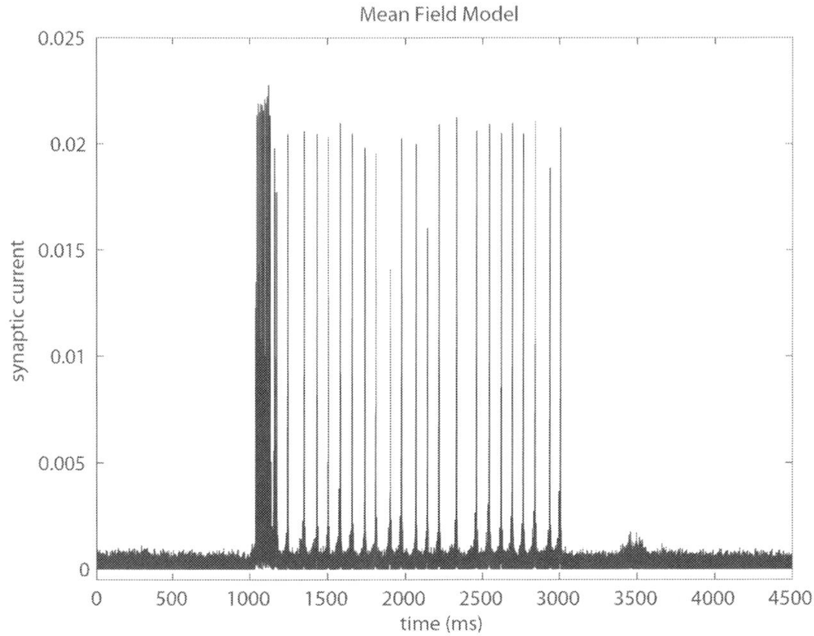

Figure 9. Mesoscopic neural mass model with sensory evoked synaptic currents from *t*=1,000 ms to t=3,000 ms. Prior to the stimulus, the system is in a stable fixed point regimen. The stochastic inputs act as perturbations around this point. Although individual neurons exhibit nonlinear dynamics, the ensemble mean dynamics are (linearly) stable to the stochastic inputs until the background current is increased. Then the fixed point state is rendered unstable by the stimulus current and large amplitude oscillations occur. These cease following stimulus termination.

In the mesoscopic neural mass, the fixed point state is rendered unstable by the stimulus current and large amplitude oscillations occur. These cease following stimulus termination. This accords with the appearance of stimulus-evoked nonlinear oscillations in the ensemble-averaged response of the microscopic system. In both models, such oscillations abate following stimulus termination. Hence, at a first pass, this neural mass model captures the mean-field response of the microscopic ensemble to a simulated sensory stimulus.

What is lost in the neural mass model? In this model, activity transits quickly from a noise-perturbed fixed point to large amplitude nonlinear oscillations. A brief, rapid periodic transient is evident at the stimulus onset (1,000 ms). The system subsequently remains in the same dynamic state until the stimulus termination. This hence fails to capture some of the cardinal properties of the microscopic ensemble, namely the coupling between the first and second moments (mean and variance). As discussed above, this process underscores the dynamical growth in the mean-field oscillations and the interdependent contraction of the interneuron spike timing variance shown in Figures 6 and 7. Because of this process the system is far more synchronized than prior to the stimulus. This synchronization leads to the damped mean-field oscillations evident in the ensemble system after the stimulus termination (3,200 ms→4,500 ms), because there is a more coherent ensemble-wide response. What is gained in the neural mass model? The addition of a third dimension (i.e., the inhibitory mean activity) to the dynamics enables the expression of chaotic dynamics [52],[112]. Hence the flow terms in the neural mass model contribute to the expression of aperiodic dynamics in addition to the stochastic inputs. This is not possible in the (planar) single neural dynamics of the microscopic system because chaotic dynamics require at least three degrees of freedom. Thus the dimension reduction afforded by the neural mass approximation allows the introduction of more complex intrinsic dynamics, permitting dynamical chaos. Whilst additional dimensions could be added to the microscopic neurons, this would add to an already significant computational burden. The massive reduction in the computational load of the neural mass approximation also allows extension of the spatial scale of the model by an array of neural masses, coupled to form a small patch of cortical tissue. Such a mesoscopic system can be endowed with additional structure, such as hierarchical [108], scale-free [113], multiscale [114], or small world [115]properties. For the present purposes, we couple a single input neural mass, as modeled above, hierarchically to a sheet with internal hyperbolic (i.e., scale-free) coupling. Intersystem coupling is purely excitatory-to-excitatory. Within the sheet, the coupling drops in proportion to spatial separation and is hence scale-free:

$$\frac{d\mu_e^j}{dt} = F\left(\mu_e^j, \mu_i^j, I_{noise}\right) + \sum_{k \neq j} \frac{C_{sheet} G\left(\varsigma_e^k\right)}{\left|x^j - x^k\right|} + \frac{C_{sens} G\left(\varsigma_e^{sens}\right)}{\left|x^j - x^{sens}\right|}. \tag{91}$$

where F incorporates all intrasystem dynamics as per Equation 89 and the indices numerate either the sensory node {sens} or the nodes within the sheet {sheet}. As above, synaptic currents are induced by the pulse density of the presynaptic neurons, rather than directly via individual presynaptic depolarization. The sensory node receives the only direct stimulus-induced currents,

$$\frac{d\mu_e^{sens}}{dt} = F\left(\mu_e^{sens}, \mu_i^{sens}, I_{noise}, I_{sens}\right) + \sum_{k \neq sens} \frac{C_{sheet} G\left(\varsigma_e^k\right)}{\left|x^{sens} - x^k\right|}. \tag{92}$$

The hierarchical nature of the system is embodied by the targeted nature of the sensory inputs and the separate parameterization of parameters that couple masses to or within the sheet, C_{sens} and C_{sheet}, respectively. It would also be possible to increase the degree of forward and backward asymmetry by incorporating purely AMPA-like kinetics for the former and NMDA-like kinetics for the latter, as has been proposed as a mechanism for perceptual inference [116],[117]. Figure 10 shows the response of the system when $C_{sens} > C_{sheet}$. Figure 10A shows the individual mean synaptic currents of all nonsensory nodes. Figure 10B shows the total synaptic currents averaged across the nonsensory sheet. Several features can be noted. For a start, despite use of the same parameters, the coupling of neural masses into an array or sheet leads to the appearance of spontaneous prestimulus activity. Stimulus-evoked activity from the sensory node reorganizes this activity from spatially incoherent to synchronized. Thus the array-averaged synaptic currents increase during the stimulus period. Hence the dynamics at this scale mirror those within the microscopic ensemble, which each node in this simulation is constructed to represent. However, at least in this simulation, the array does not exhibit a dynamic growth in the system-wide currents during the stimulus period. Presumably, if this did occur within the individual mesoscopic nodes, then the array-wide current may also grow dynamically. That is, one would anticipate coupling between moments across ensembles, as discussed in [38].

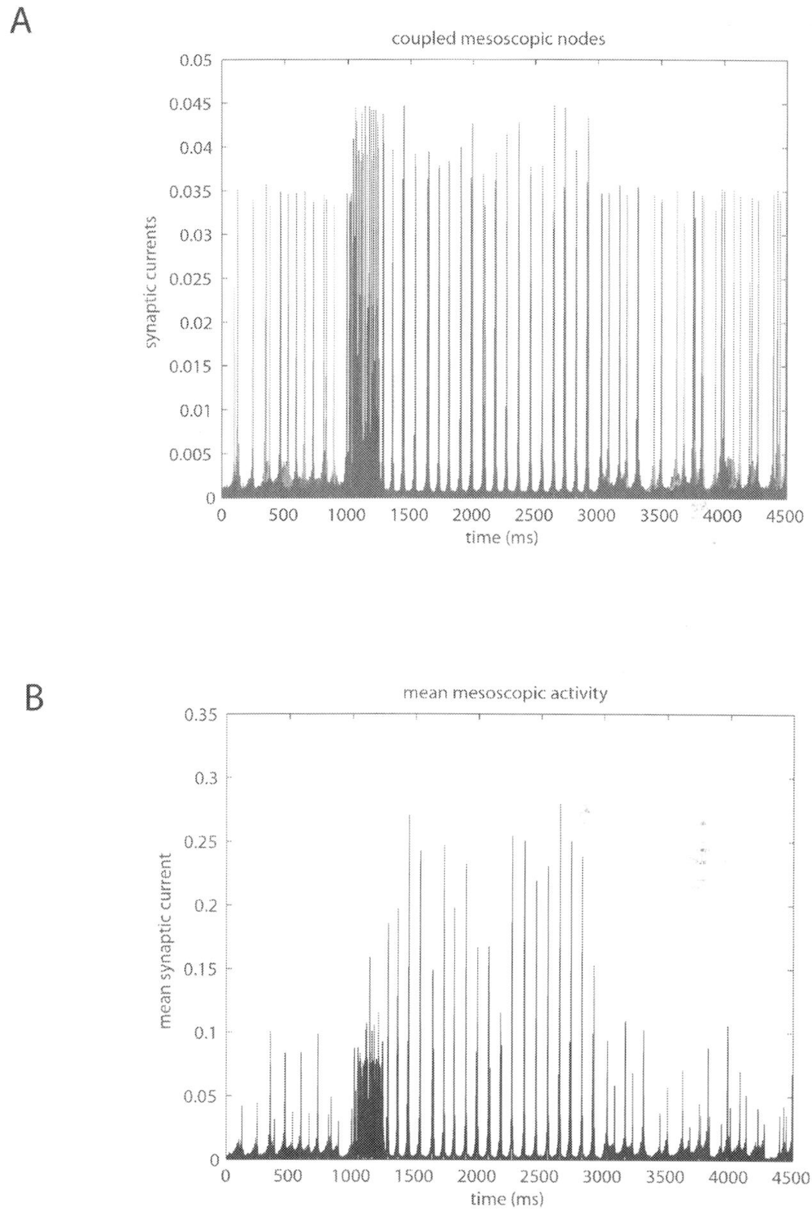

Figure 10. Coupled mesoscopic neural masses with sensory evoked synaptic currents into single sensory node from *t*=1,000 ms to t=3,000 ms $C_{sens}>C_{sheet}$. (A) Individual mean synaptic currents of all nonsensory nodes. (B) Total synaptic currents averaged across the nonsensory sheet. Stimulus-evoked activity from the sensory node reorganizes

Figure 11 shows the simulated activity following an increase in the intrasheet coupling such that $C_{sens} \approx C_{sheet}$. All other parameters are unchanged. Spontaneous prestimulus activity is clearly more coherent; consistent with a stronger internally determined dynamical state. The injection of the externally evoked sensory currents into this prior activity actually has a slightly desynchronizing effect, as evident as a decrease in the array-wide average response (Figure 11B). That is, the temporal mismatch between the within-sheet dynamics and the externally induced activity leads to more spatially complex dynamics (an increase in the spatial entropy and hence the information content of the system).

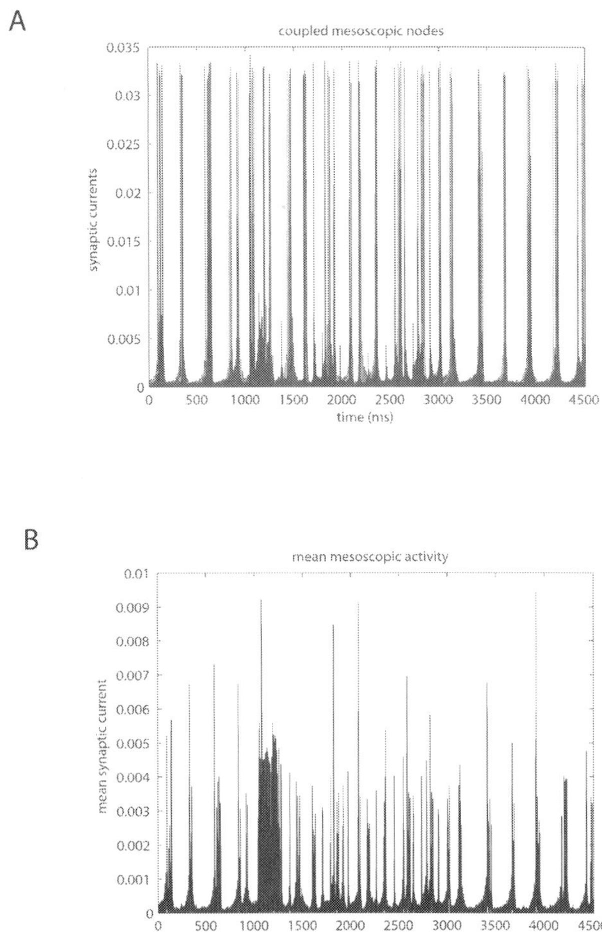

Figure 11. Coupled mesoscopic neural masses with sensory evoked synaptic currents into single sensory node from t=1,000 ms to t=3,000 ms $C_{sens} \approx C_{sheet}$.

All other parameters as in Figure 10. (A) Individual mean synaptic currents of all nonsensory nodes. (B) Total synaptic currents averaged across the nonsensory sheet. The injection of the externally evoked sensory currents into the prior activity actually has a slightly desynchronizing effect.

If the stochastic inputs, I_{noise}, are decreased below a threshold, then the spontaneous activity in the nonsensory array diminishes. The feedback effect of this quiescent activity is to suppress the stimulus-evoked activity in the sensory node. Hence there is a top-down mechanism for the complete suppression of sensory-evoked activity. Presumably, more subtle feedback effects may be possible if more forward versus backward receptor detail was modeled. In summary, these mesoscopic simulations impress a view of sensory-evoked effects as a reorganization of ongoing activity. Depending upon the ratio of internal to sensory-related coupling, this reorganization may lead to an increase or a decrease in the information content of the system dynamics.

Neural Field Dynamics at the Whole-Brain Scale

We now provide brief illustrations of sensory evoked and nonlinear activity as modeled by macroscopic field equations. As discussed in the section entitled Recent Developments in Neural Field Models, these incorporate synaptic filtering and axonal conduction delays, in addition to the population-wide conversion of membrane potentials into firing densities [50]. Significantly, they also permit the incorporation of subcortical systems, such as the thalamus [91]. Recent developments (see the Heterogeneous Connectivity in Neural Fields section) now allow elucidation of the impact of biologically relevant connection heterogeneities on the stability and conduction of cortical activity. The equations, their derivation, and relevant references are provided in the Recent Developments in Neural Field Models section.

Macroscopic Dynamics

Two crucial differences occur when moving to the macroscopic scale of the corticothalamic field model. Firstly, sensory inputs are modeled as entering the specific nuclei of the thalamus rather than directly into a cortical sensory node. The ensuing evoked corticothalamic activity can then be studied in a biologically informed framework. Second, while prestimulus activity is modeled as a noise-perturbed steady state, the system is not destabilized by sensory inputs. Instead, inputs evoke damped oscillations in the corticothalamic loop [87]. Figure 12 illustrates an example of sensory-evoked activity (Aquino et. al., unpublished data). Evoked afferent pulse densities are shown because they reflect more accurately the expected synaptic currents, through their action on postsynaptic neurons. The smooth spatiotemporal dispersion of the evoked cortical response and its time delayed corticothalamic volley are evident.

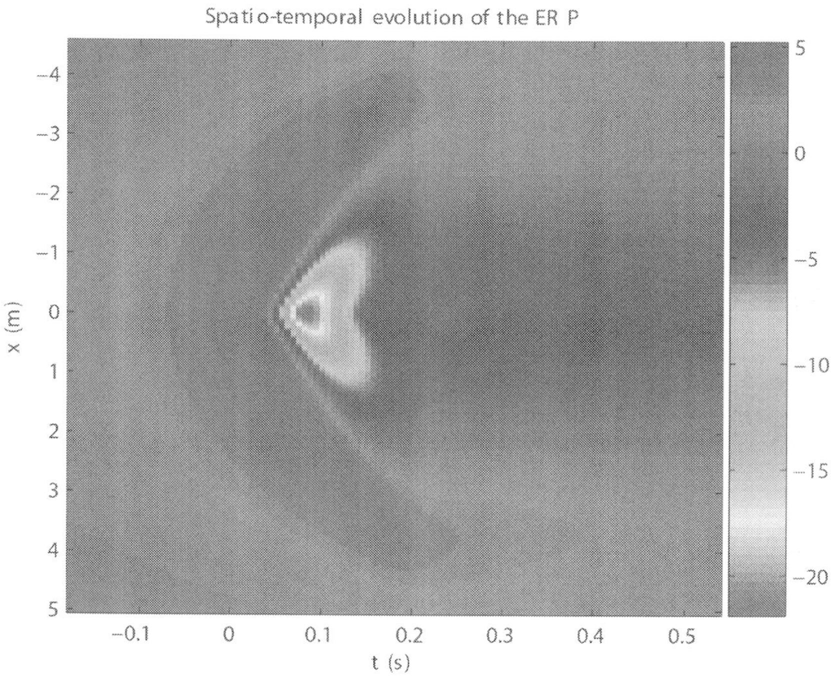

Figure 12. The spatiotemporal evolution of the evoked response of excitatory pulse densities $\mu(r,t)$ in an example of sensory-evoked activity (Aquino et. al., unpublished data).

The smooth spatiotemporal dispersion of the evoked cortical response and its time delayed corticothalamic volley are evident.

Summary of Numerical Simulations

These simulations give insight into the rich neural ensemble dynamics at different spatial scales that arise spontaneously, are evoked by sensory inputs, or follow changes in state parameters. The intention is to demonstrate concrete examples of ensemble dynamics under varying influences of flow and dispersion. The resulting dynamics can be seen to emerge from the interplay of stochastic dispersion and flow-determined ensemble contraction. The view of stimulus-evoked synaptic currents as evoking a bifurcation in neural ensemble activity derives largely from the formative work of Freeman, following detailed physiological studies of the olfactory bulb. One of the key outstanding problems is to reconcile the apparent discrepancy between proposals involving a key role of nonlinear dynamics (see also [118]) and the apparent success of mean-field models to predict measured evoked responses, without recourse to nonlinear dynamics. One approach is to construct a multiscale hierarchy, with self-consistent evolution equations at each scale and to couple the emergent dynamics from fine scales into the activity at coarser scales [114]. Although this permits small scale nonlinear activity to coincide with and influence stochastic macroscopic activity, it requires a somewhat elaborate framework. An alternative approach is to recursively enslave micro- and mesoscopic activity to predicted macroscopic field oscillations by driving them with the predicted mean-field synaptic currents. A problem here concerns the resulting emergence of sustained oscillations within mesoscopic activity and the possible causal inconsistency that this may entail.

The nature and strength of neuronal connectivity varies markedly when considered across the heirarchy of spatial scales. At the microscopic scale, connectivity is dense, concentrated equally in vertical and horizontal directions and, more or less isotropic when considered across different cortical regions. At mesoscopic scales, connectivity has a patchy, colmunar-dominated structure. At macroscopic scales, connectivity is sparser, can be considered exclusively horizontal, and is predominantly excitatory in nature. It is also characterized by heterogenous connections (large fiber tracts) which fulfill functionally defined roles. These rules are reflected in the abstractions and refinements of the models which address the different scales.

COGNITIVE AND CLINICAL APPLICATIONS

In this section, we present three distinct applications of neural ensemble modeling. We first illustrate a computational example, namely decision-making, as implemented in a mean-field model. We then illustrate healthy and pathological activity in neural field models. The healthy example is of the well-known psychophysical phenomenon of auditory streaming—the balance of segmentation versus integration in auditory perception. We then illustrate examples of spatiotemporal dynamics occurring in corticothalamic loops during Absence seizures.

Spiking Dynamics Underlying Decision-Making

Decision-making is a key brain function of intelligent behavior. A number of neurophysiological experiments on decision-making reveal the neural mechanisms underlying perceptual comparison, by characterising the neuronal correlates of behavior [119]–[121]. In particular, [119]–[124] have studied the neural mechanisms underlying perceptual comparison by measuring single-neuron responses in monkeys trained to compare two mechanical vibrations applied sequentially to the tip of a finger; the subjects have to report which of the two stimuli has the higher frequency. They found neurons in the ventral premotor cortex (VPC) whose firing rate depended only on the difference between the two applied frequencies, the sign of that difference being the determining factor for correct task performance [121]. These neurons reflect the implementation of the perceptual comparison process and may underlie the process of decision-making.

Figure 13 shows a biophysically realistic computational model for a probabilistic decision-making network that compares two mechanical vibrations applied sequentially (f1 and f2). The model implements a dynamical competition between neurons: The model enables a formal description of the transients (nonstationary) and probabilistic character of behavior (performance) by the explicit use, at the microscopic level, of spiking and synaptic dynamics of one-compartment IF neuron models. The network contains excitatory pyramidal cells and inhibitory interneurons. The excitatory recurrent postsynaptic currents (EPSCs) are mediated by AMPA (fast) and NMDA-glutamate (slow) receptors, whereas external EPSCs imposed on the network are driven by AMPA receptors only. Inhibitory postsynaptic currents (IPSCs) to both excitatory and inhibitory neurons are mediated by GABA receptors. Neurons are clustered into populations. There are two subtypes of excitatory population: namely, specific and nonselective. Specific populations encode the result of the comparison process in the two-interval vibrotactile discrimination task, i.e., if f1>f2 or f1<f2. The neurons in the two specific populations additionally receive external inputs encoding stimulus specific information. They are assumed to originate from the somatosensory area S2 and from the PFC, encoding the frequency of both stimuli f1 (stored) and f2 (present) to be compared during the comparison period, i.e., when the second stimuli is applied (see [125] for details).

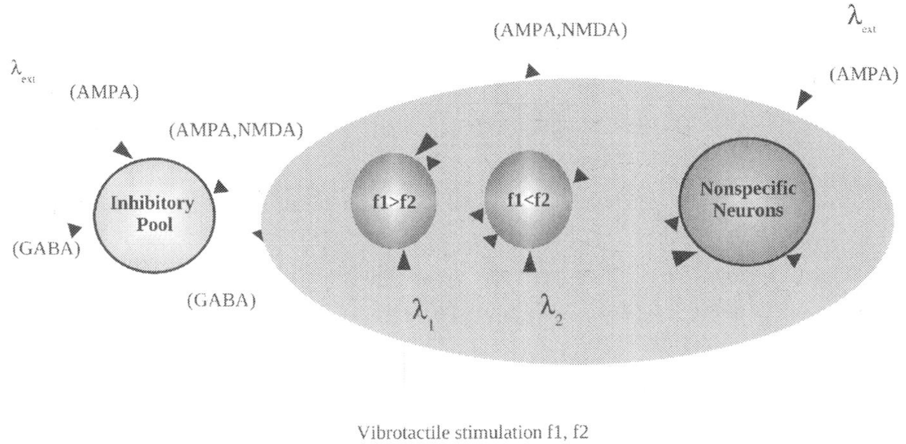

Figure 13. Decision-making neuronal network.

Minimal neurodynamical model for a probabilistic decision-making network that performs the comparison of two mechanical vibrations applied sequentially (f1 and f2). The model implements a dynamical competition between different neurons. The network contains excitatory pyramidal cells and inhibitory interneurons. The neurons are fully connected (with synaptic strengths as specified in the text). Neurons are clustered into populations. There are two different types of population: excitatory and inhibitory. There are two subtypes of excitatory population, namely: specific and nonselective. Specific populations encode the result of the comparison process in the two-interval vibrotactile discrimination task, i.e., if f1>f2 or f1<f2. The recurrent arrows indicate recurrent connections between the different neurons in a population.

The attractors of the network of IF neurons can be studied exhaustively by using the associated reduced mean-field equations. The set of stationary, self-reproducing rates, ν_i, for the different populations, i, can be found by solving a set of coupled self-consistency equations. This enables a posteriori selection of parameter regions that contain desired behaviors. In the present case, the essential requirement is that, for the stationary conditions, different attractors are stable. The attractors of interest for our task correspond to the activation (high spiking rates) or inactivation (low spiking rates) of the neurons in the specific populations $f1>f2$ and $f1<f2$. The activation of the specific population $f1>f2$ ($f1<f2$) and the simultaneous lack of activation of the complementary population $f1<f2$ ($f1>f2$), corresponds to an encoding single state associated with a motor response reporting the categorical decision $f1>f2$ ($f1<f2$). The lack of activation of both specific populations (spontaneous state) would correspond to an encoding state that cannot lead to a behavioral decision; i.e., there is no answer, or a motor response is generated randomly. The same happens if both specific populations are activated to the same degree (pair state). Because responses in animals are probabilistic in nature, the operating point of the network should be such that both categorical decisions, i.e., both states, are bistable. In addition, we have also shown that the model predicts a behavior consistent with Weber's law if, and only if, the spontaneous state is also a stable state, i.e., when the dynamical operating point of the network is in a regime of multistability. In this way, Weber's law informs the operating point of the network.

Figure 14 shows numerical simulations corresponding to the response of VPC neurons during the comparison period (to be contrasted with the experimental results shown in Figure 2 of [121]). This figure shows the average firing rate as a function of f1 and f2, obtained with the spiking simulations (diamond points correspond to the average values over 200 trials, and the error bars to the standard deviation). The lines correspond to the mean-field calculations. Black indicates $f1<f2$ ($f2=f1+8$ Hz) and gray indicates $f1>f2$ ($f2=f1-8$ Hz). The average firing rate of the population $f1<f2$ depends only on the sign of $f2-f1$ and magnitude of the difference, $|f2-f1|$; confirming again that Weber›s law cannot be encoded in the firing rate, but only in the probability with which that firing rate can be reached (that depends on the sign and magnitude of the difference between f1 and f2).

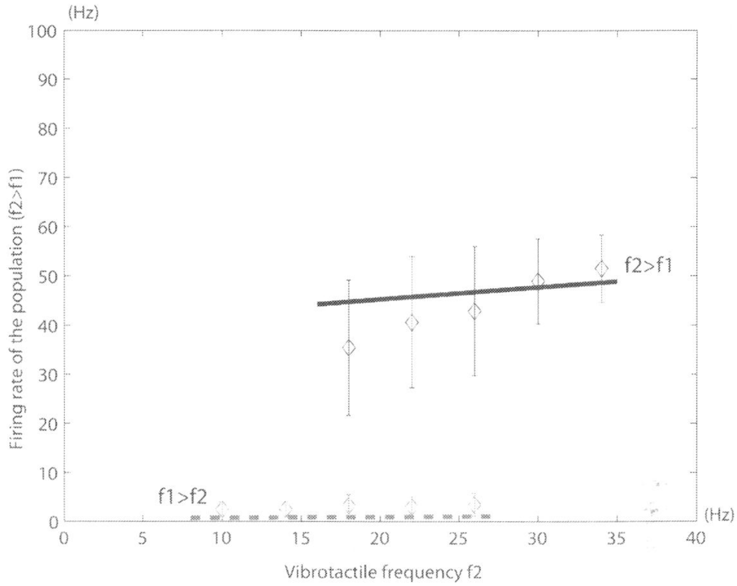

Figure 14. Average firing rate of a neuron as a function of f1 and f2, obtained with the spiking simulations of the response of VPC neurons during the comparison period (to be contrasted with the experimental results shown in Figure 2 of [121]). Diamond points correspond to the average values over 200 trials, and the error bars to the standard deviation. The lines correspond to the mean-field calculations: the black line indicates f1<f2 (f2=f1+8 Hz) and the red dashed line f1>f2 (f2=f1−8 Hz). The average firing rate of the population f1<f2 depends only on the sign of f2−f1 and magnitude of the difference, |f2−f1|.

Auditory Streaming

One of the applications of neural fields in cognitive processing is found in auditory scene analysis [126], particularly auditory streaming. Intuitively, auditory streaming or stream segregation is like listening to bass and soprano vocalists singing simultaneously. Although the two voices overlap in time, they clearly form two distinct percepts. In the laboratory, a similar effect can be created using sequences of tones. In a typical streaming experiment, two sequences are created using sets of high and low tones. Sequences vary in presentation rate and the frequency difference between the tones. The basic finding (see e.g., [127],[128]) is: (i) when the frequency separation is relatively small and/or the rate is relatively slow, listeners perceive a single integrated melody (or stream) and can accurately report the ordering of the tones, and (ii) when the frequency separation is relatively large and/or the rate relatively fast, people clearly perceive two segregated auditory streams, one with a higher pitch than the other. Essentially, there is a frequency–time boundary (known as the Fission Boundary, FB) beneath which all sequences are heard as integrated, regardless of instructions. There is a frequency–time boundary (known as the Temporal Coherence Boundary, TCB) above which all sequences are heard as segregated, regardless of instructions. In between these two boundaries exists a bistable region in which a sequence can be heard as either integrated or segregated depending upon instructions. Hysteresis phenomena are observed when traversing the bistable regime from either the FB or TCB. Many other auditory phenomena of a related nature are discussed in [128].

To capture the perceptual integration and segregation processes in the human brain, while accommodating contemporary brain theories [1],[2],[4], the authors of [126] proposed a tonotopically organized neural field for peripheral processing with projections to the higher areas that are responsible for cognitive integration. The neural field is tonotopically organized such that the frequency of the acoustic stimulus maps onto a location in neural space. The second nontonotopically organized system may either be represented by a neural field or a subnetwork. Its function is the classification of the peripheral spatiotemporal neural field dynamics. This classification process is not just a measurement (in which case the application of a simple measure to the neural field would suffice) but is itself a dynamic process. In fact, bistability and hysteresis turn out to be properties of the classification process rather than properties of the neural field dynamics. The dynamics of the neural field μ(x,t) are given by the wave Equation 39, which has been extended to accommodate auditory inputs s(x,t) as follows:

$$\left[\frac{1}{\gamma^2}\frac{\partial^2}{\partial t^2}+\frac{2}{\gamma}\frac{\partial}{\partial t}+1-r^2\nabla^2\right]\mu_v=\left(1+\frac{1}{\gamma}\frac{\partial}{\partial t}\right)\varsigma(\mu_v+s(x,t)), \tag{93}$$

where, as a reminder, γ=c/r, c is the speed of spike propagation, and r parameterizes the spatial decay of lateral interactions. The external input or stimulus to the neural sheet is $s(x,t)$: $\Re^2 \rightarrow \Re$, which contains all the spatiotemporal characteristics of the auditory input stream. Periodic boundary conditions, μ(0,t)=μ(L,t), t≥0, are used.

The second network is not tonotopically organized, hence its spatial dimension is of no relevance, when we consider only the competition of two streams. In fact, the ability to show multistable pattern formation is the only relevant property of the network and can be realized in multiple network architectures as discussed in previous sections. A simple multistable subsystem with its scalar state variable y(t) is given by the equation

$$\dot{y} = \varepsilon y - y^3 - I_0 + I(t), \tag{94}$$

where ε is a constant that captures all linear contributions. I_0 contains all constant contributions given rise to the rest state activity. The functional I(t) is specified as

$$I(t) = \int_0^L h(\mu(x,t))dx \quad h(n) = \begin{cases} 0, & n \leq \Omega \\ n, & n > \Omega. \end{cases} \tag{95}$$

where Ω is a neural activity threshold. Equations 93, 94, and 95 define the dynamics of a stream classification model in one of its simplest forms. Figure 15 illustrates the architecture of the model.

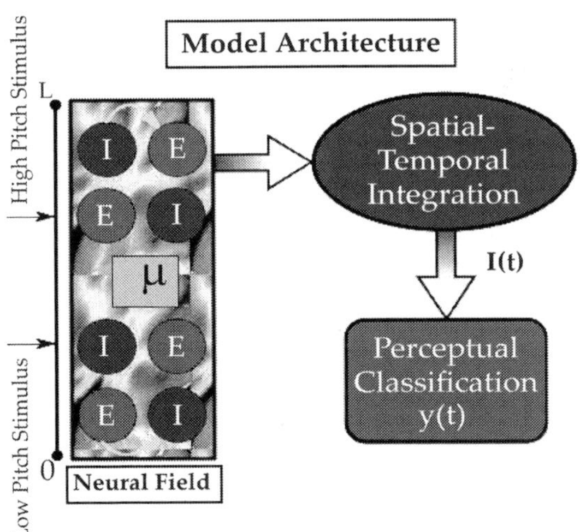

Figure 15. Cortical architecture of the model. The neural field is illustrated by the rectangular box showing the neural activity μ(x,t) composed of inhibitory and excitatory neurons. The input s(x,t) is provided at locations x_i via the Gaussian localization function $e^{-(x-x_i)^2/\delta_i}$ with width $\sqrt{\delta_i}$. The explicit model parameters used in the simulations are given in [126].

To understand van Noorden's results, we parametrize a sequence of consecutive tones by their frequency difference, Δf, and their interonset interval, IOI. As the neural field evolves, it is integrated across space and time yielding the time-dependent, but scalar, activity, I(t), driving the second system. I(t) represents the relevant information from the neural field, μ, as a spatiotemporally integrated activity measure, which depends on the amount of dispersion over space and time. The greater the dispersion, the greater will be the value of I(t) at a given time point. Figure 16 shows the contour lines of neural field activity over space x and time t for the bistable situation.

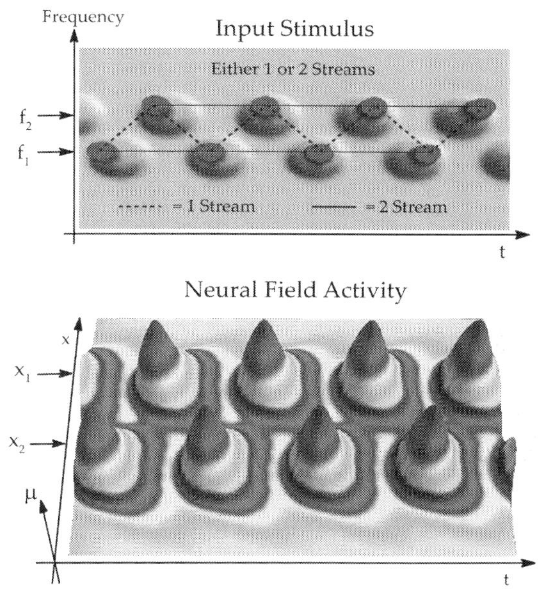

Figure 16. Bistable regime of auditory streaming. The stimulus sequences (top) and its resulting neural field dynamics (bottom).

The final state reached by the second system defined in Equation 94 with activity y will depend on I(t) and its own intrinsic dynamics. The curve of the flow is shifted up or down depending on I(t), creating either one positive or one negative fixed point. For an intermediate value of I(t), there is a bistable regime in which y can assume either one of the fixed points. The negative fixed point is identified with perceiving one stream and the positive fixed point with perceiving two streams. The time series for y are shown in Figure 17 for several different initial conditions of the activity y. After a transient the activity becomes stationary, displaying three possible scenarios (see Figure 17 from top to bottom): one stream only, or the bistable situation, in which either one integrated stream or two separate streams may be perceived, or finally two streams only. For each choice of Δf and IOI, the model Equations 93 and 94 are solved numerically and their stationary states determined. The results are plotted in the 2-D parameter space in Figure 18. TCB and the FB are reproduced in a manner that corresponds nicely to van Noorden's (1975) results including a bistable region [127]. Note that the exact experimental numerical values at which the boundaries occur vary from subject to subject and depend on the experimental methods employed [128].

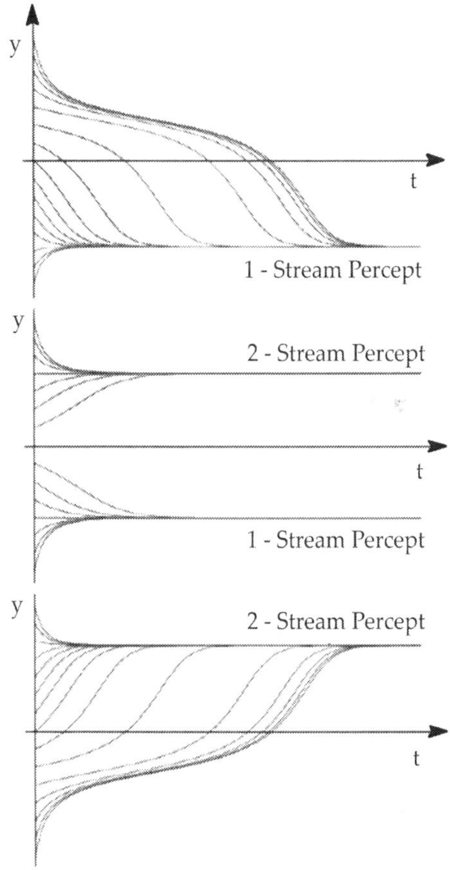

Figure 17. Percept formation. For multiple initial conditions, the time series of y(t) are plotted for the three regimes, one stream only (top), bistable (middle), and two streams only (bottom).

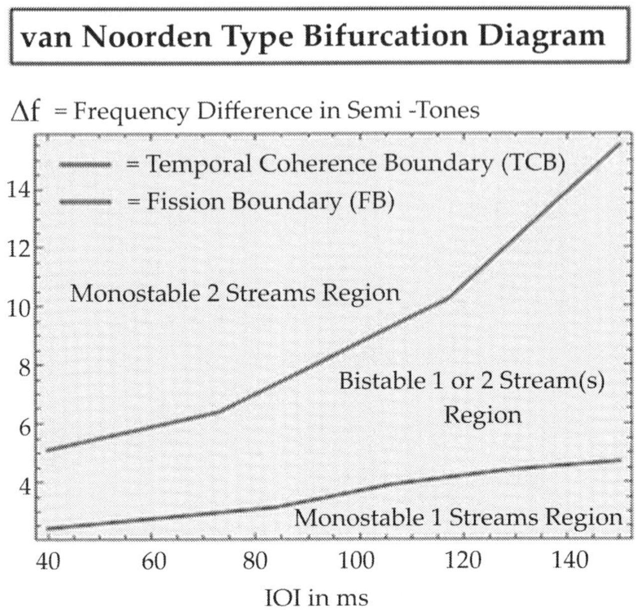

Figure 18. van Noorden's bifurcation diagram. Computational simulations yield the van Noorden's bifurcation diagram as a function of the frequency difference Δf and the IOI. The parameter space is partitioned into three regimes, one region with the percept one stream, another region with the percept two streams and a region in between which permits both.

We will briefly illustrate another phenomenon. When two interleaved rising and falling tone sequences, as shown in Figure 19, are presented, human subjects report them to be either crossing or bouncing perceptually [128],[129]. This phenomenon is known as the crossing scales phenomenon. The implementation within the neural field model of [126] is straightforward and illustrated in Figure 19.

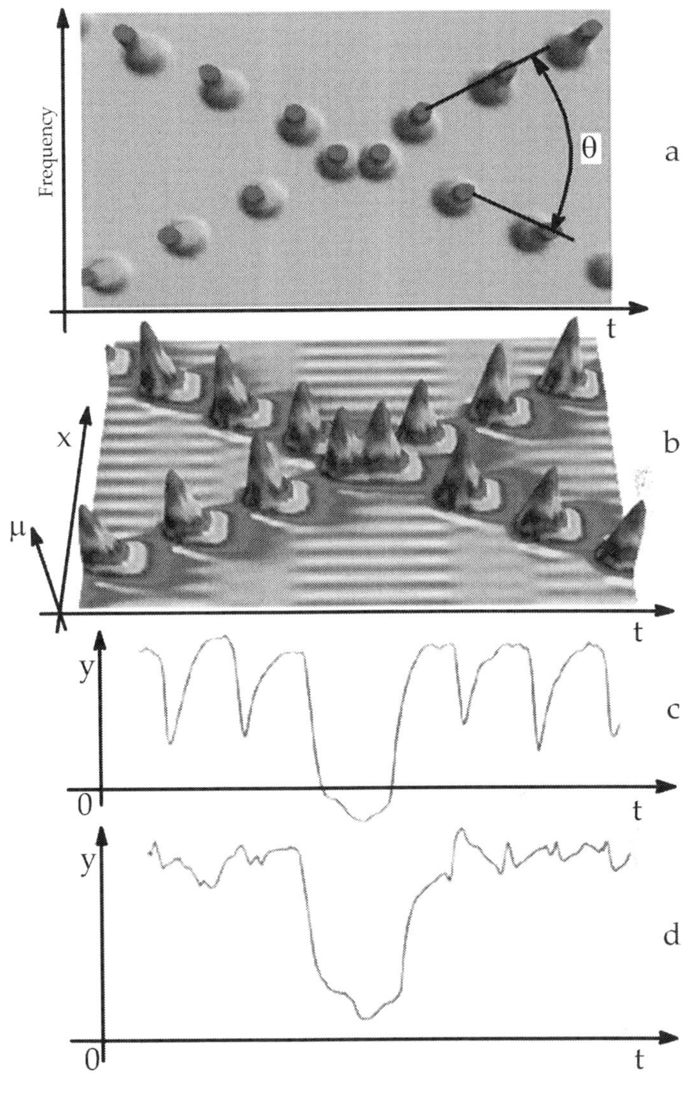

Figure 19. Crossing scales phenomenon. The input tone sequences used to form a percept of crossover are shown in (A). The resulting contours of neural field activity are plotted in (B), together with the final time series of y(t) shown in (C). In this particular case, the classification system y(t) does traverse from the positive (two streams) to the negative (one stream) fixed point and back. This trajectory is identified with the percept of crossover. In the case of the bouncing percept, the time series of y(t) will not cross the x-axis as shown in (D).

Modeling Seizures

Experimental and theoretical arguments propose that the onset of a seizure reflects a bifurcation in cortical activity from damped stochastic activity—where peaks in the power spectrum reflect damped linear resonances—to high amplitude nonlinear oscillations arising from activity on a limit cycle or chaotic attractor [86], [130]–[134]. Figure 20 presents an example of a bifurcation arising from a 3 Hz oscillatory instability in the corticothalamic neural field model of the Recent Developments in Neural Field Models section. Stochastic activity either side of the seizure can be seen, reflecting the response properties of the stable steady state mode. The large amplitude oscillations arise from a transient change in a corticothalamic state parameter from t=5 s to t=20 s. A more systematic analysis of the bifurcations in this neural field model was undertaken in [92]. It was argued that the study of these bifurcations provides a parsimonious explanation of the unique time course, symmetry, onset, and offset of both Absence and tonic clonic seizures, capturing their similarities and the differences. Further analysis of the 3 Hz (Absence) bifurcation in a reduced model argues that interactions between the reticular and specific nuclei of the thalamus contribute importantly to the Absence seizure waveform [135]. In the present simulation, the 3 Hz seizure has inherently aperiodic dynamics, as shown in the right panel of Figure 20.

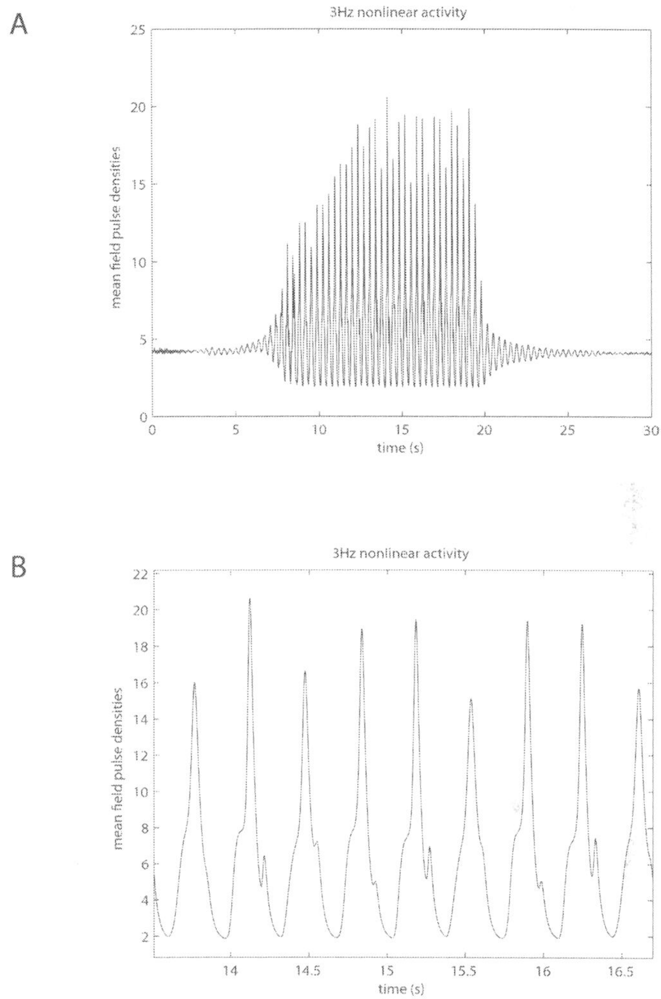

Figure 20. Simulated nonlinear oscillations arising from 3 Hz modal instability in a corticothalamic neural field model. (A) Stochastic activity either side of the seizure can be seen, reflecting the response properties of the stable steady state mode. The large amplitude oscillations arise from a transient change in a corticothalamic state parameter from t=5 s to t=20 s. (B) Shows more detail of the aperiodic oscillations.

Neural field formulations, through their implicit treatment of horizontal synaptic coupling, also lend themselves naturally to studying the spatial propagation of seizure activity, a clinically important phenomenon. An analysis of the frequency and amplitude properties of spatially extended 3 Hz seizures (Knock et. al., unpublished data) is presented in Figure 21, comparing data recorded from a young male with Absence epilepsy (left panels) to a simulated seizure in the corticothalamic model (right panels). The top row shows the temporal dynamics of the dominant frequency across a spatially extended array of (real and simulated) electrodes. The observed data (left) shows that the seizure onsets (almost) simultaneously across the scalp, although frontal electrodes lead fractionally. However, onset frequencies range from 2.7 Hz at frontal electrodes to 4 Hz over temporal regions. There follows a pattern of frequency convergence so that within 2 s of seizure onset, all cortical regions express a common frequency of 3 Hz, slowing progressively to 2.5 Hz. The seizure simulated in the corticothalamic model (right) shows a similar pattern. Peak onset frequencies in this model predominantly reflect corticothalamic conduction time lags, which have been parameterized to reflect the varying separation of cortex and thalamus. Subsequent frequency convergence in this model arises from corticortical coupling (there is no intrathalamic coupling in this simulation).

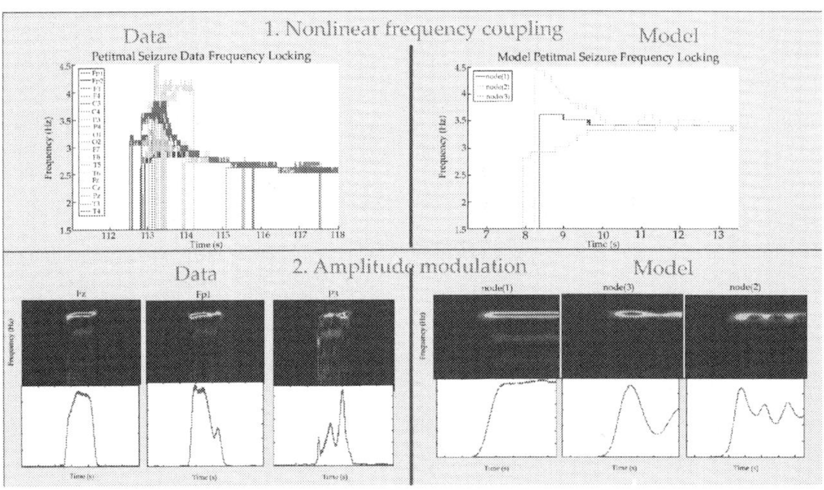

Figure 21. Spatiotemporal activity during 3 Hz spike and wave seizures. (Left) Shows analysis of an Absence seizure in an adolescent male subject. (Right) Shows the results of a simulated seizure in the corticothalamic field model described in the Recent Developments in Neural Field Models section. Top row shows the evolution of the center frequency of the dominant nonlinear mode. Lower row show the amplitude envelope of these modes. Frequencies and amplitudes were derived using a complex Morlet wavelet decomposition of the real and simulated time series.

The lower panels show the temporal evolution of the amplitude envelope of activity within the dominant mode. The principal feature of interest in the observed data (left) is the increasing modulation of the amplitude envelope as one moves from frontal electrodes, which have the strongest power, to parietal electrodes, where the onset power is weaker. These differing degrees of amplitude modulation are also present in the simulated seizure (right). Importantly, all parameters of the model are constant during the seizure. Hence the amplitude modulation is due to coupling between nonlinear modes at different spatial locations.

Whereas frequency locking is not surprising in a model with spatial coupling, the amplitude modulation is a novel, emergent property of the nonlinear dynamics.

DISCUSSION

In conclusion, we have seen that statistical descriptions of neuronal ensembles can be formulated in terms of a Fokker-Planck equation, a functional differential equation prescribing the evolution of a probability density on some phase space. The high dimensionality and complexity of these Fokker-Planck formalisms can be finessed with a mean-field approximation to give nonlinear Fokker-Planck equations, describing the evolution of separable ensembles that are coupled by mean-field effects. By parameterizing the densities in terms of basis functions or probability modes, these partial differential equations can be reduced to coupled differential equations describing their evolution. In the simplest case, we can use a single mode that can be regarded as encoding the location of a probability mass, hence neural mass models. Neural mass models can be generalized to neural field models by making the expectations a function of space, thereby furnishing wave equations that describe the spatiotemporal evolution of expected neuronal states over the cortical surface. We have tried to show the conceptual and mathematical links among the ensuing levels of description and how these models can be used to characterize key dynamical mechanisms in the brain.

Acknowledgments

VKJ thanks Felix Almonte for help with the generation of various figures. We thank the attendees of Session IV at BCW2006* for the interesting discussions, which represent the source of inspiration for the present review.

REFERENCES

1. McIntosh AR (2000) Towards a network theory of cognition. Neural Netw 13: 861–870. AR McIntosh2000Towards a network theory of cognition. Neural Netw13861870
2. Bressler SL, Kelso JAS (2001) Cortical coordination dynamics and cognition. Trends in Cognitive Science 5: 26–36. SL BresslerJAS Kelso2001Cortical coordination dynamics and cognition. Trends in Cognitive Science52636
3. Bressler SL (2002) Understanding cognition through large-scale cortical networks. Curr Dir Psychol Sci 11: 58–61. SL Bressler2002Understanding cognition through large-scale cortical networks. Curr Dir Psychol Sci115861
4. Jirsa VK (2004) Connectivity and dynamics of neural information processing. Neuroinformatics 2: 183–204. VK Jirsa2004Connectivity and dynamics of neural information processing. Neuroinformatics2183204
5. Beurle RL (1956) Properties of a mass of cells capable of regenerating pulses. Philos Trans Phys Sci Lon B 240: 55–94. RL Beurle1956Properties of a mass of cells capable of regenerating pulses. Philos Trans Phys Sci Lon B2405594
6. Rolls ET, Deco G (2002) Computational neuroscience of vision. Oxford: Oxford University Press. ET RollsG. Deco2002Computational neuroscience of visionOxfordOxford University Press
7. Tuckwell H (1988) Introduction to theoretical neurobiology. Cambridge: Cambridge University Press. H. Tuckwell1988Introduction to theoretical neurobiologyCambridgeCambridge University Press
8. Dayan P, Abbott L (2002) Theoretical neuroscience: computational and mathematical modeling of neural systems. Boston: MIT Press. P. DayanL. Abbott2002Theoretical neuroscience: computational and mathematical modeling of neural systemsBostonMIT Press

9. Jirsa VK, McIntosh AR (2007) Handbook of brain connectivity. Berlin: Springer. VK JirsaAR McIntosh2007Handbook of brain connectivityBerlinSpringer
10. Brunel N, Wang X (2001) Effects of neuromodulation in a cortical network model of object working memory dominated by recurrent inhibition. J Comput Neurosci 11: 63–85. N. BrunelX. Wang2001Effects of neuromodulation in a cortical network model of object working memory dominated by recurrent inhibition. J Comput Neurosci116385
11. De Groff D, Neelakanta P, Sudhakar R, Aalo V (1993) Stochastical aspects of neuronal dynamics: Fokker-Planck approach. Biol Cybern 69: 155–164. D. De GroffP. NeelakantaR. SudhakarV. Aalo1993Stochastical aspects of neuronal dynamics: Fokker-Planck approach. Biol Cybern69155164
12. Feller W (1951) Diffusion equations in genetics. W. Feller1951Diffusion equations in genetics. In: Proceedings of the second Berkeley symposium on mathematical statistics and probability; 31 July to 12 August, 1950; Berkeley, California, United States. Berkeley: University of California Press.
13. Ricciardi L, Sacerdote L (1979) The Ornstein-Uhlenbeck process as a model for neuronal activity. I. Mean and variance of the firing time. Biol Cybern 35: 1–9. L. RicciardiL. Sacerdote1979The Ornstein-Uhlenbeck process as a model for neuronal activity. I. Mean and variance of the firing time. Biol Cybern3519
14. Lansky P, Sacerdote L, Tomassetti F (1995) On the comparison of feller and Ornstein-Uhlenbeck models for neural activity. Biol Cybern 73: 457–465. P. LanskyL. SacerdoteF. Tomassetti1995On the comparison of feller and Ornstein-Uhlenbeck models for neural activity. Biol Cybern73457465
15. Knight B, Manin D, Sirovich L (1996) Dynamical models of interacting neuron populations. In: Gerf EC, editor. B. KnightD. ManinL. Sirovich1996Dynamical models of interacting neuron populations. EC GerfSymposium on robotics and cybernetics: computational engineering in systems applications. Lille, France: Cite Scientifique. Symposium on robotics and cybernetics: computational engineering in systems applications. Lille, France:

Cite Scientifique.

16. Omurtag A, Knight B, Sirovich L (2000) On the simulation of large populations of neurons. J Comput Neurosci 8: 51–53. A. OmurtagB. KnightL. Sirovich2000On the simulation of large populations of neurons. J Comput Neurosci85153
17. Knight B (2000) Dynamics of encoding in neuron populations: some general mathematical features. Neural Comput 12: 473–518. B. Knight2000Dynamics of encoding in neuron populations: some general mathematical features. Neural Comput12473518
18. Gerstner W (2000) Population dynamics of spiking neurons: fast transients, asynchronous states and locking. Neural Comput 12: 43–89. W. Gerstner2000Population dynamics of spiking neurons: fast transients, asynchronous states and locking. Neural Comput124389
19. Risken H (1996) The Fokker-Planck equation. Berlin: Springer. H. Risken1996The Fokker-Planck equationBerlinSpringer
20. Del Giudice P, Fusi S, Mattia M (2003) Modeling the formation of working memory with networks of integrate-and-fire neurons connected by plastic synapses. J Physiol Paris 97: 659–681. P. Del GiudiceS. FusiM. Mattia2003Modeling the formation of working memory with networks of integrate-and-fire neurons connected by plastic synapses. J Physiol Paris97659681
21. Hebb D (1949) The organization of behavior - a neurophysiological theory. New York: John Wiley. D. Hebb1949The organization of behavior - a neurophysiological theoryNew YorkJohn Wiley
22. Desimone R, Duncan J (1995) Neural mechanisms of selective visual attention. Annu Rev Neurosci 18: 193–222. R. DesimoneJ. Duncan1995Neural mechanisms of selective visual attention. Annu Rev Neurosci18193222
23. Deco G, Rolls ET (2005) Attention, short term memory, and action selection: a unifying theory. Prog Neurobiol 76: 236–256. G. DecoET Rolls2005Attention, short term memory, and action selection: a unifying theory. Prog Neurobiol76236256
24. Deco G, Lee TS (2002) A unified model of spatial and object attention based on inter-cortical biased competition. Neurocomputing 44–46: 775–781. G. DecoTS Lee2002A unified model of spatial and object attention based on inter-cortical biased competition.

Neurocomputing44–46775781

25. Corchs S, Deco G (2002) Large-scale neural model for visual attention: integration of experimental single cell and fMRI data. Cereb Cortex 12: 339–348. S. CorchsG. Deco2002Large-scale neural model for visual attention: integration of experimental single cell and fMRI data. Cereb Cortex12339348
26. Deco G, Pollatos O, Zihl J (2002) The time course of selective visual attention: theory and experiments. Vision Res 42: 2925–2945. G. DecoO. PollatosJ. Zihl2002The time course of selective visual attention: theory and experiments. Vision Res4229252945
27. Corchs S, Deco G (2004) Feature-based attention in human visual cortex: simulation of fMRI data. Neuroimage 21: 36–45. S. CorchsG. Deco2004Feature-based attention in human visual cortex: simulation of fMRI data. Neuroimage213645
28. Deco G, Rolls ET (2002) Object-based visual neglect: a computational hypothesis. Eur J Neurosci 16: 1994–2000. G. DecoET Rolls2002Object-based visual neglect: a computational hypothesis. Eur J Neurosci1619942000
29. Deco G, Rolls ET (2003) Attention and working memory: a dynamical model of neuronal activity in the prefrontal cortex. Eur J Neurosci 18: 2374–2390. G. DecoET Rolls2003Attention and working memory: a dynamical model of neuronal activity in the prefrontal cortex. Eur J Neurosci1823742390
30. Deco G, Rolls ET (2004) A neurodynamical cortical model of visual attention and invariant object recognition. Vision Res 44: 621–644. G. DecoET Rolls2004A neurodynamical cortical model of visual attention and invariant object recognition. Vision Res44621644
31. Deco G, Rolls ET, Horwitz B (2004) 'What' and 'where' in visual working memory: a computational neurodynamical perspective for integrating fMRI and single-neuron data. J Cogn Neurosci 16: 683–701. G. DecoET RollsB. Horwitz2004'What' and 'where' in visual working memory: a computational neurodynamical perspective for integrating fMRI and single-neuron data. J Cogn Neurosci16683701
32. Szabo M, Almeida R, Deco G, Stetter M (2004) Cooperation and biased competition model can explain attentional filtering in the prefrontal cortex. Eur J Neurosci 19: 1969–1977. M. SzaboR.

AlmeidaG. DecoM. Stetter2004Cooperation and biased competition model can explain attentional filtering in the prefrontal cortex. Eur J Neurosci1919691977

33. Deco G, Rolls ET (2005) Neurodynamics of biased competition and cooperation for attention: a model with spiking neurons. J Neurophysiol 94: 295–313. G. DecoET Rolls2005Neurodynamics of biased competition and cooperation for attention: a model with spiking neurons. J Neurophysiol94295313
34. Chizhov A, Graham L (2007) Population model of hippocampal pyramidal neurons, linking a refractory density approach to conductance-based neurons. Phys Rev E 75: 011924. A. ChizhovL. Graham2007Population model of hippocampal pyramidal neurons, linking a refractory density approach to conductance-based neurons. Phys Rev E75011924
35. Harrison LM, David O, Friston KJ (2005) Stochastic models of neuronal dynamics. Philos Trans R Soc Lond B Biol Sci 360: 1075–1091. LM HarrisonO. DavidKJ Friston2005Stochastic models of neuronal dynamics. Philos Trans R Soc Lond B Biol Sci36010751091
36. Griffith JS (1963) A field theory of neural nets: I. Derivation of field equations. Bull Math Biophys 25: 111–120. JS Griffith1963A field theory of neural nets: I. Derivation of field equations. Bull Math Biophys25111120
37. Griffith JS (1965) A field theory of neural nets: II. Properties of the field equations. Bull Math Biophys 27: 187–195. JS Griffith1965A field theory of neural nets: II. Properties of the field equations. Bull Math Biophys27187195
38. Freeman WJ (1975) Mass action in the nervous system. New York: Academic Press. WJ Freeman1975Mass action in the nervous systemNew YorkAcademic Press
39. Jansen B, Rit V (1995) Electroencephalogram and visual evoked potential generation in a mathematical model of coupled cortical columns. Biol Cybern 73: 357–366. B. JansenV. Rit1995Electroencephalogram and visual evoked potential generation in a mathematical model of coupled cortical columns. Biol Cybern73357366

40. Wendling F, Bellanger J, Bartolomei F, Chauvel P (2000) Relevance of nonlinear lumped parameter models in the analysis of depth-eeg epileptic signals. Biol Cybern 83: 367–378. F. WendlingJ. BellangerF. BartolomeiP. Chauvel2000Relevance of nonlinear lumped parameter models in the analysis of depth- eeg epileptic signals. Biol Cybern83367378
41. David O, Harrison L, Friston K (2005) Modelling event-related responses in the brain. Neuroimage 25: 756–770. O. DavidL. HarrisonK. Friston2005Modelling event-related responses in the brain. Neuroimage25756770
42. David O, Kiebel S, Harrison L, Mattout J, Kilner J, et al. (2006) Dynamic causal modeling of evoked responses in EEG and MEG. Neuroimage 30: 1255–1272. O. DavidS. KiebelL. HarrisonJ. MattoutJ. Kilner2006Dynamic causal modeling of evoked responses in EEG and MEG. Neuroimage3012551272
43. Wilson HR, Cowan JD (1972) Excitatory and inhibitory interactions in localized populations of model neurons. Biophys J 12: 1–24. HR WilsonJD Cowan1972Excitatory and inhibitory interactions in localized populations of model neurons. Biophys J12124
44. Wilson HR, Cowan JD (1973) A mathematical theory of the functional dynamics of cortical and thalamic nervous tissue. Kybernetik 13: 55–80. HR WilsonJD Cowan1973A mathematical theory of the functional dynamics of cortical and thalamic nervous tissue. Kybernetik135580
45. Nunez PL (1974) The brain wave equation: a model for EEG. Math Biosci 21: 279–297. PL Nunez1974The brain wave equation: a model for EEG. Math Biosci21279297
46. Amari S (1975) Homogeneous nets of neuron-like elements. Biol Cybern 17: 211–220. S. Amari1975Homogeneous nets of neuron-like elements. Biol Cybern17211220
47. Amari S (1977) Dynamics of pattern formation in lateral-inhibition type neural fields. Biol Cybern 27: 77–87. S. Amari1977Dynamics of pattern formation in lateral-inhibition type neural fields. Biol Cybern277787
48. Jirsa VK, Haken H (1996) Field theory of electromagnetic brain activity. Phys Rev Lett 77: 960–963. VK JirsaH. Haken1996Field

theory of electromagnetic brain activity. Phys Rev Lett77960963

49. Jirsa VK, Haken H (1997) A derivation of a macroscopic field theory of the brain from the quasi-microscopic neural dynamics. Physica D 99: 503–526. VK JirsaH. Haken1997A derivation of a macroscopic field theory of the brain from the quasi-microscopic neural dynamics. Physica D99503526
50. Robinson PA, Rennie CA, Wright JJ (1997) Propagation and stability of waves of electrical activity in the cerebral cortex. Phys Rev E 56: 826–840. PA RobinsonCA RennieJJ Wright1997Propagation and stability of waves of electrical activity in the cerebral cortex. Phys Rev E56826840
51. Liley D, Bojak I (2005) Understanding the transition to seizure by modeling the epileptiform activity of general anesthetic agents. J Clin Neurophysiol 22: 300–313. D. LileyI. Bojak2005Understanding the transition to seizure by modeling the epileptiform activity of general anesthetic agents. J Clin Neurophysiol22300313
52. Breakspear M, Terry J, Friston K (2003) Modulation of excitatory synaptic coupling facilitates synchronization and complex dynamics in a biophysical model of neuronal dynamics. Network: Computation in Neural Systems 14: 703–732. M. BreakspearJ. TerryK. Friston2003Modulation of excitatory synaptic coupling facilitates synchronization and complex dynamics in a biophysical model of neuronal dynamics. Network: Computation in Neural Systems14703732
53. Coombes S (2006) Neural fields. Scholarpedia 1: 1373. S. Coombes2006Neural fields. Scholarpedia11373Available: http://www. scholarpedia. org/article/Neural_fields. Accessed 7 July 2008. Available: http://www. scholarpedia. org/article/Neural_fields. Accessed 7 July 2008.
54. Ermentrout B (1998) Neural networks as spatio-temporal pattern-forming systems. Report Progress in Physics 61: 353–430. B. Ermentrout1998Neural networks as spatio-temporal pattern-forming systems. Report Progress in Physics61353430
55. Coombes S (2005) Waves, bumps, and patterns in neural field theories. Biol Cybern 93: 91–108. S. Coombes2005Waves, bumps, and patterns in neural field theories. Biol Cybern9391108

56. Colby C, Duhamel J, Goldberg M (1995) Oculocentric spatial representation in parietal cortex. Cereb Cortex 5: 470–481. C. ColbyJ. DuhamelM. Goldberg1995Oculocentric spatial representation in parietal cortex. Cereb Cortex5470481
57. Goldman-Rakic P (1995) Cellular basis of working memory. Neuron 14: 477–485. P. Goldman-Rakic1995Cellular basis of working memory. Neuron14477485
58. Durstewitz D, Seamans J, Sejnowski T (2000) Neurocomputational models of working memory. Nat Neurosci 3 (Supplement) 1184–1191. D. DurstewitzJ. SeamansT. Sejnowski2000Neurocomputational models of working memory. Nat Neurosci3 (Supplement)11841191
59. Wang X (2001) Synaptic reverberation underlying mnemonic persistent activity. Trends in Neurosciences 24: 455–463. X. Wang2001Synaptic reverberation underlying mnemonic persistent activity. Trends in Neurosciences24455463
60. Fall C, Lewis T, Rinzel J (2005) Background-activity-dependent properties of a network model for working memory that incorporates cellular bistability. Biol Cybern 93: 109–118. C. FallT. LewisJ. Rinzel2005Background-activity-dependent properties of a network model for working memory that incorporates cellular bistability. Biol Cybern93109118
61. Hahnloser R (2003) Emergence of neural integration in the head-direction system by visual supervision. Neuroscience 120: 877–891. R. Hahnloser2003Emergence of neural integration in the head-direction system by visual supervision. Neuroscience120877891
62. Redish AD, Elga AN, Touretzky DS (1996) A coupled attractor model of the rodent head direction system. Network: Computation in Neural Systems 7: 671–685. AD RedishAN ElgaDS Touretzky1996A coupled attractor model of the rodent head direction system. Network: Computation in Neural Systems7671685
63. Sharp PE, Blair HT, Cho J (2001) The anatomical and computational basis of the rat head-direction cell signal. Trends Neurosci 24: 289–294. PE SharpHT BlairJ. Cho2001The anatomical and computational basis of the rat head-direction cell signal. Trends Neurosci24289294
64. Skaggs WE, Knierim JJ, Kudrimoti HS, McNaughton BL (1995) A model of the neural basis of the rats sense of direction. Adv

Neural Inf Process Syst 7: 173–180. WE SkaggsJJ KnierimHS KudrimotiBL McNaughton1995A model of the neural basis of the rats sense of direction. Adv Neural Inf Process Syst7173180

65. Redish AD (1999) Beyond the cognitive map: from place cells to episodic memory. Cambridge (Massachusetts): MIT Press. AD Redish1999Beyond the cognitive map: from place cells to episodic memoryCambridge (Massachusetts)MIT Press
66. Redish AD, Touretzky DS (1998) The role of the hippocampus in solving the morris water maze. Neural Comput 10: 73–111. AD RedishDS Touretzky1998The role of the hippocampus in solving the morris water maze. Neural Comput1073111
67. Samsonovich A, McNaughton B (1997) Path integration and cognitive mapping in a continuous attractor neural network model. J Neurosci 17: 5900–5920. A. SamsonovichB. McNaughton1997Path integration and cognitive mapping in a continuous attractor neural network model. J Neurosci1759005920
68. Tsodyks M (1999) Attractor neural network models of spatial maps in hippocampus. Hippocampus 9: 481–489. M. Tsodyks1999Attractor neural network models of spatial maps in hippocampus. Hippocampus9481489
69. Erlhagen W, Schöner G (2002) Dynamic field theory of movement preparation. Psych Rev 109: 545–572. W. ErlhagenG. Schöner2002Dynamic field theory of movement preparation. Psych Rev109545572
70. Ben-Yishai R, Bar-Or L, Sompolinsky H (1995) Theory of orientation tuning in visual cortex. Proc Natl Acad Sci U S A 92: 3844–3848. R. Ben-YishaiL. Bar-OrH. Sompolinsky1995Theory of orientation tuning in visual cortex. Proc Natl Acad Sci U S A9238443848
71. Laing C (2003) Spiral waves in nonlocal equations. SIAM Journal on Applied Dynamical Systems 4: 588–606. C. Laing2003Spiral waves in nonlocal equations. SIAM Journal on Applied Dynamical Systems4588606
72. Ermentrout G, Cowan J (1979) A mathematical theory of visual hallucination patterns. Biol Cybern 34: 137–150. G. ErmentroutJ. Cowan1979A mathematical theory of visual hallucination patterns. Biol Cybern34137150

73. Laing C, Troy WC (2003) PDE methods for nonlocal models. SIAM Journal on Applied Dynamical Systems 2: 487–516. C. LaingWC Troy2003PDE methods for nonlocal models. SIAM Journal on Applied Dynamical Systems2487516
74. Coombes S, Owen MR (2004) Evans functions for integral neural field equations with Heaviside firing rate function. SIAM Journal on Applied Dynamical Systems 34: 574–600. S. CoombesMR Owen2004Evans functions for integral neural field equations with Heaviside firing rate function. SIAM Journal on Applied Dynamical Systems34574600
75. Chervin R, Pierce P, Connors B (1988) Periodicity and directionality in the propagation of epileptiform discharges across neortex. J Neurophysiol 60: 1695–1713. R. ChervinP. PierceB. Connors1988Periodicity and directionality in the propagation of epileptiform discharges across neortex. J Neurophysiol6016951713
76. Golomb D, Amitai Y (1997) Propagating neuronal discharges in neocortical slices: computational and experimental study. J Neurophysiol 78: 1199–1211. D. GolombY. Amitai1997Propagating neuronal discharges in neocortical slices: computational and experimental study. J Neurophysiol7811991211
77. Wu JY, Guan L, Tsau Y (1999) Propagating activation during oscillations and evoked responses in neocortical slices. J Neurosci 19: 5005–5015. JY WuL. GuanY. Tsau1999Propagating activation during oscillations and evoked responses in neocortical slices. J Neurosci1950055015
78. Miles R, Traub R, Wong R (1995) Spread of synchronous firing in longitudinal slices from the CA3 region of hippocampus. J Neurophysiol 60: 1481–1496. R. MilesR. TraubR. Wong1995Spread of synchronous firing in longitudinal slices from the CA3 region of hippocampus. J Neurophysiol6014811496
79. Kim U, Bal T, McCormick D (1995) Spindle waves are propagating synchronized oscillations in the ferret LGNd in vitro. J Neurophysiol 74: 1301–1323. U. KimT. BalD. McCormick1995Spindle waves are propagating synchronized oscillations in the ferret LGNd in vitro. J Neurophysiol7413011323
80. Connors BW, Amitai Y (1993) Generation of epileptiform

discharges by local circuits in neocortex. In: Schwartzkroin PA, editor. Epilepsy: models, mechanisms and concepts. Cambridge (United Kingdom): Cambridge University Press. pp. 388–424. BW ConnorsY. Amitai1993Generation of epileptiform discharges by local circuits in neocortex. PA SchwartzkroinEpilepsy: models, mechanisms and conceptsCambridge (United Kingdom)Cambridge University Press388424

81. Ermentrout G, Kleinfeld D (2001) Traveling electrical waves in cortex: insights from phase dynamics and speculation on a computational role. Neuron 29: 33–44. G. ErmentroutD. Kleinfeld2001Traveling electrical waves in cortex: insights from phase dynamics and speculation on a computational role. Neuron293344
82. Richardson K, Schiff S, Gluckman B (2005) Propagating activation during oscillations and evoked responses in neocortical slices. Phys Rev Lett 94: 028103. K. RichardsonS. SchiffB. Gluckman2005Propagating activation during oscillations and evoked responses in neocortical slices. Phys Rev Lett94028103
83. Nunez PL (1981) Electric Fields of the Brain. Oxford (United Kingdom): Oxford University Press. PL Nunez1981Electric Fields of the BrainOxford (United Kingdom)Oxford University Press
84. Nunez P (1995) Neocortical dynamics and human EEG rhythms. New York/Oxford: Oxford University Press. P. Nunez1995Neocortical dynamics and human EEG rhythmsNew York/OxfordOxford University Press
85. Wright JJ, Liley DTJ (1996) Dynamics of the brain at global and microscopic scales: neural networks and the EEG. Behav Brain Sci 19: 285–320. JJ WrightDTJ Liley1996Dynamics of the brain at global and microscopic scales: neural networks and the EEG. Behav Brain Sci19285320
86. Robinson P, Rennie C, Rowe D (2002) Dynamics of large-scale brain activity in normal arousal states and epileptic seizures. Phys Rev E 65: 41924. P. RobinsonC. RennieD. Rowe2002Dynamics of large-scale brain activity in normal arousal states and epileptic seizures. Phys Rev E6541924
87. Rennie C, Robinson P, Wright J (2002) Unified neurophysical model of EEG spectra and evoked potentials. Biol Cybern 86: 457–471. C.

RennieP. RobinsonJ. Wright2002Unified neurophysical model of EEG spectra and evoked potentials. Biol Cybern86457471

88. Niedermeyer E, Lopes da Silva FH (1999) Electroencephalography: basic principles, clinical applications, and related fields. Philadelphia (Pennsylvania): Lippincott Williams and Wilkins. E. NiedermeyerFH Lopes da Silva1999Electroencephalography: basic principles, clinical applications, and related fieldsPhiladelphia (Pennsylvania)Lippincott Williams and Wilkins

89. Robinson PA, Drysdale PM, der Merwe V, Kyriakou HE, Rigozzi MK, et al. (2006) Bold responses to stimuli: dependence on frequency, stimulus form, amplitude, and repetition rate. Neuroimage 31: 589–599. PA RobinsonPM DrysdaleV. der MerweHE KyriakouMK Rigozzi2006Bold responses to stimuli: dependence on frequency, stimulus form, amplitude, and repetition rate. Neuroimage31589599

90. Robinson PA, Rennie CJ, Rowe DL, O'Connor SC, Wright JJ, et al. (2003) Neurophysical modeling of brain dynamics. Neuropsychopharmacol 28: S74–S79. PA RobinsonCJ RennieDL RoweSC O'ConnorJJ Wright2003Neurophysical modeling of brain dynamics. Neuropsychopharmacol28S74S79

91. Robinson PA, Rennie CJ, Wright JJ, Bahramali H, Gordon E, et al. (2001) Prediction of electroencephalographic spectra from neurophysiology. Phys Rev E 63: 021903. PA RobinsonCJ RennieJJ WrightH. BahramaliE. Gordon2001Prediction of electroencephalographic spectra from neurophysiology. Phys Rev E63021903

92. Breakspear M, Roberts JA, Terry JR, Rodrigues S, Mahant N, et al. (2006) A unifying explanation of primary generalized seizures through nonlinear brain modeling and bifurcation analysis. Cereb Cortex 16: 1296–1313. M. BreakspearJA RobertsJR TerryS. RodriguesN. Mahant2006A unifying explanation of primary generalized seizures through nonlinear brain modeling and bifurcation analysis. Cereb Cortex1612961313

93. Robinson P, Rennie C, Rowe D, O'Connor S (2004) Estimation of multiscale neurophysiologic parameters by elctroencephalographic means. Hum Brain Mapp 23: 53–72. P. RobinsonC. RennieD. RoweS. O'Connor2004Estimation of multiscale neurophysiologic

parameters by elctroencephalographic means. Hum Brain Mapp235372

94. Rowe D, Robinson P, Rennie C (2004) Estimation of neurophysiological parameters from the waking EEG using a biophysical model of brain dynamics. J Theor Biol 231: 413–433. D. RoweP. RobinsonC. Rennie2004Estimation of neurophysiological parameters from the waking EEG using a biophysical model of brain dynamics. J Theor Biol231413433
95. Rennie CJ, Robinson PA, Wright JJ (1999) Effects of local feedback on dispersion of electrical waves in the cerebral cortex. Phys Rev E 59: 3320–3329. CJ RenniePA RobinsonJJ Wright1999Effects of local feedback on dispersion of electrical waves in the cerebral cortex. Phys Rev E5933203329
96. Koch C (1999) Biophysics of computation. Information processing in single neurons. Oxford (United Kingdom): Oxford University Press. C. Koch1999Biophysics of computation. Information processing in single neuronsOxford (United Kingdom)Oxford University Press
97. Destexhe A, Sejnowski TJ (2001) Thalamocortical assemblies: how ion channels, single neurons and large scale networks organize sleep oscillations. Oxford (United Kingdom): Oxford University Press. A. DestexheTJ Sejnowski2001Thalamocortical assemblies: how ion channels, single neurons and large scale networks organize sleep oscillationsOxford (United Kingdom)Oxford University Press
98. Sporns O (2003) Complex Neural Dynamics. In: Jirsa VK, Kelso JAS, editors. Coordination dynamics: issues and trends. Berlin: Springer. O. Sporns2003Complex Neural Dynamics. VK JirsaJAS KelsoCoordination dynamics: issues and trendsBerlinSpringer
99. Braitenberg V, Schüz A (1991) Anatomy of the cortex. Statistics and geometry. Berlin/Heidelberg/New York: Springer. V. BraitenbergA. Schüz1991Anatomy of the cortex. Statistics and geometryBerlin/Heidelberg/New YorkSpringer
100. Robinson PA (2005) Propagator theory of brain dynamics. Phys Rev E 72: 1–13. PA Robinson2005Propagator theory of brain dynamics. Phys Rev E72113
101. Robinson PA (2006) Patchy propagators, cortical dynamics, and the

generation of spatially structured gamma oscillations. Phys Rev E 73: 1–13. PA Robinson2006Patchy propagators, cortical dynamics, and the generation of spatially structured gamma oscillations. Phys Rev E73113

102. Robinson PA (2007) Visual gamma oscillations: correlations and other properties. Biol Cybern 97: 317–335. PA Robinson2007Visual gamma oscillations: correlations and other properties. Biol Cybern97317335

103. Jirsa VK, Kelso JAS (2000) Spatiotemporal pattern formation in continuous systems with heterogeneous connection topologies. Phys Rev E 62: 8462–8465. VK JirsaJAS Kelso2000Spatiotemporal pattern formation in continuous systems with heterogeneous connection topologies. Phys Rev E6284628465

104. Qubbaj MR, Jirsa VK (2007) Neural field dynamics with heterogeneous connection topology. Phys Rev Lett 93: 238102. MR QubbajVK Jirsa2007Neural field dynamics with heterogeneous connection topology. Phys Rev Lett93238102

105. Datko R (1978) A procedure for determination of the exponential stability of certain differential difference equations. Q Appl Math 36: 279–292. R. Datko1978A procedure for determination of the exponential stability of certain differential difference equations. Q Appl Math36279292

106. Freeman W (1979) Nonlinear gain mediating cortical stimulus-response relations. Biol Cybern 33: 237–247. W. Freeman1979Nonlinear gain mediating cortical stimulus-response relations. Biol Cybern33237247

107. Freeman W (1979) Nonlinear dynamics of paleocortex manifested in the olfactory EEG. Biol Cybern 35: 21–37. W. Freeman1979Nonlinear dynamics of paleocortex manifested in the olfactory EEG. Biol Cybern352137

108. Freeman W (1987) Simulation of chaotic EEG patterns with a dynamic model of the olfactory system. Biol Cybern 56: 139–150. W. Freeman1987Simulation of chaotic EEG patterns with a dynamic model of the olfactory system. Biol Cybern56139150

109. Izhikevich E (2007) Dynamical systems in neuroscience: the geometry of excitability and bursting. Cambridge (Massachusetts):

MIT Press. E. Izhikevich2007Dynamical systems in neuroscience: the geometry of excitability and burstingCambridge (Massachusetts) MIT Press

110. Breakspear M, Jirsa V (2007) Neuronal dynamics and brain connectivity. In: Jirsa VK, McIntosh AR, editors. Handbook of brain connectivity. Berlin: Springer. M. BreakspearV. Jirsa2007Neuronal dynamics and brain connectivity. VK JirsaAR McIntoshHandbook of brain connectivityBerlinSpringer
111. Morris C, Lecar H (1981) Voltage oscillations in the barnacle giant muscle fiber. Biophys J 35: 193–213. C. MorrisH. Lecar1981Voltage oscillations in the barnacle giant muscle fiber. Biophys J35193213
112. Larter R, Speelman B, Worth RM (1999) A coupled ordinary differential equation lattice model for the simulation of epileptic seizures. Chaos 9: 795–804. R. LarterB. SpeelmanRM Worth1999A coupled ordinary differential equation lattice model for the simulation of epileptic seizures. Chaos9795804
113. Freeman W, Breakspear M (2007) Scale-free neocortical dynamics. Scholarpedia 2(2): 1357. W. FreemanM. Breakspear2007Scale-free neocortical dynamics. Scholarpedia2(2)1357Available: http:// www. scholarpedia. org/article/Scale-free_neocortical_dynamics. Accessed 7 July 2008. Available: http://www. scholarpedia. org/ article/Scale-free_neocortical_dynamics. Accessed 7 July 2008.
114. Breakspear M, Stam C (2005) Dynamics of a neural system with a multiscale architecture. Philos Trans R Soc Lond B Biol Sci 360: 1051–1074. M. BreakspearC. Stam2005Dynamics of a neural system with a multiscale architecture. Philos Trans R Soc Lond B Biol Sci36010511074
115. Honey CJ, Kötter R, Breakspear M, Sporns O (2007) Network structure of cerebral cortex shapes functional connectivity on multiple time scales. Proc Natl Acad Sci U S A: 104: 10240–10245. CJ HoneyR. KötterM. BreakspearO. Sporns2007Network structure of cerebral cortex shapes functional connectivity on multiple time scales. Proc Natl Acad Sci U S A:1041024010245
116. Friston K (2003) Learning and inference in the brain. Neural Netw 16: 1325–1352. K. Friston2003Learning and inference in the brain. Neural Netw1613251352

117. Friston K (2005) A theory of cortical responses. Philos Trans R Soc Lond B Biol Sci 360: 815–836. K. Friston2005A theory of cortical responses. Philos Trans R Soc Lond B Biol Sci360815836
118. Friston K (1997) Transients, metastability and neuronal dynamics. Neuroimage 5: 164–171. K. Friston1997Transients, metastability and neuronal dynamics. Neuroimage5164171
119. Romo R, Hernandez A, Zainos A, Lemus L, Brody C (2002) Neural correlates of decision making in secondary somatosensory cortex. Nat Neurosci 5: 1217–1225. R. RomoA. HernandezA. ZainosL. LemusC. Brody2002Neural correlates of decision making in secondary somatosensory cortex. Nat Neurosci512171225
120. Romo R, Hernandez A, Zainos A, Salinas E (2003) Correlated neuronal discharges that increase coding efficiency during perceptual discrimination. Neuron 38: 649–657. R. RomoA. HernandezA. ZainosE. Salinas2003Correlated neuronal discharges that increase coding efficiency during perceptual discrimination. Neuron38649657
121. Romo R, Hernandez A, Zainos A (2004) Neuronal correlates of a perceptual decision in ventral premotor cortex. Neuron 41: 165–173. R. RomoA. HernandezA. Zainos2004Neuronal correlates of a perceptual decision in ventral premotor cortex. Neuron41165173
122. Romo R, Salinas E (2001) Touch and go: decision-making mechanisms in somatosensation. Annu Rev Neurosci 24: 107–137. R. RomoE. Salinas2001Touch and go: decision-making mechanisms in somatosensation. Annu Rev Neurosci24107137
123. Hernandez A, Zainos A, Romo R (2002) Temporal evolution of a decision-making process in medial premotor cortex. Neuron 33: 959–972. A. HernandezA. ZainosR. Romo2002Temporal evolution of a decision-making process in medial premotor cortex. Neuron33959972
124. Romo R, Salinas E (2003) Flutter discrimination: neural codes, perception, memory and decision making. Nat Rev Neurosci 4: 203–218. R. RomoE. Salinas2003Flutter discrimination: neural codes, perception, memory and decision making. Nat Rev Neurosci4203218
125. Deco G, Rolls E (2006) Decision-making and Weber's law: a

neurophysiological model. Eur J Neurosci 24: 901–916. G. DecoE. Rolls2006Decision-making and Weber's law: a neurophysiological model. Eur J Neurosci24901916

126. Almonte F, Jirsa V, Large E, Tuller B (2005) A cortical model of auditory streaming. Physica D 212: 137–159. F. AlmonteV. JirsaE. LargeB. Tuller2005A cortical model of auditory streaming. Physica D212137159
127. van Noorden LPAS (1975) Temporal coherence in the perception of tone sequences. LPAS van Noorden1975Temporal coherence in the perception of tone sequences. Ph. D. thesis, Eindhoven (The Netherlands); Eindhoven University of Technology. Ph. D. thesis, Eindhoven (The Netherlands); Eindhoven University of Technology.
128. Bregman AS (1990) Auditory scene analysis: The perceptual organization of sound. Cambridge (Massachusetts): MIT Press. AS Bregman1990Auditory scene analysis: The perceptual organization of soundCambridge (Massachusetts)MIT Press
129. Tougas Y, Bregman AS (1985a) The crossing of auditory streams. J Exp Psychol Hum Percept Perform 11: 788–798. Y. TougasAS Bregman1985aThe crossing of auditory streams. J Exp Psychol Hum Percept Perform11788798
130. Wendling F, Bartolomei F, Bellanger J, Chauvel P (2002) Epileptic fast activity can be explained by a model of impaired GABAergic dendritic inhibition. Eur J Neurosci 15: 1499–1508. F. WendlingF. BartolomeiJ. BellangerP. Chauvel2002Epileptic fast activity can be explained by a model of impaired GABAergic dendritic inhibition. Eur J Neurosci1514991508
131. da Silva F, Blanes W, Kalitzin S, Parra J, Suffczynski P, et al. (2003) Epilepsies as dynamical diseases of brain systems: basic models of the transition between normal and epileptic activity. Epilepsia 44: 72–83. F. da SilvaW. BlanesS. KalitzinJ. ParraP. Suffczynski2003Epilepsies as dynamical diseases of brain systems: basic models of the transition between normal and epileptic activity. Epilepsia447283
132. Perez Velazquez JL, Cortez MA, Snead OC, Wennberg R (2003) Dynamical regimes underlying epileptiform events: role of instabilities and bifurcations in brain activity. Physica

D 186: 205–220. JL Perez VelazquezMA CortezOC SneadR. Wennberg2003Dynamical regimes underlying epileptiform events: role of instabilities and bifurcations in brain activity. Physica D186205220

133. Kramer M, Kirsch H, Szeri A (2005) Pathological pattern formation and cortical propagation of epileptic seizures. J R Soc Interface 2: 113–127. M. KramerH. KirschA. Szeri2005Pathological pattern formation and cortical propagation of epileptic seizures. J R Soc Interface2113127
134. Wilson M, Sleigh J, Steyn-Ross D, Steyn-Ross M (2006) General anesthetic-induced seizures can be explained by a mean-field model of cortical dynamics. Anesthesiology 104: 588–593. M. WilsonJ. SleighD. Steyn-RossM. Steyn-Ross2006General anesthetic-induced seizures can be explained by a mean-field model of cortical dynamics. Anesthesiology104588593
135. Rodrigues S, Terry J, Breakspear M (2006) On the genesis of spike-wave oscillations in a mean-field model of human thalamic and corticothalamic dynamics. Phys Lett A 355: 352–357. S. RodriguesJ. TerryM. Breakspear2006On the genesis of spike-wave oscillations in a mean-field model of human thalamic and corticothalamic dynamics. Phys Lett A355352357

CITATIONS

CHAPTER 1

Daly AC, Gavaghan DJ, Holmes C, Cooper J. 2015 Hodgkin–Huxley revisited: reparametrization and identifiability analysis of the classic action potential model with approximate Bayesian methods.R. Soc. open sci.2: 150499. http://dx.doi.org/10.1098/rsos.150499.

CHAPTER 2

Michael D. Forrest, Can the Thermodynamic Hodgkin-Huxley Model of Voltage-Dependent Conductance Extrapolate for Temperature? Computation 2014, 2(2), 47-60; doi:10.3390/computation2020047.

CHAPTER 3

Aspart F, Ladenbauer J, Obermayer K (2016) Extending Integrate-and-Fire Model Neurons to Account for the Effects of Weak Electric Fields and Input Filtering Mediated by the Dendrite. PLoS Comput Biol 12(11): e1005206. https://doi.org/10.1371/journal.pcbi.1005206.

CHAPTER 4

Kornijcuk V, Lim H, Seok JY, Kim G, Kim SK, Kim I, Choi BJ and Jeong DS (2016) Leaky Integrate-and-Fire Neuron Circuit Based on Floating-Gate Integrator. *Front. Neurosci.* 10:212. doi: 10.3389/fnins.2016.00212

CHAPTER 5

Frederike Kneer, Eckehard Schöll and Markus A Dahlem, Nucleation of reaction-diffusion waves on curved surfaces, New Journal of Physics 16 (2014) 053010 doi:10.1088/1367-2630/16/5/053010.

CHAPTER 6

Ri-Qi Su, Ying-Cheng Lai and Xiao Wang, Identifying Chaotic FitzHugh–Nagumo Neurons Using Compressive Sensing, Entropy 2014, 16(7), 3889-3902; doi:10.3390/e16073889.

CHAPTER 7

André C. Marreiros, Stefan J. Kiebel, Karl J. Friston, A dynamic causal model study of neuronal population dynamics, In NeuroImage, Volume 51, Issue 1, 2010, Pages 91-101, ISSN 1053-8119, https://doi.org/10.1016/j.neuroimage.2010.01.098.

CHAPTER 8

E. C. Y. Ho, Wilson Truccolo, Interaction between synaptic inhibition and glial-potassium dynamics leads to diverse seizure transition modes in biophysical models of human focal seizures, J Comput Neurosci (2016) 41:225–244 DOI 10.1007/s10827-016-0615-7.

CHAPTER 9

Deco G, Jirsa VK, Robinson PA, Breakspear M, Friston K (2008) The Dynamic Brain: From Spiking Neurons to Neural Masses and Cortical Fields. PLoS Comput Biol 4(8): e1000092. https://doi.org/10.1371/journal.pcbi.1000092.

INDEX